MW00679374

# Automated Process Control Electronics

# Automated Process Control Electronics

**John Harrington**
**Tampa Technical Institute**

DELMAR PUBLISHERS INC.®

## NOTICE TO THE READER



*To the memory of Mr. Albert Heberlig, and for teachers everywhere.*

COVER PHOTO: Courtesy of GTE Communication Systems Corp.,
Northlake, Ill.

*Delmar Staff*

Associate Editor: Joan Gill
Assistant Managing Editor: Gerry East
Production Editor: Mary Ormsbee
Design Coordinator: Susan Mathews
Production Coordinator: Linda Helfrich

For information, address Delmar Publishers Inc.,
2 Computer Drive West, Box 15-015
Albany, New York 12212-5015



Printed in the United States of America
Published simultaneously in Canada
by Nelson Canada,
A Division of International Thomson Limited

10 9 8 7 6 5 4 3

**Library of Congress Cataloging in Publication Data**

Harrington, John, 1936–
Automated Process Control Electronics/John Harrington
p. cm.
Includes index.
ISBN 0-8273-2508-8. ISBN 0-8273-2509-6 (Instructor's guide)
1. Industrial electronics. 2. Process control. 3. Automation.
I. Title.
TK7881.H37 1989
629.8'043—dc19 88-15457
CIP

# Contents

# Preface

The wonderland of yesterday's science fiction novelist has become the reality of today's electronics technician. The pressure of a competitive world marketplace is moving American industry ever faster into automation. The area of industrial control is already a well established branch of electronics. As labor costs increase, automation will become an even larger field for the technicians of tomorrow.

This book provides the fundamentals of industrial electronic control and robotics which you will need in order to be a successful industrial electronics technician. Many different devices are detailed in order to show as much of the field as possible within the confines of a single text. The use of mathematics is minimized, except where mathematical analysis will help explain servicing techniques. The mathematics and design examples will help you understand when you can make substitutions, or why substitute parts cannot be used.

The book is written for the final course in an electronics program at a community college, vocational institute or technical college. It covers the fundamentals of process control, transducers, signal processing, feedback loops, actuators, analog and digital controllers, and the basic fundamentals of robotics. It assumes previous understanding of electricity, active solid-state devices, digital logic, and microprocessors. To help your memory, many of these subjects are reviewed briefly in the text, but these reviews should not be used as replacements for serious study in these areas: they are only reviews.

The book follows a logical path around a control loop. After a brief discussion of the control loop in the first section, it follows the system around the loop from measurement through signal processing, process controllers, to the actuators which create the process. As many examples as possible have been included to move the material from theory to practice.

Following each major section of the book you will find a section of practical troubleshooting which is based on years of field experience. Since troubleshooting is as much art as science, this view of the troubleshooter's train of thought may be the book's most beneficial feature.

Each chapter ends with a self-test and many include problems for solution. The answers to all questions are provided in an appendix. A lab manual is also available which demonstrates many of the principles found in the text.

The author hopes that his years of experience in the field and his years spent teaching electronics technicians will bear fruit in this textbook and give you a learning experience which is also a pleasure.

# Acknowledgments

The author would like to thank the following people for reviewing the manuscript at various stages and for their helpful comments:

Robert L. Arndt, Belleville Area College, Belleville, IL
Richard Aubert, Elk Grove Village, Elk Grove, IL
John Dello Russo, Porter and Chester Institute, Enfield, CT
Larren Elliott, Texas A & M University, College Station, TX
James Guyer, George Rogers Clark College, Indianapolis, IN
Jill A. Harlamert, De Vry Institute of Technology, Columbus, OH
Ted Kucharski, Waukesha County Technical Institute, Waukesha, WI
Leonard L. Place, Fox Valley Technical Institute, Oshkosh, WI
Scotty D. Richardson, Dickson State Area Vocational Technical School, Dickson, TN
Roy Roddy, Yakima Valley Community College, Yakima, WA
Sam Watson, De Vry Institute of Technology, Lombard, IL

# Section I

# Introduction

Controlling the industrial process involves a series of separate steps, each of them apparently complex. Bringing this series of steps together is the task of the manufacturing engineer; keeping the various steps working is the job of the industrial technician.

Analysis of the industrial process shows that the steps involved in controlling an industrial process are really not so complex as the first glance makes them seem. In the first section of this text, we will see an overview of the closed loop control system, from measurement through control. Other sections of the text will carry us around the loop, step by step, explaining each step in greater detail.

# 1

# An Introduction to Process Control

A robotic arm lifts a steel blank from a pallet, orients it correctly, and places it on the bed of a milling machine. The milling machine shapes the rough blank into an exactly defined shape and passes the workpiece on to a second robotic arm. The second arm places the piece on yet another milling machine which adds three precisely curved grooves to the piece. A third robotic arm moves the piece to a conveyer where it is transported to another part of the factory where still more refinements are added. Dozens of steps later, the finely machined piece of steel passes through a carburizing furnace where a hard surface is created. The part is cooled by passing it through a chamber of cold air, cleaned in a vat of solvent, and painted in a specific pattern by a robotic arm. Finally, one more robotic arm places the finished product on a shipping pallet. Not one human being has touched the workpiece since it entered the factory. One human operator sitting in an air conditioned operations center has supervised the whole process, unaffected by the high temperatures, noxious fumes, or hazardous conditions in the manufacturing plant.

The preceding scene sounds like something from the future, but the future is now. The lights in some factories in the U.S. are only turned on for the maintenance and janitorial staff. Manufacturing even a single part involves many processes. The control of those processes is the subject of this book. In this chapter, we will take a broad look at process control in its many parts. In later chapters, we will examine each of the parts of process control with an eye towards the final product.

# OBJECTIVES

You will have successfully completed this chapter when you can:

1. Define
   a. The controlled variable
   b. The manipulated variable
   c. The control setpoint
   d. The error signal
   e. The summing point
   f. The controller
2. Explain the basic ideas behind damping.

## 1.1 PROCESS CONTROL

Most industrial processes can be automated by controlling one or more variables in the manufacturing process. These process variables may include position, temperature, speed, pressure, force, thickness, weight, or volume. Whatever the process variable, the basic principles of controlling it remain the same. We will use a home heating system as our example of process control.

The first step in controlling any variable is measuring that variable. We cannot control the temperature of a process unless we can measure that temperature. A transducer for an electrical or electronic control system measures the variable and converts the measurement into an electrical signal. The thermostat of your home heating system contains a type of thermometer that measures the room temperature — the measured variable in the example.

Of course in order to control a process, we must also know the desired result — the condition of the variable that satisfies our control objective. This is the *control setpoint*; for example, the temperature setting on your thermostat.

At some point in a control system, the measured variable and the process setpoint must be brought together and compared. We call that point the *summing point*. The result of the comparison is an error signal. The error signal may be a simple on-off signal or it may include an indication of the amount of correction needed. The thermostat provides an on-off type of error signal, sending a signal to the heater when the room temperature falls below the control setpoint. The error signal is input to the controller, which makes the decision to adjust the process to more closely match the control setpoint.

The final step in process control does the work of adjusting the variable to match the control setpoint more closely. The actuator performs the work the rest of the system is controlling. When your thermostat sends an on signal to the heater, the heating element is turned on to raise the room temperature. In this example, we are not acting

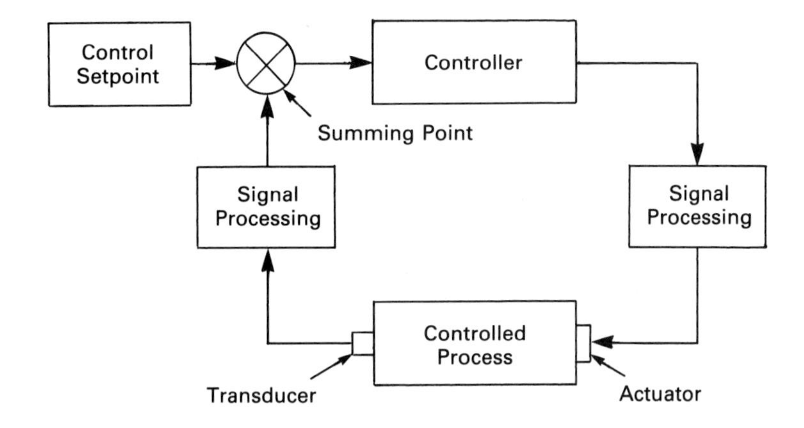


**Figure 1.1** The elements of process control

on the room temperature directly; we are acting on the temperature in the heating chamber of the furnace. This is the manipulated variable: it will, in its own turn, affect the room temperature. When the temperature reaches the control setpoint, the thermostat simply stops sending the on signal and the heating element switches off.

Figure 1.1 illustrates in block diagram form all of the elements of process control and the direction of signal flow. (All control systems may not include all of the elements shown in Figure 1.1.) The figure shows that automatic process control is a closed loop system. The transducer measures the controlled variable and produces a feedback signal. The signal processing section may include amplifiers, filters to guard against noise, voltage-to-current converters, or pulse-shaping circuits. The output of the signal processor is fed to the summing point, a comparator, where it is compared to the control setpoint. The result is an error signal which is fed to the controller. The controller makes the decisions about whether to adjust the controlled variable, how much to adjust it, and for how long. The output of the controller is a control signal, which may in turn require some processing to alter the controlled variable. This processing may include pulse shaping, voltage-to-current conversion, and switching. The processed control signal is then sent to the actuator, which alters the process variable. The transducer now reports the effects of the alteration to the controller and the process repeats.

## *1.2 DAMPING: FINE TUNING THE CONTROL PROCESS*

As you drive your car on the highway, lift your foot from the accelerator. Your car will continue to move at the same speed for some distance before slowing down. This is the principle of inertia, which states: "an object at rest tends to remain at rest — an object in motion tends to remain in motion." No physical process can respond instantly at full speed, nor can it stop instantly. Every process includes a certain amount of inertia, which can have an undesirable effect on the control process. A system that controls the speed of a large dc motor provides us with a good example.

Connected to the output shaft of the motor is a tach generator, which has an output dc voltage in direct proportion to the motor's speed. This voltage is used to control the pulse width of the drive current to the motor. The longer the pulse width, the faster the motor will run. For purposes of illustration, let's assume that the motor is running too slowly.

The low output from the tach generator will cause the controller to increase the pulse width, speeding up the motor. When the motor reaches the setpoint speed, the controller will respond by adjusting the pulse width, but the motor is still accelerating from the earlier control signals. Because of its mass, the flywheel effect (inertia) will keep the motor running too fast for a short while. The controller will respond by reducing the pulse width even more, and the motor will begin to slow down as the momentum of the motor is overcome. The motor will continue to oscillate between a speed that is too slow and a speed that is too fast for quite some time. This is an undamped or underdamped response.

A typical undamped response curve is shown in Figure 1.2. It may be used to illustrate the response of a system to changes in either the control setpoint or the controlled variable. The time between the change in the system and the correction of the change is the *response time.* The difference between the desired setpoint and the actual system results in the *steady-state error.*

Through careful design of a system, the ability of the system to reach the desired setpoint can be improved. More complex circuitry in an analog system or more complex software in a digital system can overcome the system's tendency to oscillate. Whether this is done or not is usually decided on the basis of cost versus benefits. Often the system with the fastest response time has the greatest steady-state error.

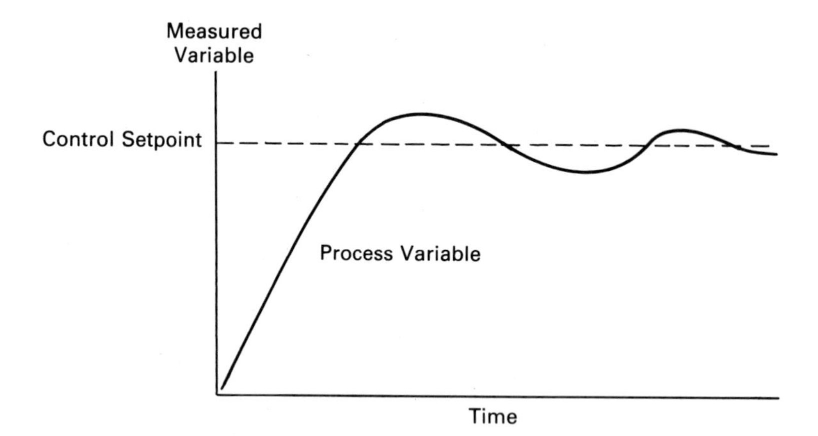


**Figure 1.2** An undamped response curve

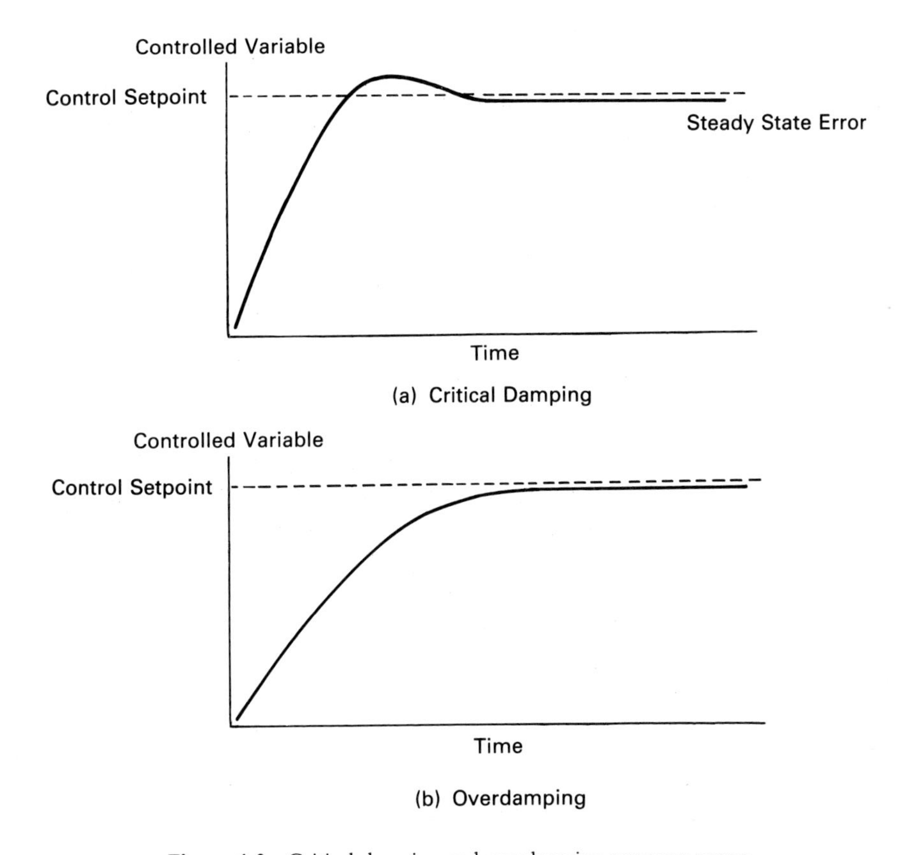


**Figure 1.3** Critical damping and overdamping response curves

The more complex (and expensive) systems have a lower error and a slower response to changes. The systems engineer must carefully evaluate the application requirements to select the most cost-effective system.

Figure 1.3 illustrates typical response curves of systems that employ damping in different degrees. The critically damped system response offers the best compromise between accuracy and response time. The overdamped system response is too slow for the slight gain in accuracy, but may be used where overshoot is unacceptable. We will deal more thoroughly with the techniques for damping in Chapter 17 and in Section III of the text.

## 1.3 SOME EXAMPLES OF PROCESS CONTROL

The old-fashioned potbellied stove provides a classic example of a process without feedback. The homeowner (our system operator) lights a fire on the grate when the air in the home becomes too chilly. Soon a fire is blazing in the stove, the chimney is roaring, and the sides of the stove are glowing a friendly cherry red. Within an hour the room is too warm and the homeowner must adjust the damper in the stovepipe to reduce the draw. The fire dwindles, the room becomes chilled again, and the damper must be adjusted again. With patience and experience our homeowner will eventually arrive at an optimum setting of the damper. When the wind outside dies away, however, the draw in the chimney changes again and the cycle of too chilly, too warm will be repeated. This change in conditions may be termed a *disturbance*.

In the system we just described the homeowner is at once the sensor, the controller, and the actuator. He or she decides that it is too cool in the room and ignites the fire. Later the homeowner decides that it is too warm in the room and adjusts the damper. An observer recording the temperature in the room over a period of time would construct the curve of an undamped system, continually oscillating above and below the control setpoint.

A more modern electric home heating system operates more efficiently. The thermostat contains a bimetal element curved into a helical coil. As the temperature in the room decreases, the coil unwinds. A mercury switch attached to the coil closes at the preset temperature, sending the heater an on signal, which operates a relay, which turns on the heating element. As the heater warms the room, the bimetal coil closes and the mercury switch turns off the heater.

If the thermostat responded instantly to temperature changes in the room, it would tend to cycle on and off continuously, not a very desirable circumstance. Because of mechanical lags in the thermostat's bimetal coil and the mass of the air in the room, there is a gap in the system's response. The system will turn the heater on at one temperature and turn it off at another, higher temperature. The difference between the on setting and the off setting is the system's *hysteresis*. Our observer in this room would plot the curve of a critically damped system.

Temperature is also the most often controlled variable in the industrial process. The temperature of many chemical reactions is very closely controlled. The temperature of steel is controlled in the tempering process, in carburization, in annealing, and in many other processes.

*Heat treating* is a commonly used process for hardening steel to a depth of a few millimeters from the surface. Cutting tools often have heat-treated edges to stay sharp longer without suffering from the brittleness that would be caused by having hardened steel throughout the whole tool. As the heat-treated parts leave the heating chamber, they are immersed in a bath of quenching oil as a final step. The temperature of that bath must be maintained within a limited range to avoid destroying the effects of the heat treatment. But the constant flow of hot steel parts into the bath causes its temperature to rise.

Figure 1.4 shows a typical circuit for controlling the temperature of a quench bath. A thermistor probe is inserted into the tank as a transducer. The transducer output is compared with the control setpoint, and when an overtemperature exists, the error voltage causes the controller to turn on a pump. The pump circulates the hot oil through a heat exchanger filled with chilled water, reducing the oil's temperature.

Another very common industrial variable is pneumatic pressure. Many industrial operations, including robotics, are performed using pneumatic, rather than electrical,

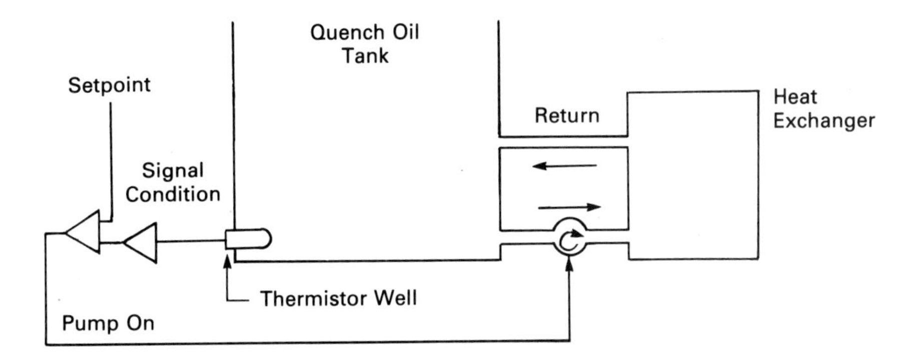


**Figure 1.4** A quench bath temperature control system

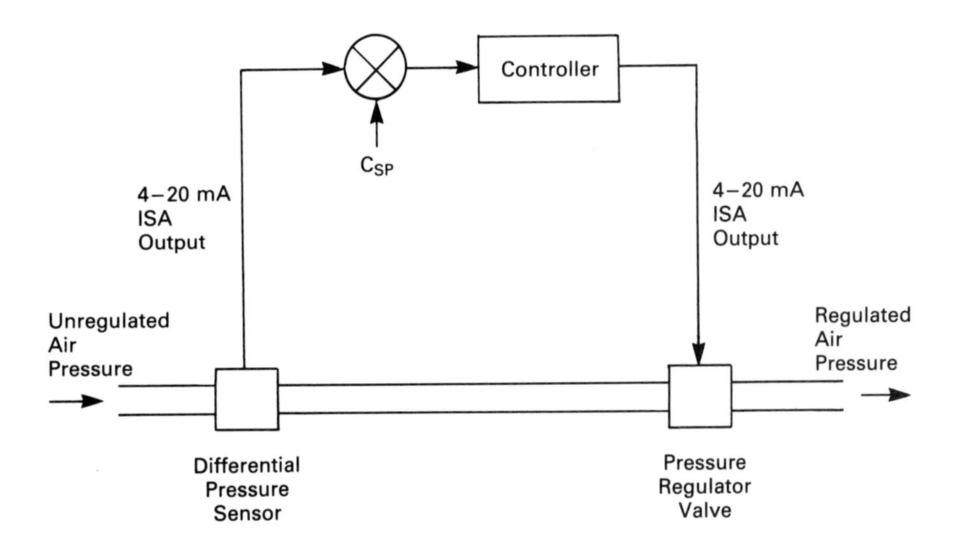


**Figure 1.5** A pneumatic pressure control system

actuators. Electronic controllers are used to regulate the pressure in pneumatic systems, as well as to control the actuators themselves.

The system illustrated in Figure 1.5 shows one approach to the control of pneumatic pressure in a single pneumatic line. The pressure transducer is a differential pressure sensor that outputs a current proportional to the pressure in the line. Its output is compared to the control setpoint and the resulting error signal is transmitted to the controller. The controller outputs a current proportional to the required amount of correction. This current controls an actuator which controls a valve. The opening and closing of the valve regulates the pressure in the output line.

## SUMMARY

We have seen in this section that process control involves many factors:

1. The process to be controlled must be measured by a transducer.
2. The desired process output is determined by the control setpoint.
3. The transducer output is compared to the control setpoint at the summing point.
4. The difference between the control setpoint and the measured value is the error signal.
5. The error signal is input to a controller.
6. The controller decides what action needs to be taken to adjust the process.
7. The output of the controller drives the actuator, which actually adjusts the process.

## SELF-TEST

1. An automatic control system adjusts variations in ________________.
2. The process variable is measured by a sensor known as a(n) __________.
3. The desired condition of the process variable is indicated by the ________.
4. The output of the transducer is compared with the control setpoint at the ______________________.
5. The result of the comparison is a(n) ______________________ signal.
6. The error signal is input to a(n) ______________________.
7. The actual adjustments to the process variable are made by the ________ ______________.
8. A system that oscillates above and below the control setpoint is ________ ______________.
9. A system that has a small overshoot and then settles to a state slightly below or above the control setpoint has ______________________ damping.
10. An overdamped system has no ______________________.

## QUESTIONS/PROBLEMS

1. Describe an example of an automatic control system (other than the heating system) in your domestic environment. Draw a block diagram of the system, identifying the parts in relationship to Figure 1.1.
2. You have been assigned the task of designing a control system that will maintain the temperature of a tank of water at 150 degrees F. The temperature in the tank may not exceed the control setpoint at any time. Draw a block diagram of the system and explain the type of damping (if any) you would select.

# Section II

# Measuring the Process Variable

In the first section of this book, we saw an overall view of the industrial control process. Section II is devoted to measuring the process variable. This is the point at which the design of a controlled process begins, and it is a logical beginning for the serious study of the control process.

In this section, we will examine many of the techniques that are used for measuring a controlled variable. Since our work is primarily electronic, we will concentrate on the methods used for converting a physical variable into an electrical signal. This study of transducers will encompass many chapters and prepare us for the work beyond: controlling the variable.

# 2

# Measuring Temperature

Without a doubt, the most measured of all process variables is temperature. Processes as widely varied as the annealing of steel and the completion of a delicate chemical process depend upon very closely controlled temperature conditions. If the steel is too cold, its surface becomes brittle and unsuited to the task for which it was designed. If the chemical process becomes too warm, the result may be a violent explosion. Clearly, an effective means of temperature control is an important facet of industry.

## OBJECTIVES

You will have successfully completed this chapter when you can:

1. Describe the basic operating principles behind the
   a. Bimetal thermostat
   b. Thermocouple
   c. Thermopile
   d. RTD
   e. Thermistor

## 2.1 THE BIMETAL THERMOSTAT — HEAT TO MECHANICAL MOTION

Probably the most familiar automation device in the world is the ubiquitous bimetal thermostat. This simple, effective device adorns walls in homes, schools, offices, and factories throughout the country, quietly monitoring room temperature and switching heaters and air conditioners on and off without notice.

The principle of the bimetal thermostat is simple: metals expand at different rates. The major element of the bimetal thermostat is two strips of different metals bonded together. As the bimetal strip is heated, the two metal strips begin to expand. Because one of the metals is expanding faster than the other, the bimetal strip bends or curves as the temperature increases, bending away from the metal with the higher coefficient of expansion. As the metal cools, the bending effect is reversed. Engineers have devised many different ways to make use of the bending effect of the bimetal strip.

Chances are that the thermostat in your home has a long coil of bimetal inside its plastic case. The longer the bimetal strip, the greater the degree of bending, and the more sensitive is the bimetal element. The manufacturer has coiled the long strip into a spiral simply to reduce the size of the controller. The center of the spiral is attached to the plastic case of the thermostat, and the outside end is free to move. A mercury switch is most likely attached to the moving end of the coil.

The spiral is arranged so that the metal with the higher rate of expansion is on the outside of the coil. Thus, when the temperature increases, the coil tightens, tilting the mercury switch in one direction. As the temperature decreases, the coil opens up, tilting the mercury switch in the opposite direction. An SPDT mercury switch can now take care of both heating and cooling.

Other types of bimetal controls use the bimetal itself as a part of the switch. It becomes the wiper, or moving contact, in a switching arrangement. The bimetal element may be arranged as a straight piece, a spiral, a helix, or even a disc for different types of control functions.

Figure 2.1 illustrates some of the different types of bimetal temperature sensors. The familiar dial thermometer is yet another, and the thermostat in your car relies on the same principle to open a valve and circulate the coolant when it reaches a predetermined temperature. Bimetal temperature sensors are rugged, inexpensive, and reliable, which explains their widely varied use, but they also have certain drawbacks. Some of their drawbacks are:

1. The bimetal thermostat is essentially a binary device. The switch controlled by a bimetal element is off or on. Thus it can tell you whether the temperature is above or below a preset point, but it cannot give you an absolute value for the temperature in a control circuit.
2. When the bimetal itself is a part of the circuit, the current must be limited to prevent self-heating from the voltage drop in the bimetal. If you fail to take self-heating into consideration, your control may well create some surprising results. Heating controls will turn off before the process is sufficiently hot, and cooling controls will not turn off until the process is cooler than desired.

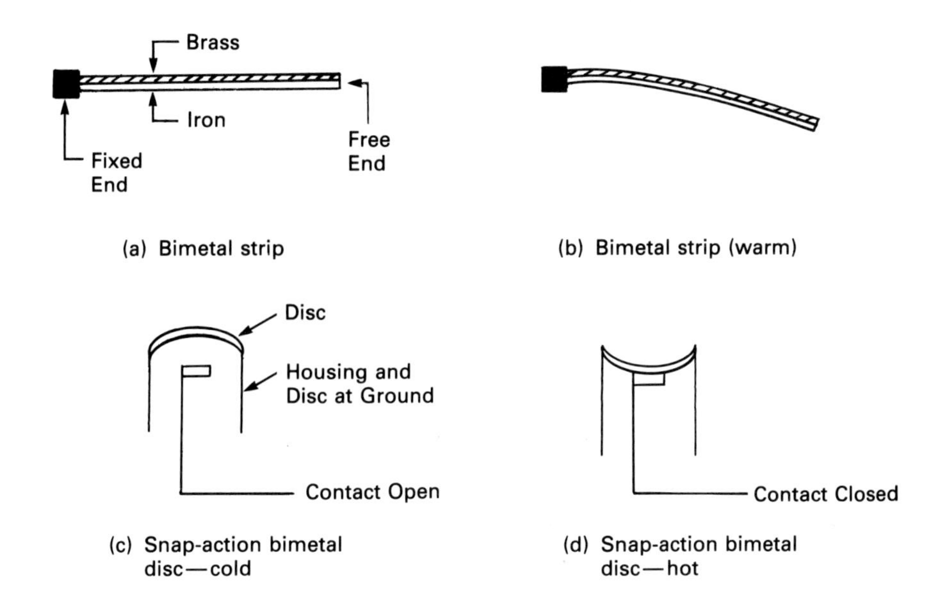


**Figure 2.1** Some examples of bimetal thermostats

3. When the bimetal is used as a part of the switching circuit, the slowly operating switch is subject to arcing and the types of damage that arcing can cause to switch contacts.

When high accuracy and repeatability are needed, the bimetal element should not be used. Still, even with its faults, you will find the bimetal in wide industrial and domestic use.

## 2.2 THE THERMOCOUPLE — HEAT TO ELECTRICITY

In 1821, Thomas Seebeck discovered that when two wires of different metals are joined at both ends and one end is heated, a small current flow depends upon the difference in temperature between the two junctions. If the circuit is opened at the center, an open circuit voltage (the Seebeck voltage) can be measured.

Figure 2.2 illustrates the Seebeck effect and the Seebeck voltage. For small changes in temperature, the Seebeck voltage is linearly proportional to the temperature. This fact forms the basis of one of industry's favorite temperature measuring devices, the *thermocouple.*

The American National Standards Institute (ANSI) has defined standard performance curves for different types of thermocouples. Various alloys of metals have been standardized, allowing consistency in thermocouples made by different manufacturers.

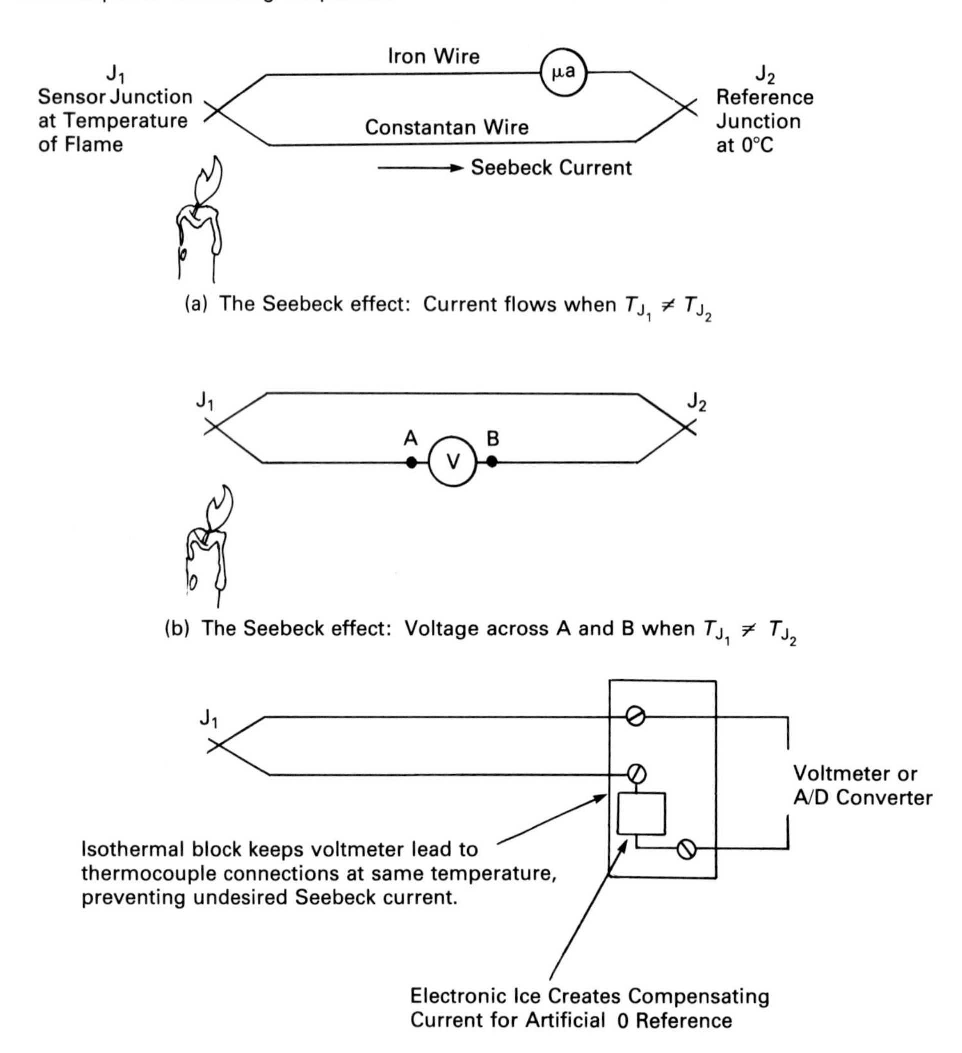


**Figure 2.2** Principles of the thermocouple

The variety of thermocouple types is important because it allows the designer to specify a thermocouple that is suited to the particular atmosphere and temperature range of a task. A thermocouple that is well suited to one application may be destroyed by the atmosphere of another. Manufacturers such as Omega Engineering Inc. supply detailed information to assist the designer in selecting the correct thermocouple for

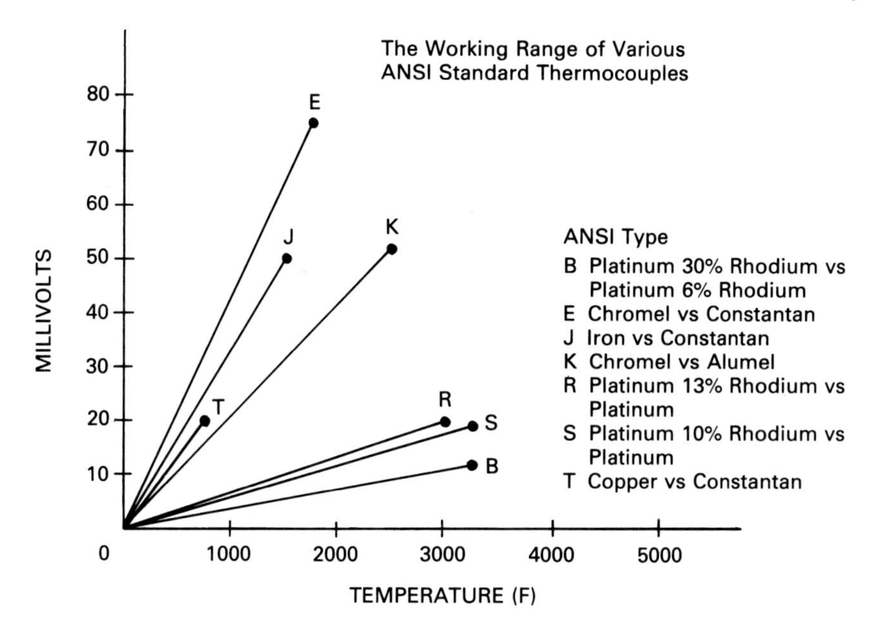


**Figure 2.3** Common thermocouple response curves

a particular job. When in doubt, the manufacturers will also provide you with engineering assistance.

The graph in Figure 2.3 shows the response of various ANSI standard thermocouples. The following table lists the most common alloys used in thermocouple design.

| *COMMON NAME* | *ALLOY* |
|---|---|
| CONSTANTAN | COPPER-NICKEL |
| CHROMEL | NICKEL-CHROMIUM |
| ALUMEL | NICKEL-ALUMINUM |

In addition to these alloys, various alloys of platinum and Rhodium are used for some atmospheric and temperature conditions, and alloys of tungsten and rhenium are used in very high temperature applications (around 3000 °C), which would destroy other temperature measuring devices.

A standard ANSI type *J* thermocouple is made up of a junction of iron wire with constantan wire. It has a useful range between 0° and 750° Celsius, or 32° to 1400° Fahrenheit. With the reference junction at 0 °C, the output voltage is 0.507 mV when the measuring junction is at 10°. Then when the junction is at 100 °C, we would expect the output to be 5.07 mV (10 × 0.507). The National Bureau of Standards publishes tables that show that the actual output will be 5.268 mV. This is an error of nearly 4%.

Over a narrower range, calculations can be more exact, and temperature control is most often used over a limited range. The output at 110°, for example, is given as 5.812 mV. We can then calculate that the change in output per degree in the range of 100° to 110° will be 0.054 mV:

$$5.812 - 5.268 = 0.544$$

$$\text{and } \frac{0.544}{10} = 0.054$$

A temperature of 105° should yield an output voltage of

$$5.268 + (5 \times 0.054) = 5.540 \text{ mV}$$

This is the exact voltage that is published in the NBS tables.

Figure 2.4 shows a few of the many commercially available thermocouple types.

Practical use of the Seebeck effect is more complicated than simply taking a voltage measurement. The Seebeck voltage is the result of the temperature difference between two dissimilar metals. When a voltmeter and its leads are added, two more junctions are created. These junctions may produce undesirable Seebeck voltages in series with the desired voltage. Fortunately, there is a simple approach that will prevent this problem from occurring. The connections to the voltmeter circuit in Figure 2.2e are made on an *isothermal* block. This is an electrically insulating terminal block made of a substance that readily conducts heat. When the isothermal block is used, there is no temperature difference between the junctions where the voltmeter leads connect to the thermocouple leads. Since the Seebeck voltage is created by the temperature difference between two junctions, the voltmeter connections made on the isothermal block will not add any undesired voltage to the desired Seebeck voltage.

Accurate measurement of temperature with thermocouples requires a known temperature — the *reference temperature* — at one of the junctions. The Seebeck voltage depends on the difference between the temperatures at the two thermocouple junctions, one known and one unknown. The traditional approach to a reference temperature is to immerse one of the thermocouple junctions in an ice bath. The 0 degrees Celsius temperature of the ice bath provides a very reliable reference, but it is inconvenient. Outside the laboratory, this technique is impractical. The field of electronics provides us with two different techniques.

Figure 2.5 illustrates the methods used for eliminating the ice bath reference temperature. A *thermistor* is a semiconductor resistance that varies according to temperature. The resistor labeled $T$ in Figure 2.5a is a thermistor. It is fed by a constant current source. The voltage drop across the thermistor is inversely proportional to the temperature of the isothermal block. By measuring the temperature of the isothermal block, we can calculate the temperature at $T_{J_1}$, and compensate for the temperature at $T_{J_2}$. When this is done under computer control, it is called software compensation.

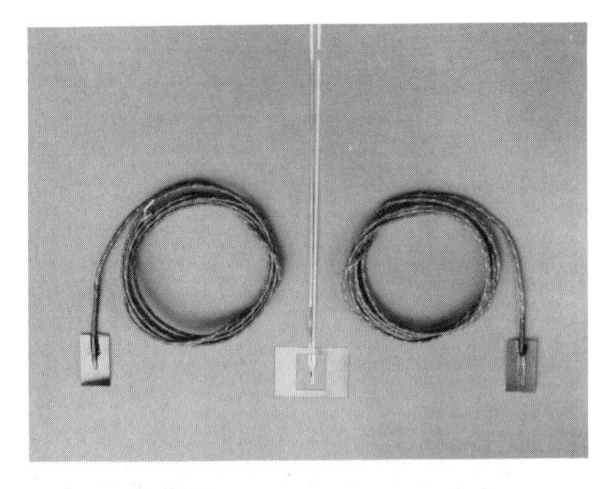

**Figure 2.4a** Typical industrial thermocouples. The metal tab can be welded directly to a metal container to ensure good thermal connection. Other types are available which may be bolted to the container. *(Courtesy of Omega Engineering)*

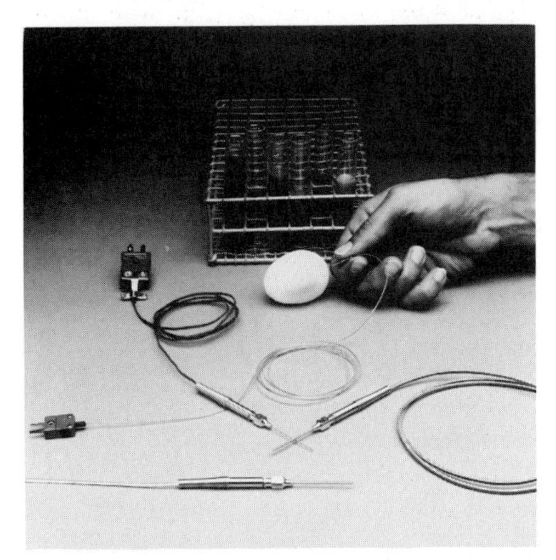

**Figure 2.4b** Veterinary thermocouples *(Courtesy of Omega Engineering)*

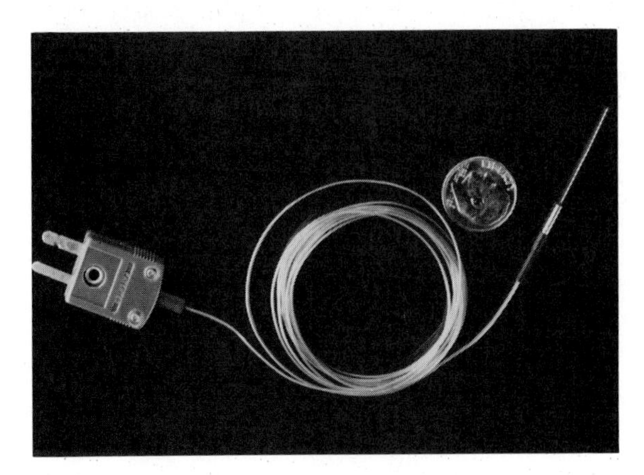

**Figure 2.4c** Closeup of veterinary thermocouple *(Courtesy of Omega Engineering)*

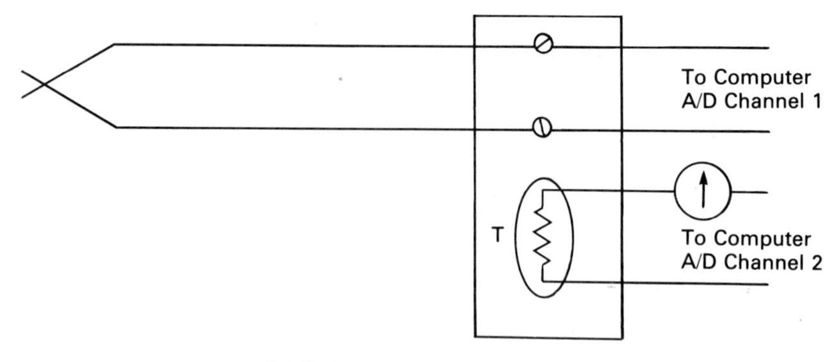


(a) Software compensation

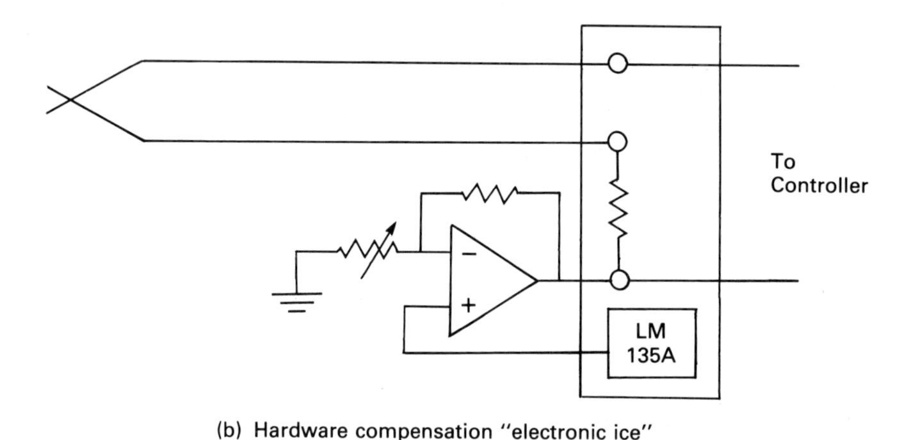


(b) Hardware compensation "electronic ice"

**Figure 2.5** Thermocouple compensation

Figure 2.5b illustrates hardware compensation. This circuit includes a precision integrated-circuit temperature sensor mounted on the isothermal block. The LM135A temperature sensor is a typical monolithic temperature sensor, which has a linear output of 10 mV/degree Kelvin. This output is multiplied by an appropriate amount by the op amp and added in series with the Seebeck voltage from the thermocouple junction. The circuit is often called the electronic-ice point reference.

Both techniques work equally well. The software compensation does require computer time with its attendant delay, but it lends itself well to servicing different types of thermocouples. The electronic-ice circuit provides almost instantaneous response, but it is limited to serving only one type of thermocouple unless switching arrangements are made to change the gain resistance for each type of thermocouple it must serve.

Thermocouples are one of the preferred temperature measuring devices in modern industrial practice. Their tolerance for very high temperatures and extremes of hostile atmospheres allows them to be used where other types of temperature sensors would be melted by the heat or rapidly corroded by the fumes. Thermocouple *wells* allow the thermocouple to be inserted into the bottoms of tanks of fluid without the danger of leaks and still provide for simple replacement of defective units. The ANSI standards allow for easy second sourcing of thermocouples without problems of recalibration. The use of isothermal blocks and either software or hardware compensation provides accuracy, ease of use, and ease of calibration and repair. All told, the thermocouple is the sensor of choice for many industrial and laboratory tasks.

A special application of the thermocouple is the *thermopile*, which is made up of a number of thermocouples in a series-aiding configuration. The small thermocouple output voltages (or currents) are additive, like batteries in series, which increases the output voltage. Thermopiles are often used in hand-held optical instruments that focus infrared energy onto the closely spaced thermocouple junctions and provide the operator with a direct reading of the temperature of the process at which the device has been aimed.

## 2.3 THE RTD — TEMPERATURE TO RESISTANCE

From your earlier studies in electricity, you are aware that the resistance of pure metals has a positive temperature coefficient. In other words, the resistance of a metal increases as the temperature increases. This simple fact is the basis of the resistive temperature detector, the RTD.

All metals produce a positive change in resistance for a positive change in temperature. Because the amount of error in a system is less when the resistance value of the RTD is large, metals with high resistivity are preferred for construction of practical RTDs. The most common RTDs are made of platinum, nickel, or nickel alloys. Tungsten, with its relatively high resistivity, is used for very high temperature applications, but its brittleness makes it difficult to work. The resultant cost limits its use to special applications. Of the most commonly used RTD materials, the most linear response is achieved with platinum. The RTD exhibits much greater linearity over its entire range of usefulness than does the thermocouple. The effect is especially pronounced at lower temperatures.

The resistivity of platinum is 59 ohms per circular mil foot (CMF). A standard wire table informs us that a 30 gage wire (the fine wire used for wire wrapping is 30 gage wire) has a cross-sectional area of 100 circular mils. Thus, a one-foot length of 30 gage platinum wire would have a resistance of less than one ohm. Clearly, we must use either very small wire or large lengths of wire to obtain a useful resistance to measure.

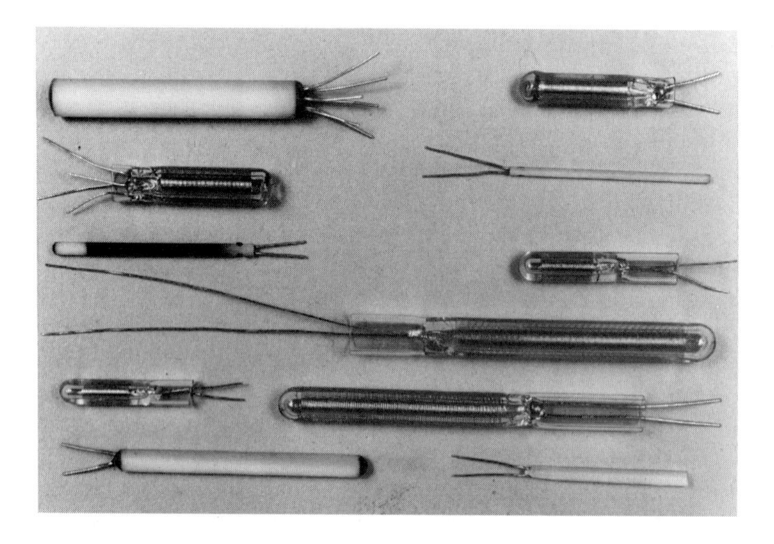

**Figure 2.6a** Platinum wire RTDs *(Courtesy of Omega Engineering)*

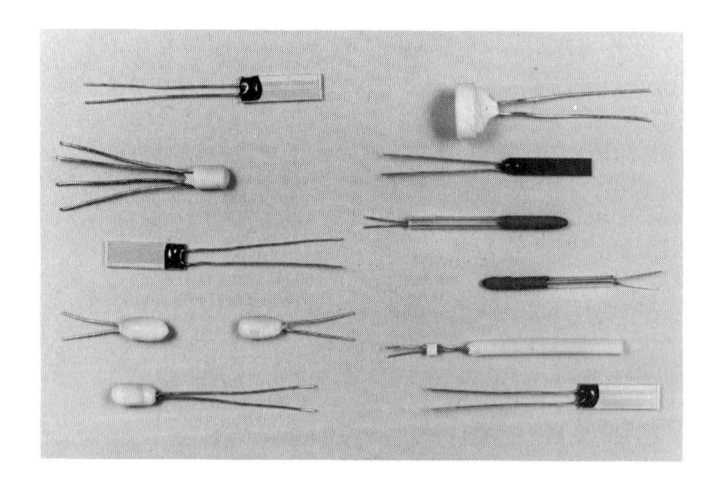

**Figure 2.6b** Platinum film RTDs *(Courtesy of Omega Engineering)*

While laboratory-grade RTDs do use relatively long lengths of platinum wire coiled around a mica form and encased in a glass tube, this method of construction suffers from both fragility and a slow response time. Commercial RTDs, such as those shown in figures 2.6a and b, must be more rugged to withstand shock and vibration, and must respond rapidly to changes in temperature. They are often wound as a bifiler coil (noninductive winding) around a ceramic bobbin, then sealed in molten glass.

This ensures the integrity of the RTD under extremes of vibration, but the bobbin and the platinum wire must have almost perfect matching in thermal expansion characteristics. If this is not so, the resistance of the wire will change not only with temperature but with strain from the difference in expansion between the wire and the bobbin.

The latest technique is to deposit a thin film of platinum metal on a small ceramic substrate, etch it with a laser trimming system, and seal it in glass. The result is an RTD with a fairly high resistance, good resistance to vibration, and low cost because the manufacturing technique is more easily automated.

Typical resistance values for platinum RTDs vary from about 10 ohms for the hand-wired laboratory RTD to several thousand ohms for thin film RTDs. The single most common value is 100 ohms at 0°C. The temperature coefficient (alpha) of platinum is 0.00385 ohms/degree at 0°C, which is the average slope from 0°C to 100°C. For a 100-ohm RTD, this means that the resistance will change 0.385 ohm for each degree change in temperature.

Both the small change and the small initial resistance mean that the resistance of the meter leads between the sensor and the measuring instrument may play a significant role in a measurement. A 5-ohm lead resistance will contribute an error of 5/0.385, or about 13° in our system. The classic method of avoiding this problem is the Wheatstone bridge.

Figure 2.7 illustrates the use of a bridge for measuring the resistance of the RTD. In Figure 2.7a, the bridge resistors other than the RTD must have a zero temperature coefficient to prevent them from contributing their own resistance changes to the system. Figure 2.7b shows the RTD separated from the bridge by extension wires to protect the bridge resistors from the temperatures that the RTD is measuring. This unfortunately returns us to our original problem — the resistance of the lead wires can contribute errors to the system. Figure 2.7c solves this problem by adding a third wire. Wires *A* and *B* are carefully measured to be the same length so that their resistance effects will cancel themselves out. Wire *C* carries no appreciable current because it is only a sense lead.

The three-wire bridge, however, adds its own nonlinearity to that of the RTD, and requires calculations to compensate for the lead resistance. As the output voltage of the circuit increases, the accuracy decreases. When the bridge is close to balance, however, the three-wire technique can be quite effective.

Figure 2.8 shows another, more accurate technique for measuring the resistance of the RTD. $I_{S_1}$ is a constant current source; that is, the output current is constant over a wide range of load resistance. The output voltage measured by the DVM is then directly proportional to the RTD's resistance, and is insensitive to lead length. The disadvantage of having to use four wires rather than three is a small price to pay for the increased accuracy of the system.

Unlike the thermocouple, the RTD is not a self-powered device. The current that passes through the RTD for measurement causes the phenomenon of self-heating, which injects an error into the system. Use the lowest excitation current possible con-

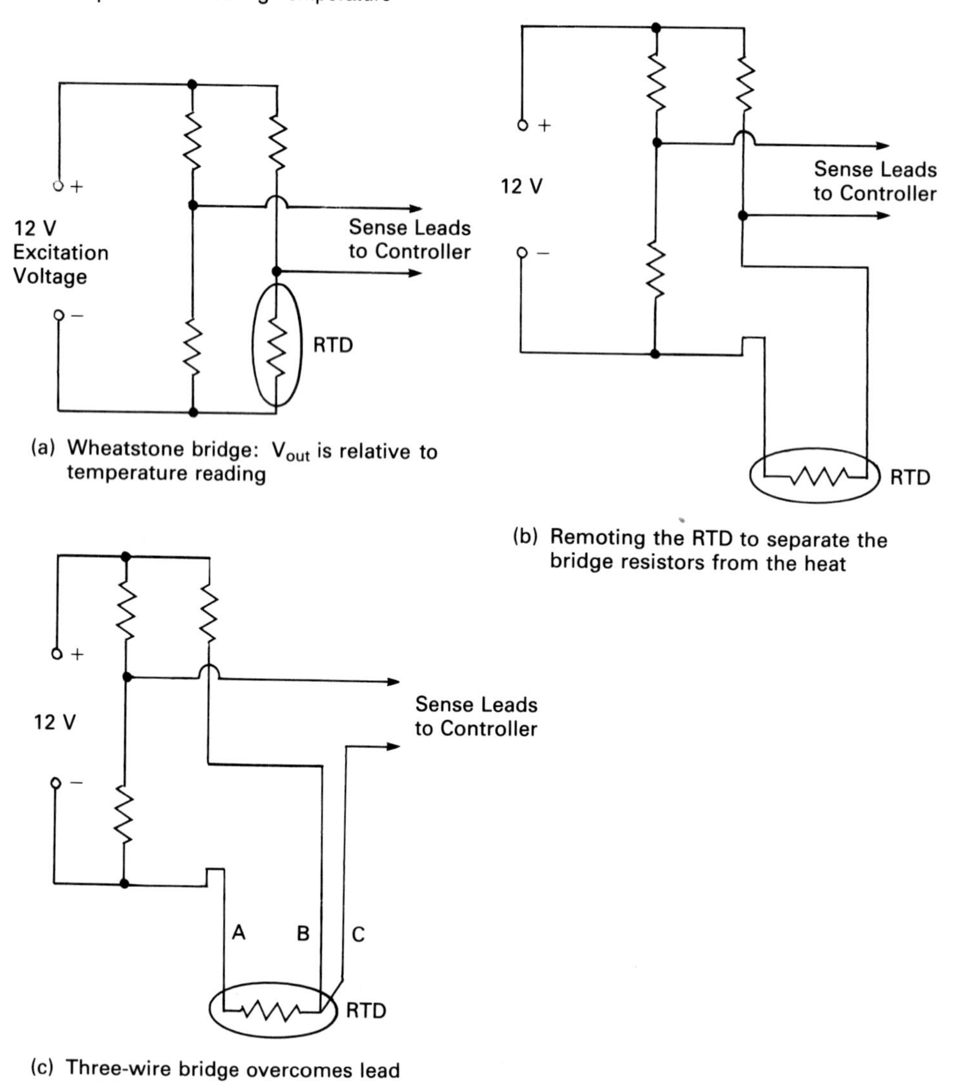


**Figure 2.7** Using the Wheatstone bridge to measure the temperature of an RTD

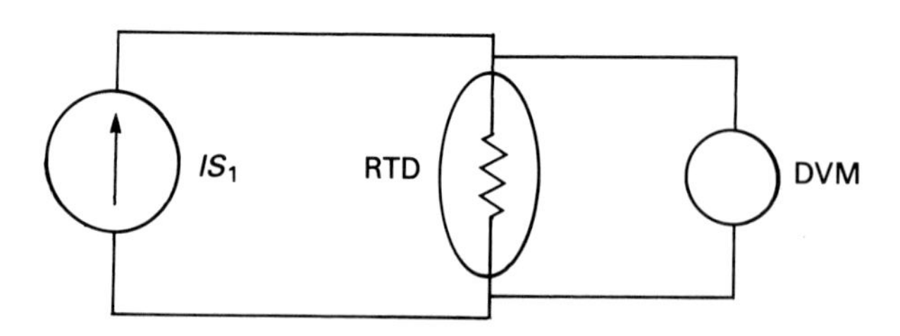


**Figure 2.8** Constant current source provides excellent accuracy without lead length problems.

sistent with accurate measurement. Physically large RTDs suffer less from self-heating, but their response time is slower than smaller units, and they exhibit thermal shunting, a tendency to shunt heat away from the process being measured. The selection of the correct RTD generally involves compromises in the areas of self-heating, response time, and thermal shunting.

## 2.4 THE THERMISTOR — A SEMICONDUCTOR APPROACH

Semiconductors exhibit a temperature coefficient that is the opposite of the temperature coefficient of metals: as the temperature increases, their resistance decreases. This phenomenon, which causes thermal runaway in poorly designed transistor circuits, is the basis of the thermistor.

The thermistor, like the RTD, is a temperature-sensitive resistance. In fact, it is considerably more sensitive than the RTD, with temperature coefficients as large as four percent or more. In other words, while a 100-ohm platinum RTD will have a resistance change of only 0.385 ohms/degree, a 100-ohm thermistor will have a resistance change of 4 ohms/degree. Also, thermistors typically have much higher resistance than RTDs, further increasing their sensitivity. A common thermistor value is 5 kilohms at 25°C. This thermistor will change by 200 ohms ($0.04 \times 5000$) for each degree change in temperature. Thus, we can use the thermistor to detect very small temperature changes that would be unmeasurable with either thermocouples or RTDs. However, we pay for this sensitivity in linearity.

Figure 2.9 shows the linearity of the thermistor compared with that of the RTD and the thermocouple. As you can see, the thermistor comes in a poor third. However, for measurements over a short portion of any curve, the segment will approximate a straight line. Over a limited portion of its measuring range, a few relatively simple calculations by the controlling computer can compensate for the nonlinearity of an individual thermistor. The use of computer controllers has largely supplanted efforts to linearize the output of thermistors.

The higher resistance of the thermistor means that the resistance of connecting leads will contribute a much smaller margin of error than with RTDs, facilitating their use in two-wire bridge circuits. Their relatively small size also means that they will exhibit a fast response time and contribute less thermal shunting to a system. On the other hand, their use is limited to a range of only a few hundred degrees, and they are subject to calibration changes if they are exposed to temperatures near the upper limit for extended periods of time. They are also typically less rugged than either thermocouples or RTDs and require care in mounting to avoid damage.

## 2.5 THE MONOLITHIC TEMPERATURE SENSOR — THE INTEGRATED APPROACH

Recent advances in the design of semiconductor integrated circuits have led to monolithic IC temperature sensors and controllers.

The National Semiconductor LM135, LM235, LM335 series is typical of the monolithic temperature sensors on the market. This interesting device appears to the

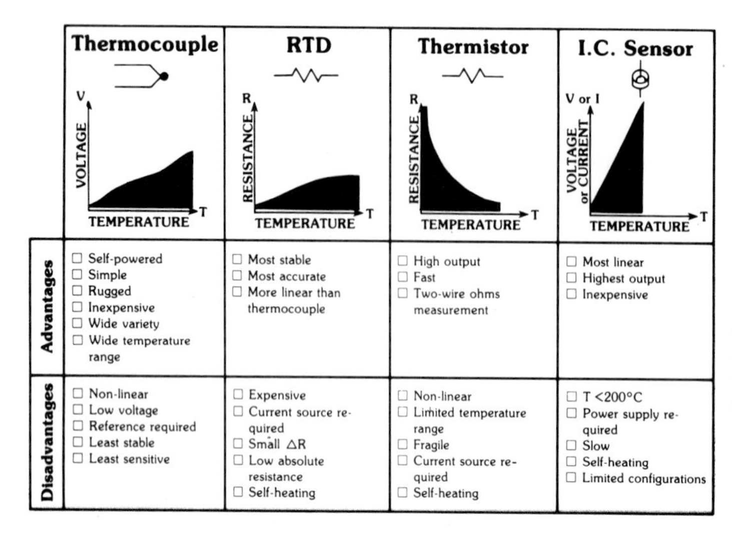

| | Thermocouple | RTD | Thermistor | I.C. Sensor |
|---|---|---|---|---|
| | V / VOLTAGE — TEMPERATURE — T | R / RESISTANCE — TEMPERATURE — T | R / RESISTANCE — TEMPERATURE — T | V or I / VOLTAGE or CURRENT — TEMPERATURE — T |
| Advantages | ☐ Self-powered<br>☐ Simple<br>☐ Rugged<br>☐ Inexpensive<br>☐ Wide variety<br>☐ Wide temperature range | ☐ Most stable<br>☐ Most accurate<br>☐ More linear than thermocouple | ☐ High output<br>☐ Fast<br>☐ Two-wire ohms measurement | ☐ Most linear<br>☐ Highest output<br>☐ Inexpensive |
| Disadvantages | ☐ Non-linear<br>☐ Low voltage<br>☐ Reference required<br>☐ Least stable<br>☐ Least sensitive | ☐ Expensive<br>☐ Current source required<br>☐ Small ΔR<br>☐ Low absolute resistance<br>☐ Self-heating | ☐ Non-linear<br>☐ Limited temperature range<br>☐ Fragile<br>☐ Current source required<br>☐ Self-heating | ☐ T <200°C<br>☐ Power supply required<br>☐ Slow<br>☐ Self-heating<br>☐ Limited configurations |

**Figure 2.9a** Comparison chart for the different types of temperature transducers *(Courtesy of Omega Engineering)*

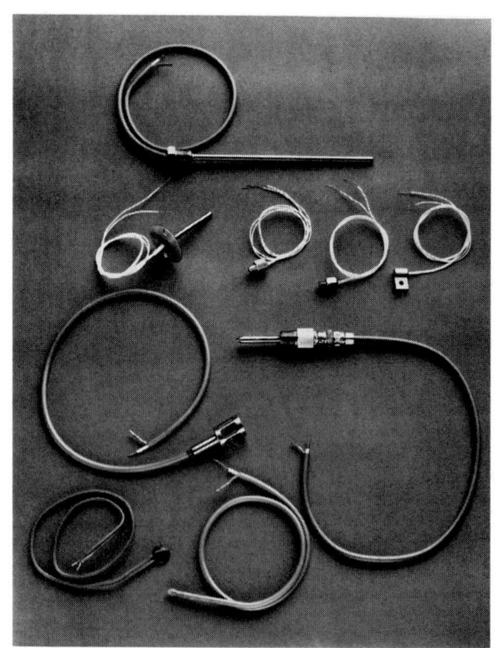

**Figure 2.9b** Industrial thermistors *(Courtesy of Omega Engineering)*

circuit as a Zener diode whose breakdown voltage is proportional to temperature. The output voltage of the LM135 is 10 mV/degree K, and it is extremely linear over its entire range of measurement temperatures. Since 0 K is $-273\,^\circ C$, the output of the LM135 at room temperature (25 °C) would be:

$$(25\,^\circ C + 273\,^\circ C)\,(10\text{ mV}) = 2.98\text{ V}$$

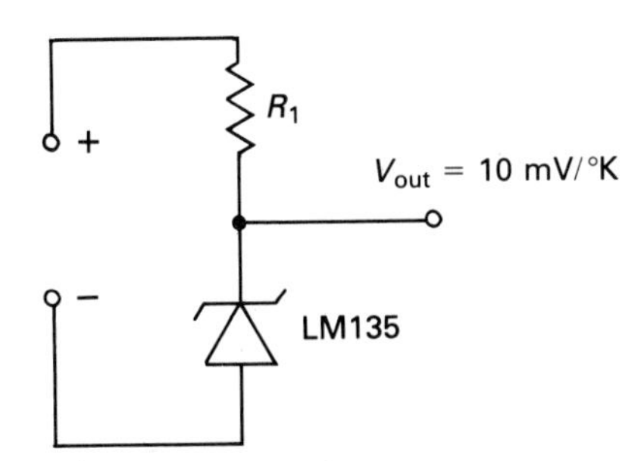


(a) The basic LM135 circuit.
The IC appears to the circuit as a temperature-sensitive Zener diode.

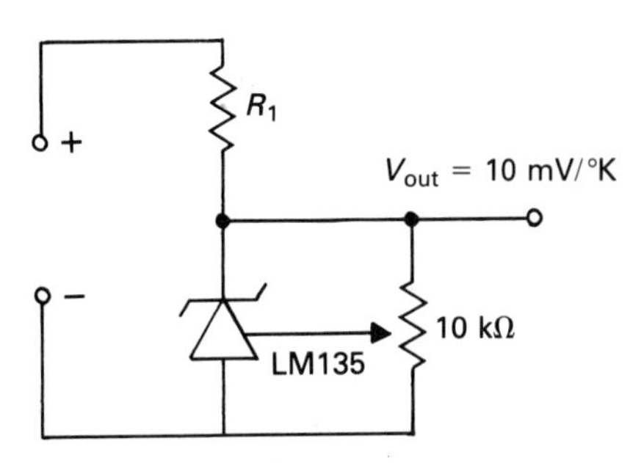


(b) The LM135 as a calibrated sensor.
Adjust potentiometer for $V_{out}$ = 2.982 V at 25°C. Calibration is true for all temperatures.

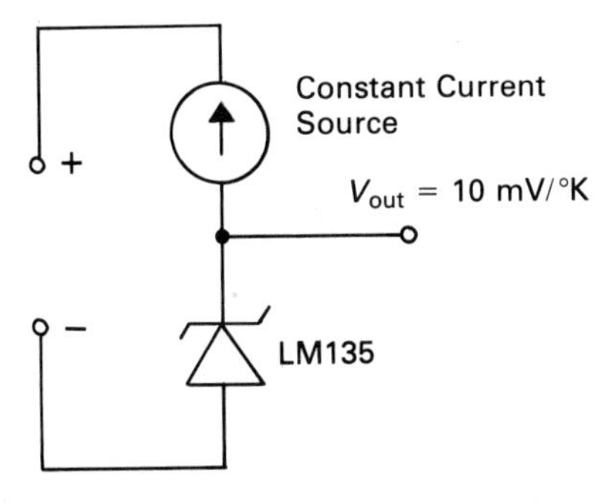


(c) The LM135 powered by a constant current source. May be mounted remotely if IR drop in line is accounted for.

**Figure 2.10** The LM135 Series IC in three different circuit configurations

A third lead is available on the device for calibration. When it is calibrated to read correctly at one temperature, it will read correctly at all temperatures.

Figure 2.10 shows the LM135 series IC in three different circuit configurations. The circuit in Figure 2.10a is the basic circuit. Resistor $R_1$ must be sufficiently large to appear as a current source to the circuit, and $V_{CC}$ must be kept small enough to limit the current through the IC to less than 5 mA. The operating current may be as low as 400 microamps. It should be kept as low as possible to reduce self-heating. The circuit in Figure 2.10b illustrates the use of the calibration adjustment terminal. When the potentiometer is adjusted for the correct output voltage at a known calibration temperature, the temperature measurement will be correct over the entire operating range of the device. Figure 2.10c illustrates a technique for remote temperature sensing with the LM135 using a constant current source.

The LM3911 is an IC that includes both a temperature sensor and an operational amplifier on one chip. Like the LM135, the temperature sensor has a 10 mV per degree

Kelvin output. This output can be coupled to the op amp to serve as either a measuring device or a full-featured temperature controller, depending on the circuit configuration.

Like the thermistor, monolithic temperature sensors may be damaged by continued use near or above their upper limits. The LM135 has a useful range between −55 °C and +150 °C. The manufacturer states that continued use above 150 °C will shorten the life expectancy of the device. For applications within these temperature limits, however, monolithic temperature sensors seem to be the best choice. One obvious application, which has already been mentioned, is to provide an electronic-ice point for compensating thermocouples.

## SUMMARY

In this chapter, we have examined different devices used for measuring temperature.

1. The bimetal element is rugged and reliable, but its accuracy and repeatability are not good. It is useful primarily as a binary device for detecting a specific temperature, rather than for reporting intermediate temperatures.
2. The thermocouple element provides a means for accurately reporting temperatures over a broad range. The Seebeck effect allows the thermocouple to convert temperature directly into current or voltage. Thermocouples are reliable and fairly rugged, but they are slightly nonlinear.
3. The RTD is a very linear, temperature-sensitive resistance element. Platinum RTDs offer high repeatability and excellent accuracy, but they require very accurate measuring techniques.
4. The thermistor is a semiconductor version of an RTD. Thermistors have a much higher temperature coefficient than metal RTDs, which makes them much more sensitive. They are, however, only useful over a narrow range of temperatures, are fairly fragile, and are highly nonlinear.
5. Monolithic temperature sensors are very linear, semiconductor, temperature-measuring devices. Like thermistors, they are useful over a narrow range of temperatures and are somewhat fragile.

## SELF-TEST

1. A bimetal temperature sensor works because different metals ___________ at different rates when they are heated.
2. A long bimetal element is ________________________ sensitive than a short bimetal element.
3. Bimetal elements are rugged and reliable, but they have poor ___________.

4. When wires made of different types of metal are joined at their ends, an electrical current will flow when the temperatures at the two ends are ______.
5. This phenomenon is called the ____________________ effect.
6. When one end of the loop is open, we can measure a(n) ____________.
7. The temperature measuring device that uses the Seebeck effect is called a(n) ____________________.
8. When using a thermocouple, the connections made to the voltmeter should be made on a(n) ____________________ block to prevent the generation of undesired Seebeck voltages where the thermocouple wires make connections with the voltmeter wires.
9. For accurate temperature measurement with a thermocouple, we must know the temperature at one end of the thermocouple loop. This is the ______ ____________ temperature.
10. In a laboratory a(n) ____________________ is used as a reference temperature.
11. In an industrial environment, the ice bath is not practical. One approach is to measure the temperature of the isothermal connecting block with a thermistor or monolithic temperature sensor and calculate the measured temperature in a computer program. This is the ____________________ approach.
12. An electronic circuit may also be used to electronically add a compensating current in series with the Seebeck current of the thermocouple junction. This circuit is called ____________________.
13. Metals have a(n) ____________________ temperature coefficient.
14. A temperature sensing device that uses this positive temperature coefficient is called a(n) ____________________.
15. One of the most commonly used metals for RTDs is ____________.
16. The temperature coefficient of platinum is 0.00385. If a particular platinum RTD has a resistance of 120 ohms at 25° Celsius, what is its resistance at 100° Celsius? At 0° Celsius? ____________________
17. Semiconductor materials have a(n) ____________________ temperature coefficient.
18. The ____________________ uses this negative coefficient for sensing temperature.
19. A particular thermistor has a temperature coefficient of 4%. At 100° Celsius, it has a resistance of 2000 ohms. What is its resistance at 50° Celsius? At 0° Celsius? ____________________
20. An LM135 monolithic temperature sensor has been calibrated for 10 mV/degree Kelvin. What is the voltage output at 25° Celsius? ____________

## QUESTIONS/PROBLEMS

1. You have been assigned the task of designing a temperature controller for an environmentally controlled room. The temperature in the room is to be 72 °F, within +/−2 °F. After considering the needs of the task, you have elected to use an LM135 monolithic temperature sensor. The output of the sensor at 72 °F will be ______________________________.

# 3

# Sensing Position

Robots, milling machines, drill presses, and palletizers all share one common need — they must know where the workpiece is located. Position sensors provide the feedback to systems controllers that require this vital information. In this chapter, we will explore some of the many devices used for position sensing.

## OBJECTIVES

You will have successfully completed this chapter when you can:

1. Explain the importance of accurate position sensing in automation.
2. Explain the use and adjustment of limit switches.
3. Explain the operating principles of photoelectric position sensors.
4. Explain the operating principles of magnetic proximity detectors.
5. Explain the operating principles of the linear variable differential transformer.
6. Explain the operating principles of absolute position encoders.
7. Explain the techniques used in encoding angular or rotary position.

## 3.1 AN OVERVIEW OF POSITION SENSING

Position sensing covers a wide variety of variables, ranging from simply sensing the presence of a workpiece to locating its position with absolute accuracy, from measuring the precise depth of a drill bit to measuring the exact angle of rotation of a shaft. Position sensors themselves show the same wide variety, ranging from simple switches to precision potentiometers to variable transformers. The accuracy of the position sensing equipment determines the precision of the finished work.

The natural urge when developing position sensing circuits is to strive for maximum accuracy at the cost of complicated circuitry. This conflicts with the need for simplicity, which increases reliability. Clearly, we must study the compromises in order to make an intelligent decision where position sensing is concerned.

## 3.2 LIMIT SWITCHES, DETECTING PRESENCE

The simplest and most familiar form of position detection is the limit switch. Limit switches keep your garage door opener from running the door off its track, detect the presence of a write protect notch in floppy disks, and ring the bells in schoolhouses across the country. They are so ubiquitous that they are almost invisible. It is a limit switch, for example, that turns on the dome light when your car door opens. Another sounds a buzzer to remind you to fasten your seat belt, and still another may remind you not to leave your key in the ignition.

In Figure 3.1, we can see several different varieties of limit switches. Although they vary in size and in the means of actuation, they are all essentially the same basic thing: a momentary contact push-button. The switch mechanism in most of the switches shown in Figure 3.1 is a snap-action spring-loaded switch. The mechanism assures position action with a quick switching time by causing the switch to snap closed once the actuating lever has reached a given point. The resulting switch closure has fewer bounces than would a simple push button, and protects the switch contacts from arcing.

There are two different ways to adjust the lever-actuated snap-action limit switch. When limited accuracy is acceptable, they may be adjusted by making careful bending adjustments to the lever. Care must be exercised, however, to assure that the workpiece will not bend the lever on contact, which would ruin the adjustment. The more precise method allows the position of the switch itself to be adjusted. In these cases, the switch housing has oblong screw holes, allowing minute adjustments to be made. In some cases, the switch is mounted to a metal plate, which can itself be adjusted for positioning the switch and its actuator.

Limit switches are available in many contact styles, but the most common is the standard SPDT arrangement. One terminal of the switch is marked *C* or *COM* for common, another is marked *NO* for normally open, and the third is marked *NC* for normally closed. The *normal* mode of the switch is not actuated — with no pressure on the actuating lever or button.

Limit switches are used to detect the presence of a workpiece on an assembly line, allowing automated machinery to work on it. For example, a limit switch may

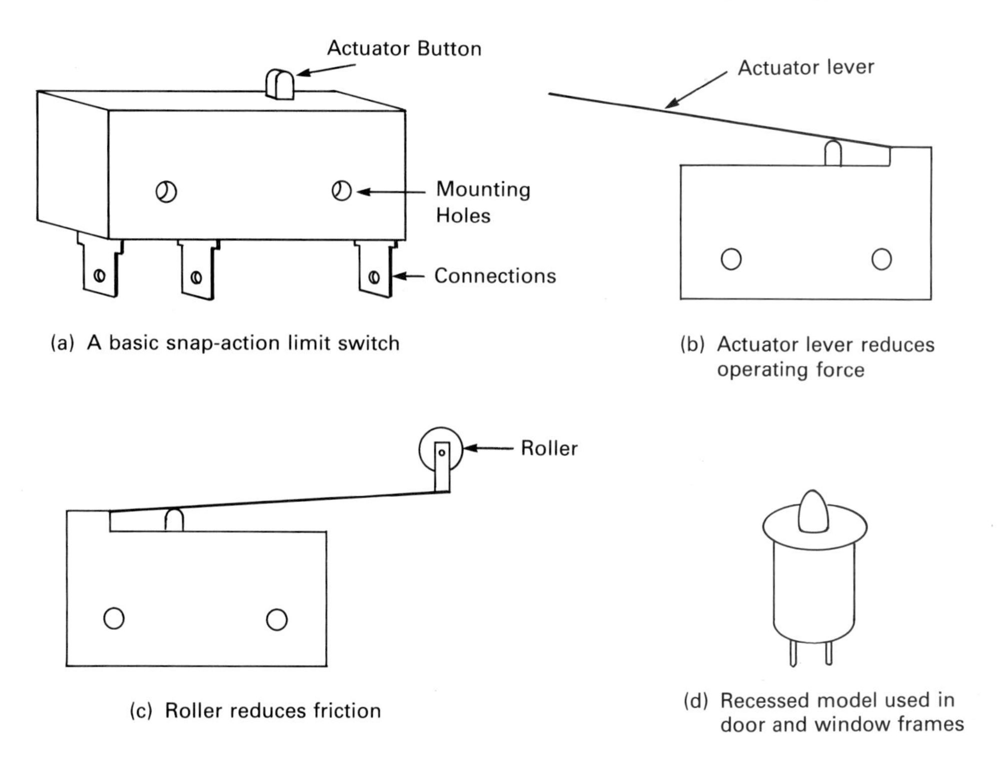


**Figure 3.1** Limit switch variations

determine that a refrigerator door has reached a particular spot on the assembly line. The switch closure signals a painter robot to begin painting the door. Because the line moves at a constant speed, the robot can follow a *blind path* in order to complete the painting with no additional sensors required. By including the limit switch in the system, it is possible to have the workpieces spaced irregularly on the line, without having the robot paint the air when it expects a door but doesn't get one.

In the office, limit switches may control the operation of the copy machine. The drum controller illustrated in Figure 3.2 shows an example of how precise timing can be achieved in such machines. The drum is rotated by a synchronous clock motor running at low RPMs. The limit switch leaf springs, which ride on the drum, are in their actuated position. When the leaf actuator of the switch encounters a notch, the NO contacts will open and the NC contacts will close. At the end of the notch, the switch will be actuated again by the drum, providing a time-controlled sequence of activities in the machine. The popular timer switches that are used to control your house lights, your water heater, and the automatic lighting of illuminated signs operate on a similar concept. The switches, however, are closed by a protrusion on the timer disk, rather than opened by a notch.

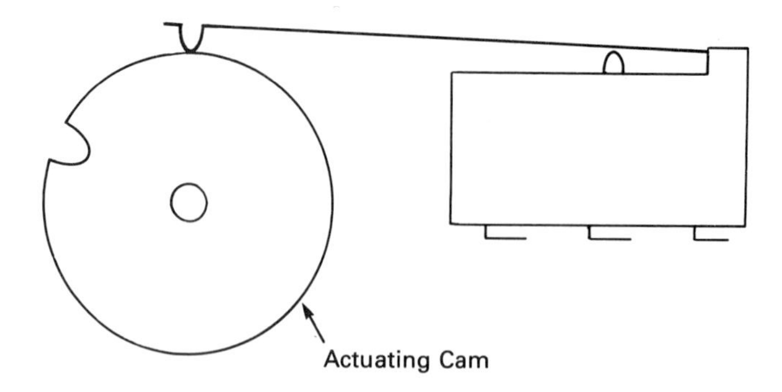


**Figure 3.2** A basic drum controller. Large drum controllers may have 20 actuating cams and switches, each individually adjustable.

## 3.3 PHOTOELECTRIC LIMIT SWITCHES

Another form of limit switch utilizes photosensitive cells. There are many different types of photosensitive devices available to industry.

In Figure 3.3, we can see some of the photosensitive devices that are used in modern industry. Some of the devices detect transmitted light. An opaque solid must pass between the photocell and the light source to break the light beam. Other devices rely on reflected light. They are used to detect the presence of light-colored and reflective materials.

Figure 3.4 illustrates the three photosensors most used in industry today. The phototransistor, Figure 3.4a, is a very popular device. Phototransistors are also available as photo Darlingtons, Figure 3.4b, which increase the sensitivity of the circuit. The phototransistor operates in the cutoff region of its characteristic curve with the base dark. When light strikes the base, the transistor turns on and collector current flows in proportion to the amount of light striking the base. For use as a limit switch, the light level is made bright enough that the transistor will go from cutoff to saturation. Some phototransistors have external base connections that can be voltage-divider biased to a level just below turn-on, which increases their sensitivity.

Phototransistors provide gain as well as light sensitivity, but the phenomenon of *charge time* slows their reaction and limits the useful frequency at which they can be actuated. Charge time is the result of the time the current carriers in the base of the transistor take to complete their journey through the base after the transistor is switched off. It is the cause of propagation delay in TTL logic and it reduces the on-off switching time of phototransistors. This is rarely a problem when we are dealing with

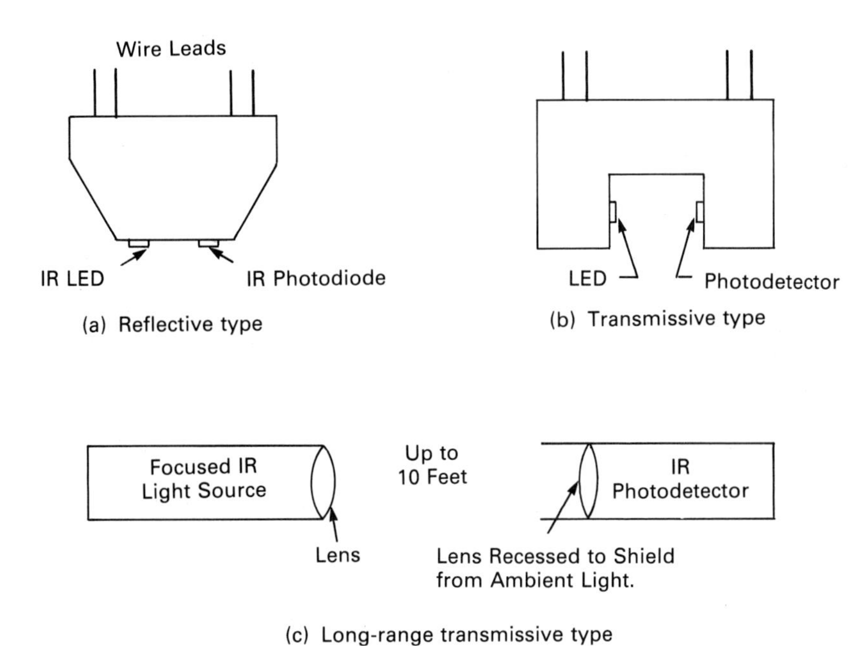


**Figure 3.3** Common photodetectors

mechanical phenomena, but it must be considered when high-speed counting is needed — as in the use of optical discs for motor speed control.

Photodiodes, Figure 3.4c, provide the same reliable response as phototransistors, but they can operate at higher speeds. They do not offer amplification. They are often connected to switching transistors to increase their sensitivity. Photodiodes can be made with thinner layers of semiconductor materials than phototransistors; therefore, they can switch off faster because their charge time is less. Photodiodes are wired into a circuit in reverse bias, with the cathode connected to the positive power supply terminal and the anode connected to the negative power supply terminal. When light strikes the junction, the reverse breakdown voltage of the diode is reduced to the point where the diode conducts in reverse, signaling the presence of light.

The cadmium sulfide, or CdS, photoresistor, Figure 3.4d, offers yet another source for detecting and measuring light. Many of today's automatic cameras make use of CdS cells. The resistance of the CdS cell varies, depending upon the amount of light striking the surface. By measuring the resistance of the cell, we can detect the presence or absence of light. The CdS cell, however, suffers from "memory." When the cell is exposed to very bright light, it becomes blinded for a period up to several seconds. During this blind time, it outputs a signal proportional to the light that blinded it. For this reason, we must be cautious in using CdS photocells.

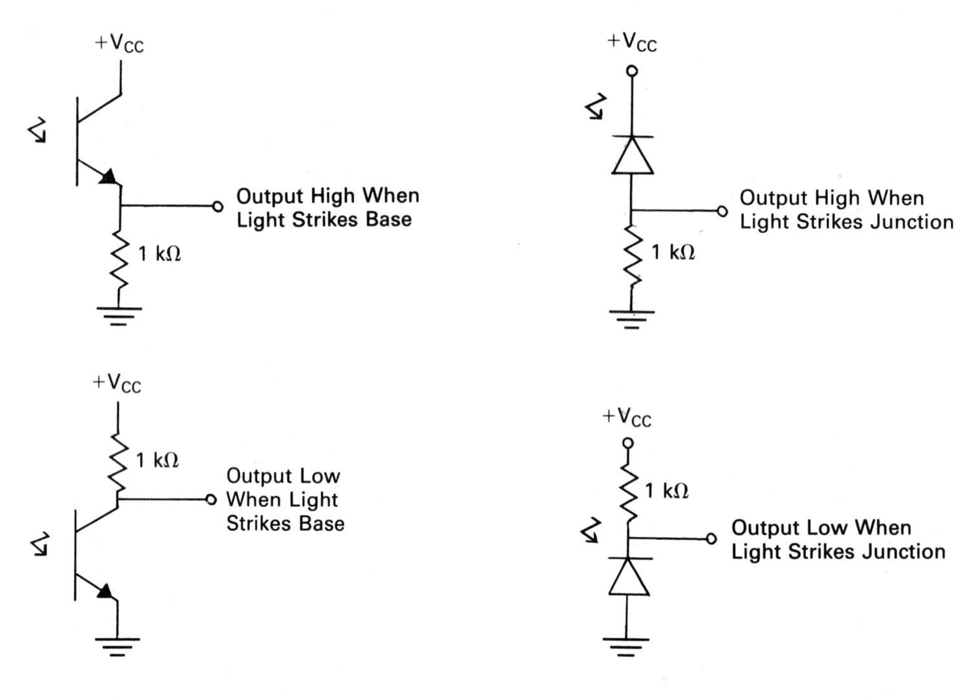


**Figure 3.4a** The phototransistor

**Figure 3.4c** The photodiode

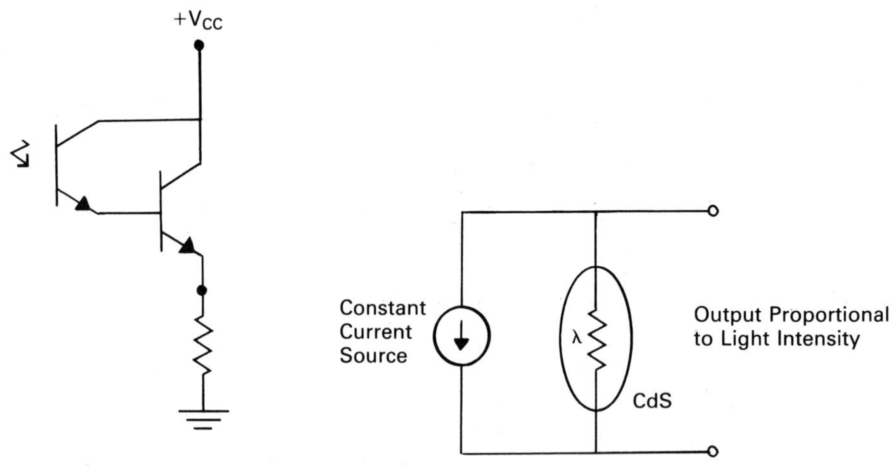


**Figure 3.4b** A photo Darlington increases sensitivity

**Figure 3.4d** The Cadmium sulphide (CdS) photoresistor

Various photovoltaic devices are also finding some industrial use. These photocells convert light directly into electricity. They are most often found in circuitry that measures and controls light intensity, rather than in limit switch applications.

Another consideration when using photodetectors is that when they are interfaced with digital circuits, it becomes almost mandatory to use Schmitt trigger circuits to "clean up" their output. Otherwise, the analog output of a photodetector may hover just at the level needed to allow the digital circuit to *race*. Racing occurs when a digital circuit is presented with a signal that is just at the threshold level of a logic high or a logic low. Rather than counting one pulse under these conditions, the circuit oscillates.

Photodetectors, however, offer a significant improvement over mechanical switches in certain applications. There are no switch leaves or rollers to adjust or wear out, there are no contacts to bounce, arc, or burn, or get dirty and corroded, and they offer much higher operating speeds. They do need an unobstructed light path, and the lenses on the lamp and the photocell need to be kept clean. It is also important to shield the photodetector from ambient light levels, but this is rarely a problem.

## 3.4 MAGNETIC DETECTION OF POSITION

The magnetic properties of ferrous metals can also be used in position sensing. The magnetic reed switch is made up of two magnetic switch contacts sealed in a glass tube. When a magnetic field is present, the magnetic reeds are attracted to each other and close the switch contact. Ferrous metals will "conduct" the lines of force from a magnet to the reed switch, causing the switch to close when the metal is in position. Figure 3.5 illustrates this principle.

Because the switch contacts are sealed in the tube, reed switches are not sensitive to environmental dust, oil, or corrosive atmospheres. Some precautions are necessary

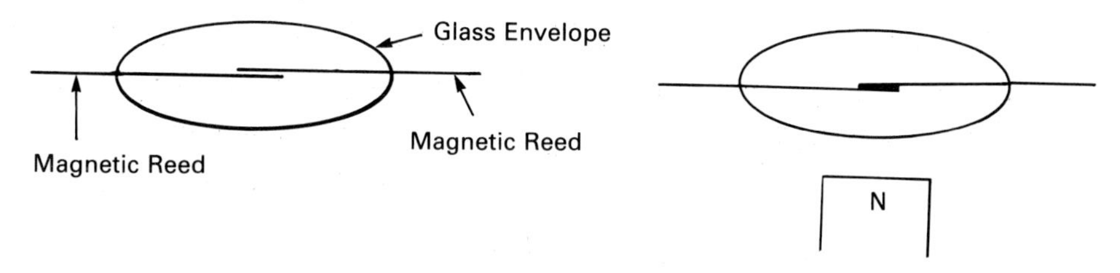


**Figure 3.5** The magnetic reed switch

when installing them, however. The seals at the ends of the glass tube are rather fragile: when bending, cutting, or soldering these leads, it is important to shield the seals from shock and stress. Using a pair of long nosed pliers or hemostats, grasp the leads next to the seals. Apply bending or cutting force or soldering heat beyond this protective grip.

**Figure 3.6a** The Maxi-Mag magnetic motion detector. Available in many sizes and styles for almost any conceivable task. *(Courtesy of Electro Corporation)*

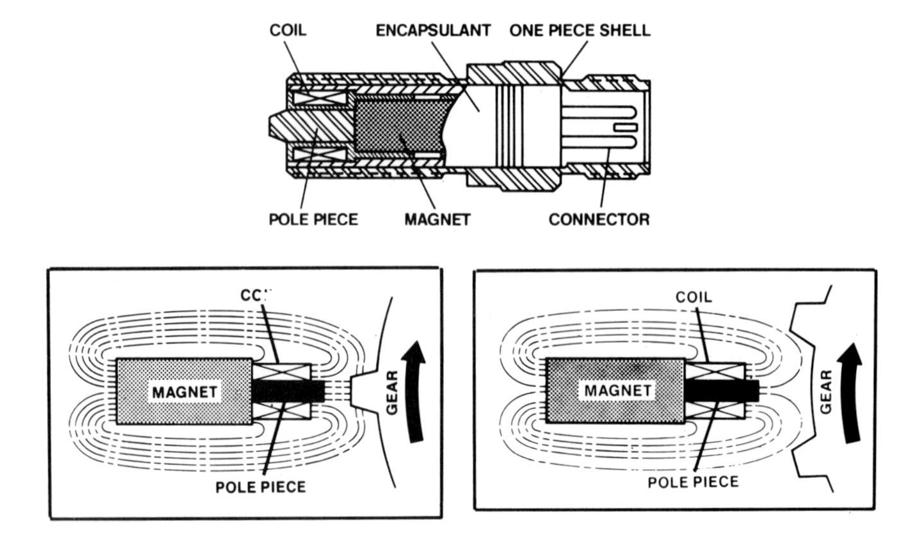


**Figure 3.6b** The operating principles of the Maxi-Mag motion detector. *(Courtesy of Electro Corporation)*

The magnetic motion detector shown in Figure 3.6a can detect moving ferrous metals. The operating principles of this device are shown in Figure 3.6b. The permanent magnet provides a magnetic field through the coil. The magnetic field is changed by the passing of a discontinuity — for example, a gear tooth or the edge of a metal plate. This change in the magnetic field induces a current in the coil. The size of the current depends on the speed of the workpiece as it passes. This device outputs a sine wave when excited by a conventional gear. The frequency of the sine wave is directly proportional to the speed of the gear and its *pitch* (the number of teeth on the gear). In that application, it is used in motor speed sensing and control circuits. It is limited by its ability to detect only discontinuity, but within these limits, its lack of moving parts and its overall reliability make it an excellent choice in many applications.

## 3.5 INDUCTIVE PROXIMITY DETECTORS

Inductive proximity sensors offer a noncontact proximity detector for all conductive materials. They operate by projecting a narrow high-frequency field (150–250 kHz) in front of the detector. When a conductive material is present, the eddy currents induced by this field load down the oscillator, either reducing the oscillator level to nearly zero or to about half of the unloaded level, depending on the type of detection. The *killed oscillator* type of detector is more common, but its maximum operating speed is reduced because the oscillator has to restart each time. The detection of eddy currents gives these devices a second name: eddy current detectors. Eddy current detectors are available in a wide variety of shapes and sizes for use in almost any imaginable job. Several of these are illustrated in Figure 3.7.

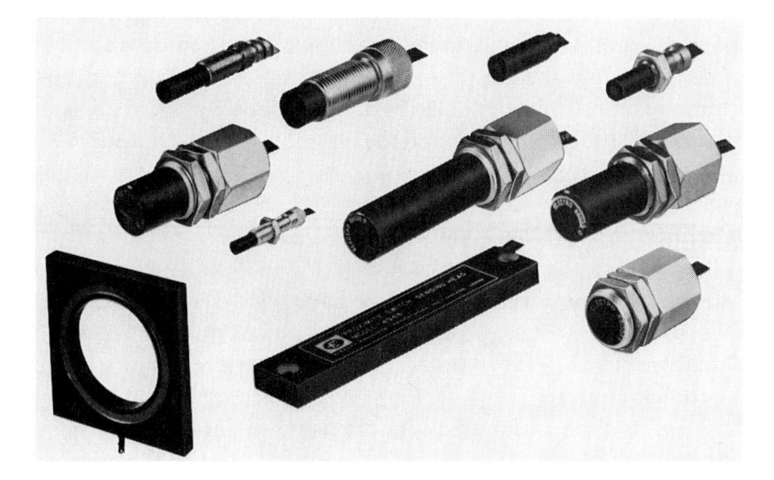

**Figure 3.7** Some examples of inductive proximity detectors. Material must pass through the circular model in the lower left, used for measuring diameters. The threaded models provide easy and almost infinite fine adjustments. *(Courtesy of Electro Corporation)*

## 3.6 ABSOLUTE POSITION DETECTORS

Most of the devices mentioned previously are primarily used to detect the presence or absence of a workpiece or tool. Their application is limited to fixed positions and they must usually be rearranged, or *retooled*, to accommodate design changes or production variations. Absolute position detectors offer more flexibility by reporting the exact position of the tool or workpiece.

In Figure 3.8, we can see that many of the common electronic parts may be used as position detectors. In Figure 3.8a, a potentiometer provides resistive feedback to the controller. By measuring the resistance, we can determine the exact position of the work. Combining precision multiple-turn linear-taper pots and rack and pinion gearing provides very accurate position sensing.

In Figure 3.8b, we see a linear variable capacitor used in the same way, while in Figure 3.8c, a variable inductance is used to detect position. In both cases, the detector is used in the frequency-determining portion of an oscillator. A frequency counter can now report position with amazing accuracy.

Figure 3.9 illustrates the operating principles of a very interesting type of absolute position encoder, the linear variable differential transformer, or LVDT. The primary of the LVDT is powered by an ac excitation voltage, usually 12 V, 60 Hz, which is readily available from most controller circuits. When the movable core is exactly centered between the two identical secondaries, the output voltages of $S_1$ and $S_2$ are the same. These output voltages are usually rectified by precision rectifiers and combined at a summing point. The output of $S_1$ would be rectified to a positive output voltage, and the output of $S_2$ would be rectified to a negative output voltage. Thus, when $V_{S_1} = V_{S_2}$, the output of the summing point is 0 volts.

When the movable core is to the left of center, $S_1$ has the higher output and the summing point reports with a positive output voltage. When the movable core is to the right of center, $S_2$ has the higher output and the summing point reports with a negative output voltage. Thus, the polarity of the output voltage reports whether the tool is right of center or left of center, and the magnitude of the output voltage reports exactly how far left or right it is.

## 3.7 SHAFT ENCODERS: REPORTING ANGULAR POSITION

All of the absolute position encoders we have discussed are available in designs that can report angular or rotary position. Rotary potentiometers report the position of small robotic arms with great accuracy. Select a linear-taper pot for the job to avoid calculation problems.

When measuring the position of a continuously rotating shaft, an ordinary potentiometer fails because there is usually a dead spot as it passes through its extremes. While special premium pots that can help avoid the difficulty are available, optical discs and resolvers provide more popular means of measuring this type of rotation.

An incremental encoding optical disc consists of a disc made of glass or plastic

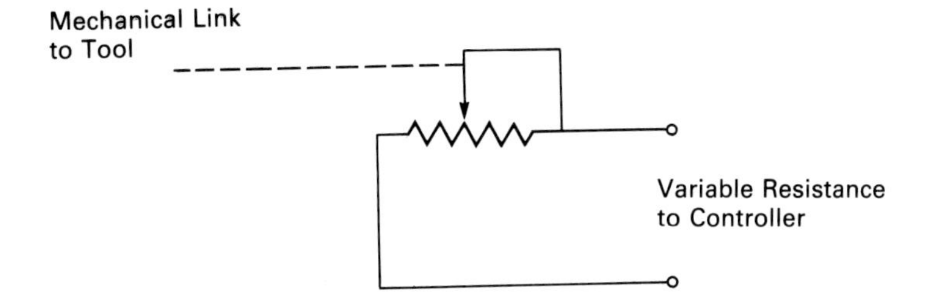


**Figure 3.8a** Resistive position sensor

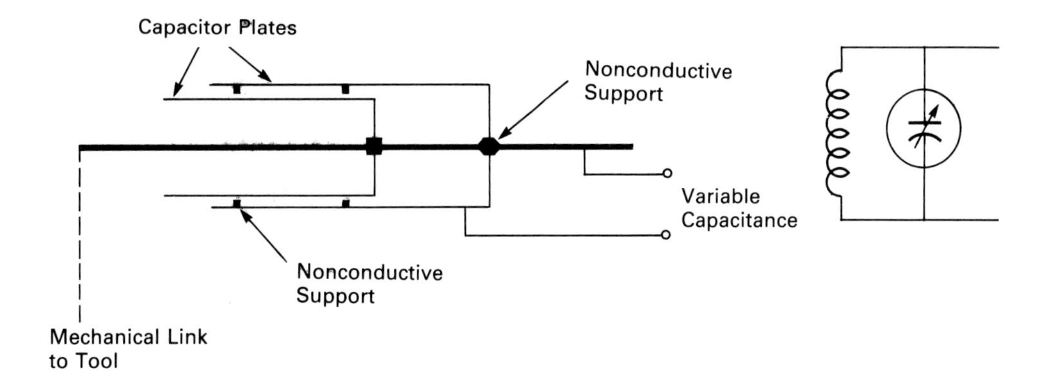


**Figure 3.8b** Capacitive position sensor

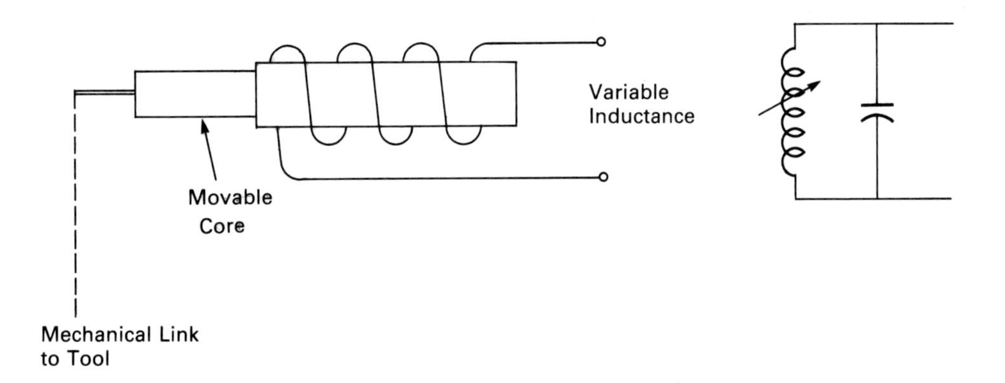


**Figure 3.8c** Variable inductance position sensor

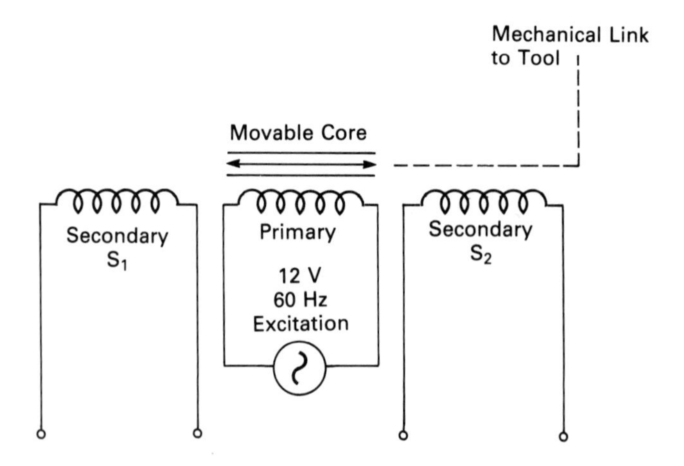


**Figure 3.9** The linear variable differential transformer, LVDT

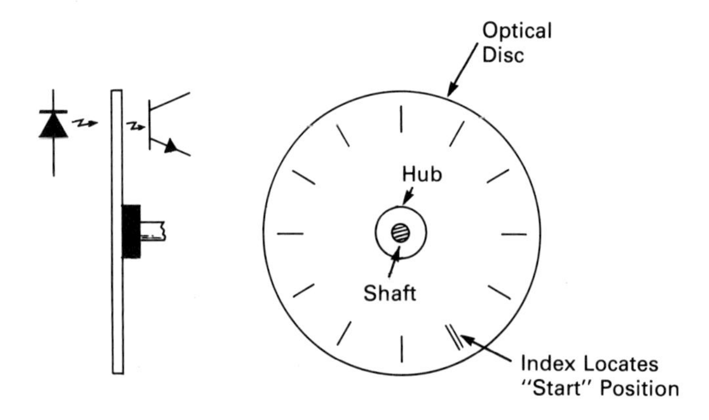


**Figure 3.10a** Incremental optical shaft encoder

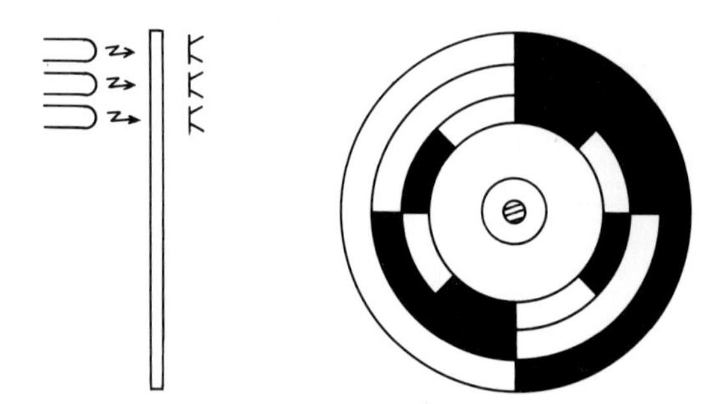

**Figure 3.10b** A three-bit optical position encoder

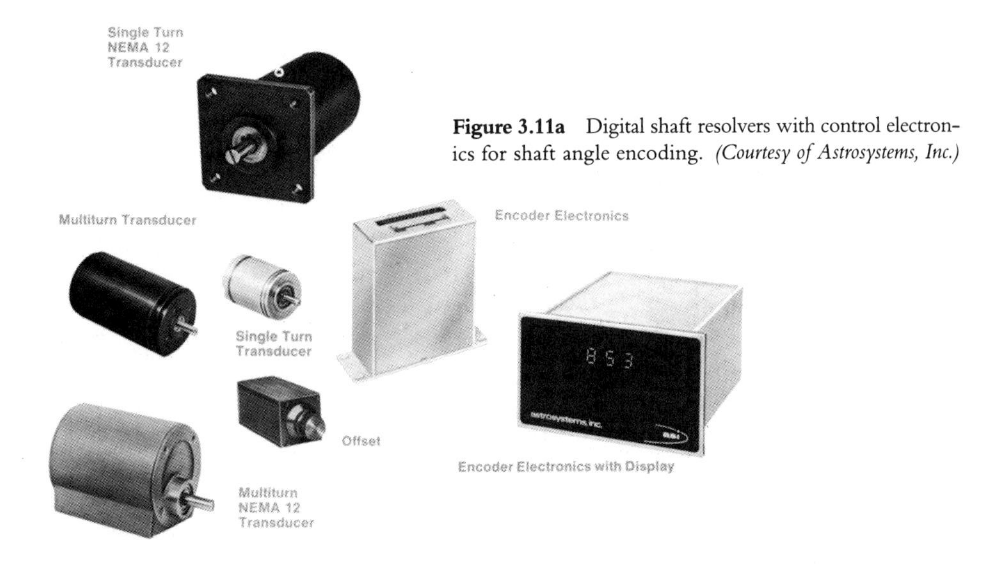


**Figure 3.11a** Digital shaft resolvers with control electronics for shaft angle encoding. *(Courtesy of Astrosystems, Inc.)*

**Figure 3.11b** The Durapot "wiperless potentiometer" resolver provides noise-free analog dc output voltage, which is proportional to the shaft angle. *(Courtesy of Astrosystems, Inc.)*

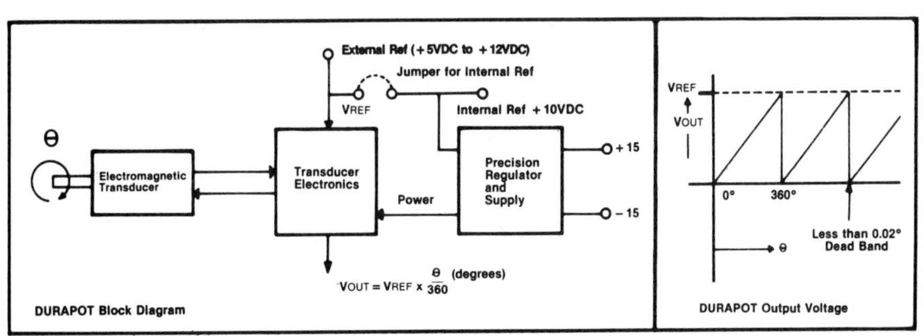


**Figure 3.11c** Block diagram of Durapot hookup illustrates the simplicity of the system. *(Courtesy of Astrosystems, Inc.)*

with very accurately etched lines spaced regularly about its circumference. A light source on one side and a photodiode on the other side complete the sensor. This is depicted in Figure 3.10a. Digital circuitry or computers keep track of the number of passing pulses and calculate the angular position with simple up-down counting. In fact, this same proven device is used for controlling the motor speed of data recorders that can record data at megabaud rates. A metal disc with carefully drilled holes can be substituted for the glass disc if less resolution can be accepted.

Another optical approach uses a disc encoded with concentric rings of alternately transparent and opaque segments. This is illustrated in Figure 3.10b. A 12-bit absolute optical encoder can provide positional feedback to one part in 4096, or about 0.09° (365/4096). This same approach may also be used to report linear position. Unfortunately, this highly accurate approach requires 13 wires between the light sensors and the digital circuitry for 12-bit accuracy. In addition, we need extra wiring for the light source. Optical encoders also suffer from environmental problems with dirt, oil mists, and lubricants, all of which are present in the industrial environment.

A resolver is essentially a rotary version of the LVDT. The primary of the transformer is wound in the case of the resolver and the single secondary is wound on the shaft. As the secondary rotates, the position of the secondary changes relative to the primary, and the output varies as the sine of the shaft angle. Since each position of the shaft has a unique output, resolution of a resolver is limited only by the hardware or software used to calculate position. Resolvers usually require a six-wire cable, but the cabling requirements are much less stringent than for digital position encoders because they enjoy a far greater immunity to noise. Figure 3.11 illustrates several styles of resolvers.

## SUMMARY

In this chapter, we have seen that many of the common electrical and electronic components can be used for measuring or detecting position. We have also looked at a few of the devices that are specially designed for position measurement.

1. The limit switch is a simple push-button switch that is actuated by contact with the tool or workpiece. Its on-off binary output is useful for detecting presence.
2. The photosensitive detectors can provide noncontact detection of presence. They include:
   a. Phototransistors and photo Darlingtons, which include amplification as well as photodetection.
   b. Photodiodes, which operate on the same principles, but are useful at higher speeds.
   c. Photoresistors, which find only limited use in position detection.

3. Magnetic reed switches can provide useful detection of ferrous metals.
4. Magnetic proximity detectors can be used to detect a discontinuity in magnetic materials. These devices are often used to measure rotational speed, as well.
5. Eddy current detectors report the presence of conductive materials.
6. Absolute position encoders provide exact positioning information.
7. The LVDT can report very accurate linear position with its differential output.
8. Angular or rotary shaft position reporting can also be accomplished using conventional electronic components such as potentiometers, variable capacitors, or variable inductors.
9. Optical disc encoders offer an alternative angular encoder. They include:
   a. The incremental disc, which outputs pulses to up-down counters for position locating.
   b. The digital disc, which can report with high accuracy and doesn't forget its position after a power failure.
10. Resolvers work like rotary LVDTs and can give very accurate rotary position information.

## SELF-TEST

1. A limit switch is used to detect ______________________.
2. In its simplest form, a limit switch is a(n) ______________________.
3. The ______________________-activated limit switch is more sensitive than a button-activated limit switch.
4. The snap-action limit switch helps avoid problems with ______________.
5. A drum controller is used to ______________________ a series of events in time.
6. A phototransistor acts like a(n) ______________________ when light strikes the base.
7. Phototransistors operate in the ______________________ of their curve when the base is dark.
8. A photodiode is superior to a phototransistor in terms of ______________.
9. When photosensors are interfaced to TTL logic, they must be coupled through a ______________________ to clean up the signals.
10. A(n) ______________________ detects the presence of a magnetic field with a switch closure.
11. A magnetic motion detector is sensitive to ______________________ in the magnetic field.

12. Magnetic motion detectors are particularly useful for reporting ________ as a frequency.
13. The inductive proximity detector generates a high-frequency field that creates ____________________________ in any conductive material within the field.
14. When a conductive surface passes near an inductive proximity detector, the oscillator level ____________________________.
15. A(n) ____________________________ detects the exact position of a tool, rather than its presence.
16. A(n) ____________________________ reports tool position to the controller as a variable resistance.
17. Both the ________________ and the ________________ report position as a frequency.
18. A linear variable differential transformer reports position as the difference of two ____________________________.
19. The rotary position of a shaft is also called its ________________________ position.
20. The optical disc that has precisely spaced lines around its circumference is a(n) ____________________________ encoder.
21. The digital optical encoder can provide great accuracy, but at the cost of ____________________________.
22. A(n) ____________________________________ is an electromagnetic shaft angle encoder.

# 4

# Measuring Strain

When we think of the word *strain*, we tend to think of the effort we put forth when we lift something heavy. Certainly, the strained back is one of modern man's most common ailments.

In a way, that first thought can stay with us, for mechanical strain is directly related to force. The same device that you use to measure weight in one application will be used by someone else to measure vibration.

The modern world demands energy efficiency, which means that machinery must be light in weight. The same world also demands product liability, which means that the lightweight machinery must do no harm. Because of these two contradictory demands, product engineers must know more about what is happening in the parts of a machine than ever before. By measuring the strain on the machine parts, the engineer can calculate the stresses within the parts and predict reliability, based on the properties of the material.

This is possibly one of the biggest uses of strain measurement today, but it is by no means the only one. We also use strain measurement to measure weight and vibration. In this chapter, we will examine the devices used for these important measurements.

## OBJECTIVES

You will have successfully completed this chapter when you can:

1. Define strain.
2. Describe the unbonded strain gage and explain its general principles of operation.
3. Describe the bonded foil strain gage and explain its general principles of operation.
4. Describe the semiconductor strain gage and explain its principles of operation.

## 4.1 STRAIN

When force is applied to a body, the body deforms. The deformation of the body is called *strain*.

In practical terms, we will speak here primarily of the change in the length of the body, or the fractional change per unit of length. Figure 4.1a illustrates this idea. The body is shown in solid lines before the force is applied, and in dotted lines after the force is applied. Strain may be either tensile (positive) or compressive (negative). An example of compressive change is shown in Figure 4.1b.

We can determine the strain on the body in Figure 4.1a by taking two micrometer measurements, one before the force is applied and one after. The strain will then be

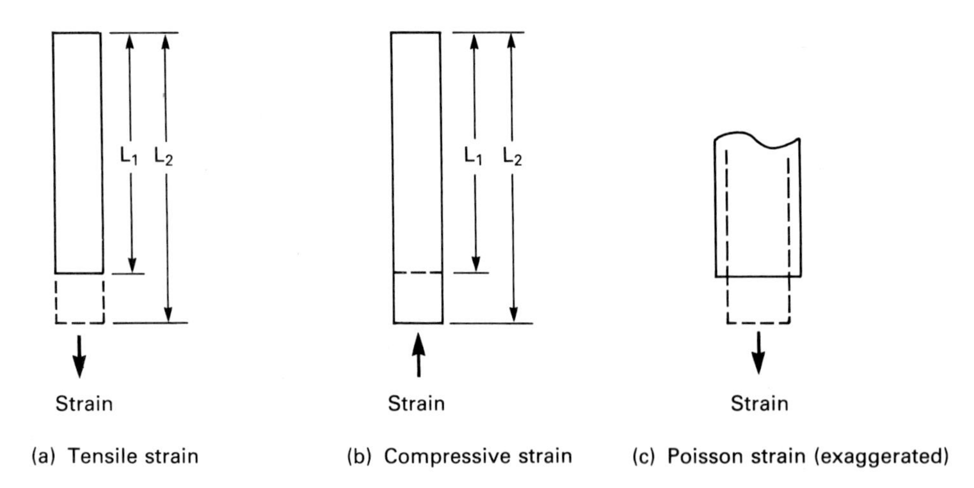


**Figure 4.1** The effects of strain. The percentage of strain may be calculated:

$$\frac{L_1 - L_2}{L_1} \times 100$$

expressed as a percentage of change in unit length. For example, if the steel bar in Figure 4.1a measured 4 inches long before the force was applied and 4.01 inches long after the force was applied, then the percentage of change was

$$\frac{0.01''}{4''} \times 100 = 0.25\%$$

The change in length in this example is typical of those found with most structural metals.

Another change occurs in materials that are put under strain this way. Figure 4.1c points out that when we stretch the metal bar, it not only gets longer, but its width decreases. If you've ever pulled taffy, you have seen this phenomenon. It is called the Poisson strain.

## 4.2 THE STRAIN GAGE

Of course, measuring such changes with a micrometer in operating machinery is not very practical. The solution is the strain gage, which converts these strain changes into an electrical parameter that can be monitored by a meter movement or sampled by a computer.

There are three common types of strain gages in current use:

- the unbonded strain gage
- the bonded foil strain gage
- the semiconductor strain gage

Of these, the first two work on the same principle: when a conductor is stretched, its length increases and its diameter or width decreases, resulting in an increase in its resistance. By measuring the change in resistance, we can determine how far the wire was stretched. If the wire is in some way attached to the source of strain we want to measure, the wire will undergo the same strain as the member.

## 4.3 THE UNBONDED STRAIN GAGE

The unbonded strain gage is used primarily in *load cells*, as shown in Figure 4.2. The wire is strung tightly around the support posts so that strain in the support bar is transferred to the wire. Notice that the wire runs are more or less parallel to the direction of the expected strain. Strain in the direction at right angles to the expected strain will be ignored by the strain gage wires.

Load cells are used to measure weight. If the load cell is used to suspend the weight, the strain will be tensile, or positive, indicated by an increase in resistance. If the load cell is used to support the weight, the strain will be compressive, or negative, as indicated by a decrease in resistance. That is why the wire is wound tightly onto the load

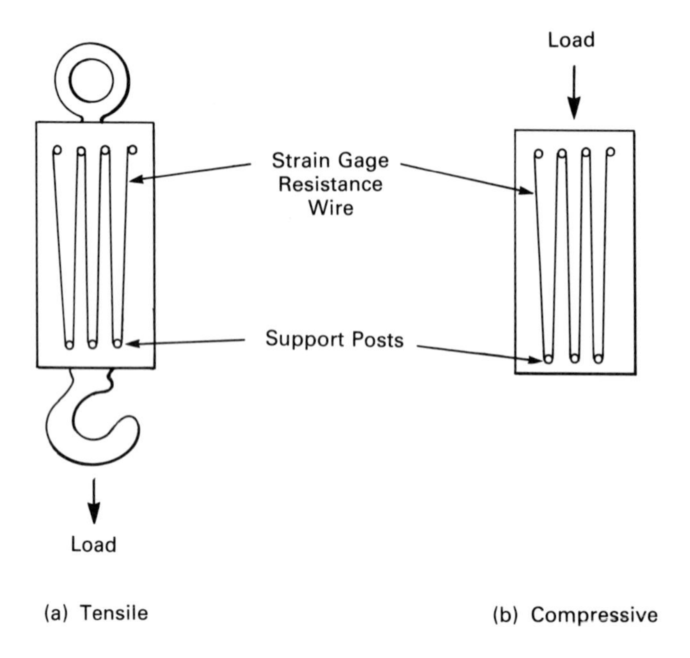


**Figure 4.2** Load cells — an example of the unbonded strain gage.

cell body, under constant and predetermined strain. It is not practical for a technician to rewire a load cell in the field, unless there is special equipment available.

## 4.4 THE BONDED STRAIN GAGE

The bonded strain gage is made of similar wire that is tightly bonded to a plastic or paper support. The support is then cemented to the member under strain, and it undergoes the same strain as the member. Until recently, the bonded strain gage was made of wire bonded to the support, but that approach has given way to the foil strain gage. Bonded foil strain gages are made with a printed circuit process, which uses conductive alloys that have been rolled into a thin foil and bonded to the backing. Constantan, a copper-nickel alloy, is one of the most commonly used materials. Figure 4.3 shows several different types of bonded foil gages.

The gages shown in Figures 4.3a, 4.3b, and 4.3c are single bonded strain gages. They are used in straightforward strain measurements. The gages shown in Figures 4.3g and 4.3h actually have two gages mounted at right angles to each other. The second gage is used for temperature compensation in the bridge circuits that are used

**Figure 4.3** Several different styles of the bonded foil strain gage.

to read the gage. The gage shown in Figure 4.3k is made up of four separate gages, which are ready to be wired into a bridge configuration.

The gages shown in Figures 4.3e, 4.3i, and 4.3j are specially designed for measuring strain on machine parts. For example, several gages, such as the one shown in Figure 4.3e, are used to monitor strain in jet aircraft engines. A small microcomputer stores samples of the strain data on cassette tape, and a master computer in the maintenance headquarters reads and analyzes the taped data and warns of impending failure long before it can happen.

Bonded foil strain gages are available in several standard resistance values. The most common is 120 ohms, followed by 350 ohms. Also available, but seen less often, are 600-ohm and 300-ohm bonded foil strain gages.

Bonded foil strain gages have several inherent problems. The resistance of the metal foil increases with temperature, and compensation must be provided. One solution to this problem was suggested in the description of the gages in Figures 4.3f and 4.3g. A second gage, which is used solely for temperature compensation, is mounted at right angles to the measuring gage.

Temperature also causes the gage and the member to which it is cemented to expand. Today, you can buy gages with a carefully controlled thermal expansion factor to match the material to which it will be cemented. You must be careful to attain this match to prevent the gage from giving false readings from thermal strain.

The cement used to fasten the gage to a machine member or load cell is a very important part of the accuracy of the gage. Only cement that has been approved by the gage manufacturer should be used. There are two types of cement in general use. *Cold setting cement* sets at room temperature. *Thermal setting cement* requires an elevated temperature to cause it to set. When conditions do not permit the use of cements, special gages that can be welded to the machine member are available.

Strain gages also need protection from moisture and physical damage. Special putties and polyurethane varnishes are supplied by all gage manufacturers for this purpose.

Another important problem with either bonded or unbonded wire strain gages and bonded foil strain gages is their small change in resistance when under strain. Recall that in the section on strain, we mentioned that a part under strain will most likely change only a fraction of one percent in length. A typical 120-ohm strain gage will have a change in resistance of less than one ohm under such conditions. The measuring circuits for such a small change must be very accurate indeed! Typically, strain gages are used in one of the many bridge configurations. We saw a few of these in Chapter 2. We will study bridge networks in detail in a later chapter.

## 4.5 THE SEMICONDUCTOR STRAIN GAGE

The semiconductor strain gage is much more sensitive than the foil strain gage. The semiconductor strain gage makes use of the phenomenon of *piezoresistance*, the change in resistance caused by a strain applied to the semiconductor. Generally speaking, a semiconductor strain gage is 100 times more sensitive than either a foil or a wire gage. On the other hand, semiconductor gages are more fragile and temperature sensitive than foil and wire gages. Semiconductor gages are currently used in special applications where they will be protected from shock and extremes of temperature. The piezoresistive strain gage is also used in many pressure measuring transducers. This application will be discussed in Chapter 5.

## 4.6 GAGE FACTOR

The *gage factor*, or GF, of a strain gage is the measure of resistance change with strain. Gage factor is an indication of the sensitivity of the gage.

$$\frac{\Delta R \div R}{\Delta L \div L}$$

In other words, it is the fractional change in resistance per the fractional change in length. For most bonded foil strain gages, the gage factor is approximately equal to two. Closer figures are supplied with each strain gage, because the gage factor varies from lot to lot as they are manufactured.

A gage factor of two means that if there is a 0.5% change in the length of a 120-ohm strain gage, its resistance will change by

$$2 \times 120 \times 0.005 = 1.2 \text{ ohms}$$

Such a small change obviously requires sensitive techniques to be measured.

The gage factor of a semiconductor strain gage may be 200. Then a change of 0.5% in length will cause a change of

$$200 \times 120 \times 0.005 = 120 \text{ ohms}$$

This is a 100% change in resistance. Under this strain, the gage will measure 240 ohms. The semiconductor strain gage is clearly much more sensitive.

## SUMMARY

In this chapter, we have seen the techniques used for measuring strain.

1. Strain is the change in length of a body when a force is applied to the body.
2. Strain that increases the length of a body is called *tensile strain*.
3. Strain that decreases the length of a body is called *compressive strain*.
4. A strain gage converts the change in a body's length into electrical signals.
5. When a conductor is stretched, its length increases, its diameter decreases, and its resistance increases as a result.
6. The unbonded strain gage is used mainly in load cells, which are used to measure weight.
7. The bonded foil strain gage is used in measuring strain.
8. Strain gages are also sensitive to temperature changes, and compensation must be included in the circuit to allow for temperature change.
9. A semiconductor strain gage is much more sensitive than a foil strain gage.
10. The gage factor is a measure of the sensitivity of a strain gage. Gage factor is the ratio of change in resistance to change in length.

## SELF-TEST

1. The deformation of a body by a force is called ______________________.
2. This deformation can be measured by a(n) ______________________.
3. When wire is stretched, its resistance will ______________________.
4. The unbonded strain gage is primarily used in measuring ____________.
5. Load cells are used for measuring ______________________.

6. The bonded foil strain gage must be ______________________________ to the member undergoing strain.
7. A typical bonded foil strain gage will increase its resistance about ________ ______________________________ with a 1% increase in length.
8. A semiconductor strain gage is ______________________________ sensitive than a bonded foil strain gage.
9. The semiconductor strain gage is ______________________________ fragile than a bonded foil strain gage.
10. A typical semiconductor strain gage will increase its resistance about _____ ______________________________ with a 1% increase in length.

## QUESTIONS/PROBLEMS

1. A 3-inch bonded foil strain gage is stretched by 0.5%. The nominal resistance of the gage is 120 ohms and the gage factor is 1.96. What is the resistance of the gage while it is being stretched?
2. A bonded foil strain gage is cemented to a bridge member. The nominal length of the gage is 2 inches. When a large truck crosses the bridge, the strain gage length becomes 2.02 inches. What is the resistance of the gage at this time if the gage factor is 2?

# 5

# Measuring Fluid Pressure

The machines that drive modern industry are powered in many different ways. We are all familiar with the electrically powered machines, but we may be surprised to know that much of the work that takes place in a factory is powered by fluid power — pneumatics and hydraulics. Pneumatic power tools range from drills to impact wrenches to medium-duty robots, while hydraulic power raises heavy loads in forklifts, operates presses, and powers heavy-duty high-precision robotics.

Webster defines a *fluid* as any substance that conforms to the shape of its container, such as a gas or a liquid. A gas expands to fill its container and may be compressed. The molecules in a liquid are more tightly bonded, and liquids do not usually expand or compress. Pneumatically powered machines operate on compressed gas, and hydraulically powered machines operate on liquid under pressure from a pump. In both cases, it is vital to measure and control the pressures in the system. In this chapter, we will explore some of the devices used to measure fluid pressure.

## OBJECTIVES

You will have successfully completed this chapter when you can:

1. Describe the operating principles of a bellows pressure sensor.
2. Describe the operating principles of a diaphragm pressure sensor.

3. Describe the operating principles of a piston pressure sensor.
4. Describe the operating principles of a Bourdon tube pressure sensor.
5. Describe the operating principles of a solid-state pressure sensor.

## 5.1 PRESSURE SENSORS — A FIRST LOOK

Pressure sensors can be divided into two broad classes: pressure switches and pressure transducers. We may define a pressure switch as one that actuates at a preset pressure level. A pressure switch resembles a limit switch in that it does not report any pressure other than the setpoint pressure. Most pressure switches contain two main parts. One of these parts is the pressure-sensing element, which responds to pressure by moving or expanding. The second part is a switch that is actuated when the pressure-sensing element presses on the switch. Pressure switches may be preset to respond to either increasing or decreasing pressure.

A pressure transducer, on the other hand, reports the system pressure directly at all times. The system controller (the brain of the system) determines when or if to close a contact or make other adjustments to the system.

Both types of pressure-sensing devices have their optimum uses. Pressure switches are generally simpler in design and fill most needs quite well. Pressure transducers are widely used in medicine, aerospace, and other applications where close, accurate, and continuous monitoring of pressure is important.

## 5.2 THE BELLOWS PRESSURE SENSOR

A bellows pressure sensor is made of a resilient (springy) material that is formed in the shape of a round bellows. Modern bellows made of sheet polyurethane materials are effective at pressures up to 100 pounds per square inch (PSI) of pressure. Bellows made of metallic materials are available for higher pressures.

Parts a and b of Figure 5.1 illustrate the operation of the bellows pressure sensor. One end of the bellows is closed, while the other end is connected through a fitting to the system pressure. As pressure increases, the bellows expands, causing a limit switch to act as a pressure switch, or causing a position detector to act as a pressure transducer. The operating range of the bellows can be increased by working it against a spring as is shown in Figure 5.1c. By operating the bellows in a sealed chamber that is connected to a second source of pressure, the bellows also works reliably as a differential pressure sensor.

The bellows pressure sensor is a reliable, long-life device that has been an industry standard for many years. In some systems, in fact, the bellows may act directly on a valve to control system pressure, eliminating the need for a switch.

## 5.3 THE DIAPHRAGM PRESSURE SENSOR

A diaphragm is a thin disk made of metal, silicon rubber, or polyurethane. The disk is secured in a sealed housing, which is connected to system pressure. As pressure

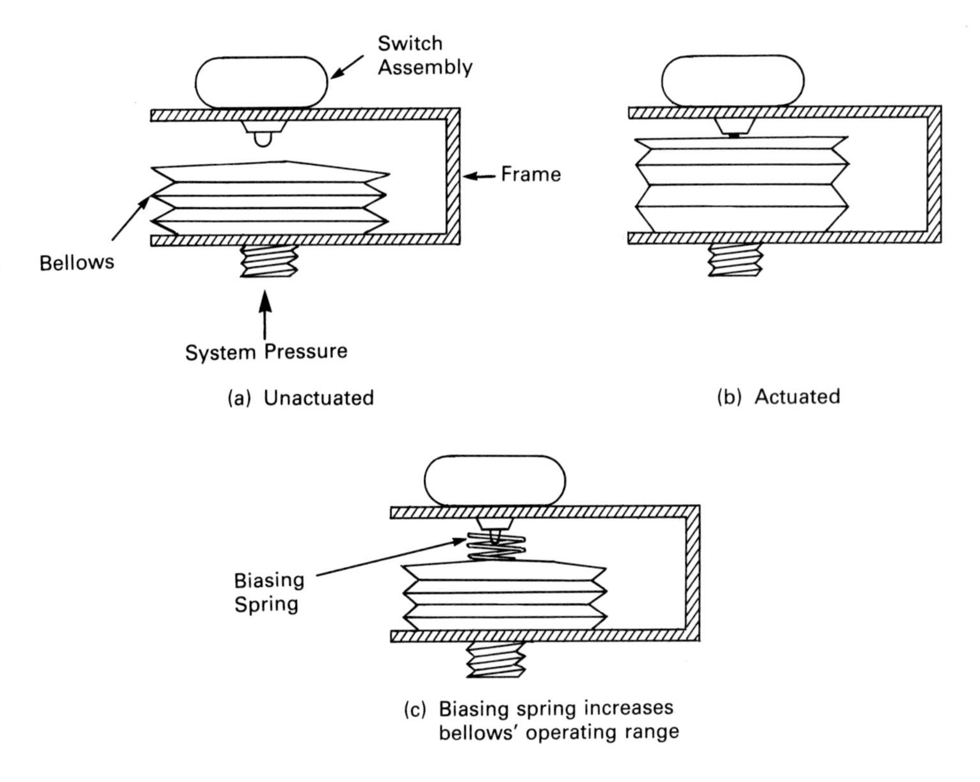


**Figure 5.1** The bellows pressure sensor

increases, the diaphragm expands upward at the center. The motion may be used to actuate a switch or an absolute position encoder, which reports back to the controller. Figures 5.2a and 5.2b illustrate the operating principles of the diaphragm element.

Diaphragms are somewhat less sensitive than bellows pressure sensors, but they can usually withstand greater system pressure — up to about 150 PSI. The sensitivity can be increased by "rippling" the diaphragm, as shown in Figure 5.2c. Another approach connects two diaphragms back to back, as shown in Figure 5.2d, to form a *capsule*. Diaphragm and capsule sensors also have the advantage of being able to measure vacuum.

## 5.4 THE PISTON PRESSURE SENSOR

When system pressures are too high for either the bellows or the diaphragm sensors, piston pressure sensors provide reliable pressure sensing at pressures up to 20,000 PSI. The piston is housed in a cylinder much like the cylinders found in an automobile engine. One end of the cylinder is connected to system pressure through

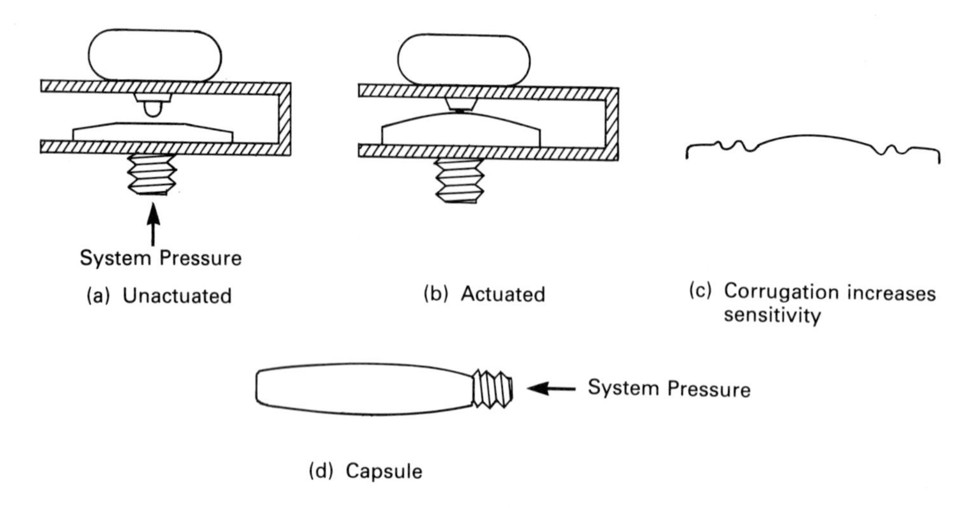


**Figure 5.2** The diaphragm pressure sensor

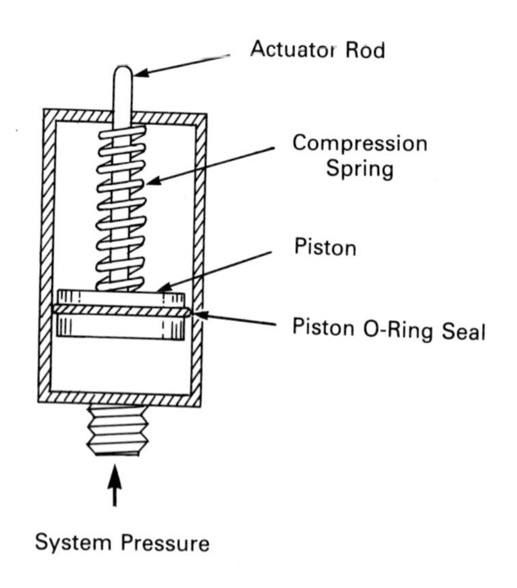


**Figure 5.3** The piston pressure sensor

a fitting, allowing system pressure to act on the piston. The system pressure forces the piston against a calibrated compression spring, Figure 5.3, and the piston movement actuates a limit switch.

The piston pressure sensor does not have the same degree of accuracy as the other sensors discussed in this chapter. Tolerances of 1% to 2% are common in premium units, while the other sensors in this chapter are readily available in tolerances of less than 1%. The piston pressure sensor, however, does provide reliable operation over a long life span. It also allows millions of on-off cycles and frequent large pressure changes without risk of failure.

## 5.5 THE BOURDON TUBE PRESSURE SENSOR

Probably the most used of all mechanical pressure sensors, the Bourdon tube operates reliably at pressures up to 18,000 PSI. A Bourdon tube is made of a flattened or oval tube of metal that has one end sealed and the other end connected to system pressure. The tube is usually bent into the shape of a *C*. When pressure is increased, the tube straightens, much as a curved balloon straightens when you inflate it. The movement of the tube can operate a limit switch or an absolute position encoder to report pressure to a controller. This operation is shown in Figure 5.4. Most direct-reading dial-type pressure gages are Bourdon tubes.

Bourdon tubes operate well at high pressure and offer accuracy better than 0.5%. They are not generally recommended for systems that have high cycling rates because too frequent flexing causes the tubes to become brittle from metal fatigue. This makes them more suitable for use in systems that do not undergo frequent large changes in system pressure.

## 5.6 THE SOLID-STATE PRESSURE SENSOR

All of the preceding pressure sensors are basically mechanical pressure-sensing devices that may be used to operate a dial, a switch, or some type of absolute position

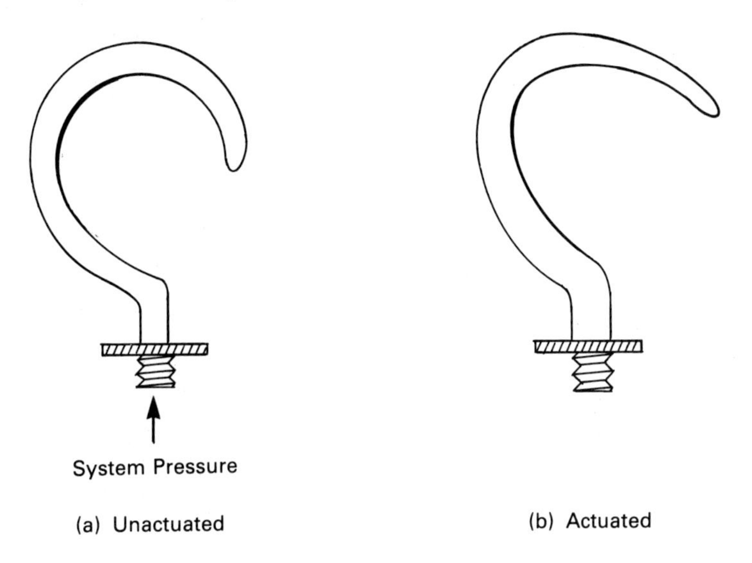


**Figure 5.4** The Bourdon tube pressure sensor

encoder — they are not of themselves electronic sensors. The solid-state pressure sensor is a purely electronic sensor designed to report system pressure as either a proportional voltage or current.

Solid-state pressure sensors may be found in a wide variety of equipment — from sensitive units used for measuring blood pressure to units that operate up to 10,000 PSI in a hostile atmosphere and under extreme temperature.

Most solid-state pressure sensors make use of the phenomenon of piezoresistance. The piezoresistance of a semiconductor is the change in resistance of the semiconductor material that is caused by a strain applied to the material. This is the same characteristic that makes the semiconductor strain gage work. In fact, a solid-state pressure sensor is a strain gage applied to the pressure-sensing application.

The sensing element of a semiconductor pressure sensor is made of four identical piezoresistors that are diffused into the surface of a thin disk of silicon. Electrical connections are made through gold pads that are bonded to the surface of the silicon disk. The side of the disk opposite the resistors is etched to form a diaphragm. The outer edges are left unetched to serve as a rigid support structure. Figure 5.5 shows the general construction details of a silicon solid-state pressure sensor.

The diaphragm with its piezoresistive elements is mounted in a housing and fitted with connections to system pressure. The reverse side of the diaphragm may be open to atmospheric pressure, sealed in a vacuum, sealed with a reference pressure, or open to a second source of pressure for differential pressure measurement.

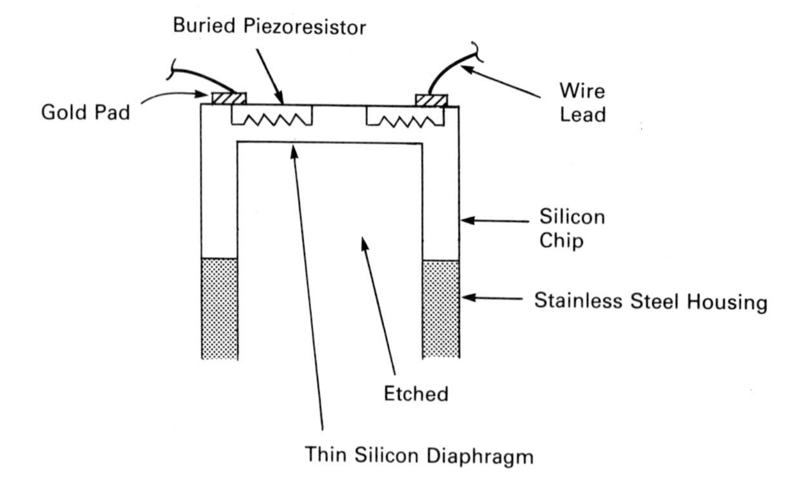


**Figure 5.5** Construction details of the silicon diaphragm.

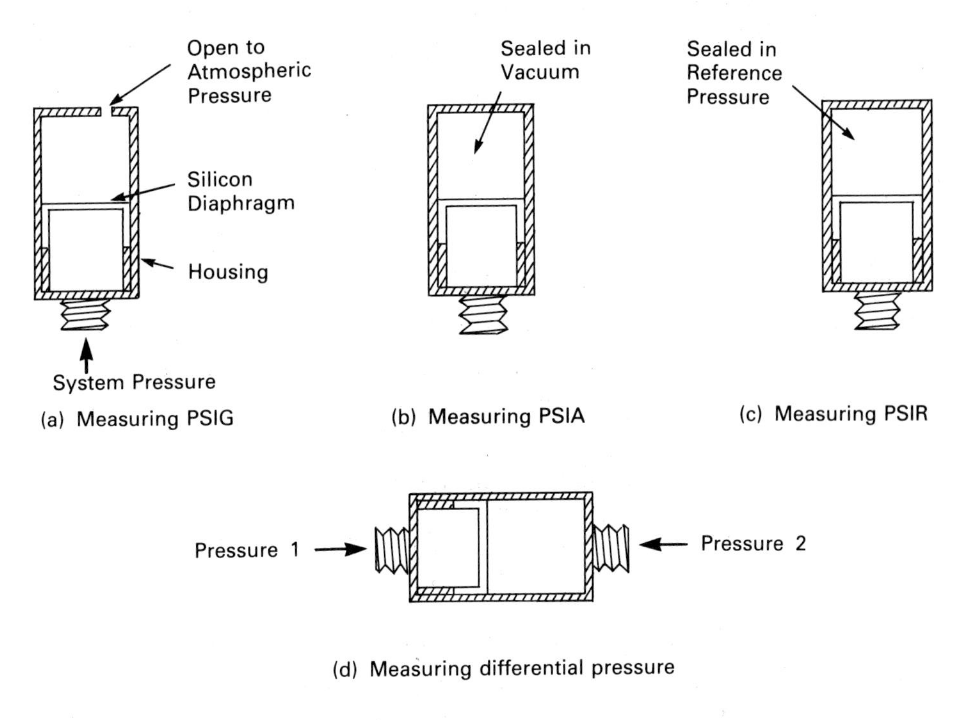


**Figure 5.6** Measuring PSIG, PSIA, PSIR, and differential pressure.

When the reverse side is open to ambient atmospheric pressure, the unit will measure *gage pressure* (PSIG). This is pressure relative to the atmospheric pressure, which is 14.7 PSI at sea level. When the reverse side is sealed with a vacuum, the unit measures *absolute pressure* (PSIA), which is pressure relative to zero. To convert PSIG to PSIA, add the current atmospheric pressure.

When the reverse side is sealed with a reference pressure, the unit will read pressure relative to that specific pressure (PSIR). When the reverse side is connected to a second pressure source, it will read the difference between the two pressures. In that application, the pressure source on the reverse side is usually limited to dry, noncaustic gases. For other differential pressure measurements, two gages must be used. Figure 5.6 illustrates these different applications.

The sensing elements in the solid-state pressure sensor are arranged as a resistive bridge. When there is no stress applied to the diaphragm, the output voltage is 0 V because the bridge is balanced. Under strain, the resistors change values as illustrated in Figure 5.7. For details on how to calculate the output voltage of the bridge, you may wish to consult Chapter 11.

The resistance of any substance changes as temperature changes. This will affect the output voltage of any resistive network. Many pressure sensors are available as

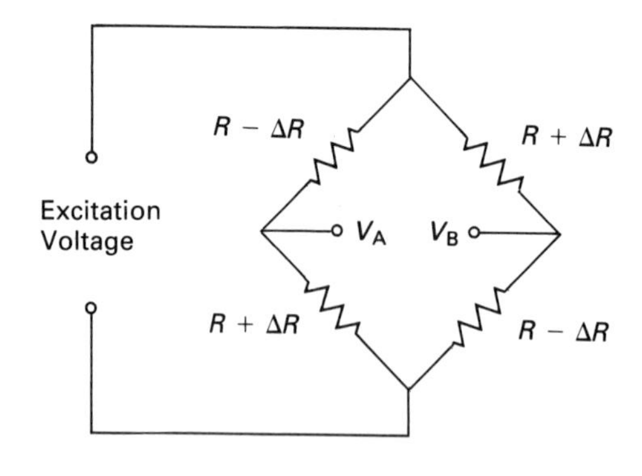


All four resistors are equal under zero pressure

$\Delta R$ is the change that results from change in pressure.

$$V_A = \frac{V_{EXC.}}{(R + \Delta R) + (R - \Delta R)} \times (R + \Delta R)$$

$$V_B = \frac{V_{EXC.}}{(R + \Delta R) + (R - \Delta R)} \times (R - \Delta R)$$

$$V_{out} = V_A - V_B$$

**Figure 5.7** The solid-state pressure-sensor bridge network

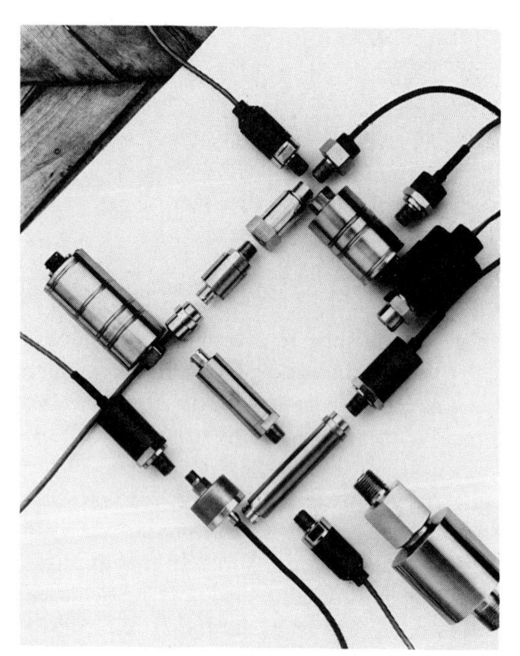

**Figure 5.8** Solid-state pressure sensors *(Courtesy of Omega Engineering)*

pressure transducers, which have built-in temperature compensation and allow accurate measurement under a wide range of operating conditions. Unless special reasons preclude their use, these units are preferable to those that require external compensation or calculations.

For high-pressure work and under conditions that may cause deterioration of the silicon diaphragm, semiconductor strain gages are bonded to stainless steel diaphragms. The basic principles of operation are the same as for those with silicon diaphragms. In addition, some units for high-temperature work may use foil strain gages bonded to the stainless steel diaphragm. The foil strain gages are less sensitive to temperature changes, but they provide low output voltage, on the order of 3 mV/V of excitation voltage.

Solid-state pressure sensors, Figure 5.8, offer a wide range of sensitivity, accuracy of 0.5%, reliable operation through tens of millions of cycles, and excellent vibration resistance. They can be used to provide electronic readout of pressure from remote locations, eliminating the need for expensive piping, valves, and fittings. With the rise in electronic and digital control circuitry, they are rapidly moving to the forefront in pressure-measuring technology.

## SUMMARY

In this chapter, we have seen the major devices used for measuring pressure. Pressure sensors are divided into two classes: those that operate a switch at a specified pressure, and those that report a pressure measurement to a controller, which then determines what action should be taken.

1. The bellows pressure sensor is useful at pressures up to 100 PSI. System pressure expands a bellows, which may be used to operate a snap-action switch or a position detector.
2. The diaphragm pressure sensor is useful at pressures up to 150 PSI. System pressure deforms the diaphragm, which may be used to operate a snap-action switch or a position detector. The diaphragm may be described as a flat bellows. Two diaphragms may be mounted back to back to form a capsule. This increases the sensitivity of the diaphragm.
3. The piston pressure sensor is useful at pressures up to 20,000 PSI. It is a good choice for systems that have frequent on-off cycling. System pressure forces the piston against a calibrated spring. A piston rod may then be used to operate a snap-action switch or a position detector. The piston sensor is the least accurate of the sensors, typically offering accuracies of +/− 2%, as contrasted to an accuracy of 0.5% for most other sensors.
4. The Bourdon tube pressure sensor is useful at pressures up to 18,000 PSI. The sensor is a curved, flat metal tube that is straightened by increasing system pressure. The motion of the tube may be used to operate a snap-action switch or a position detector. Bourdon tubes are not well-suited to systems with rapid on-off cycling.
5. The solid-state pressure sensor is useful at pressures up to 10,000 PSI. The solid-state pressure sensor provides an electrical output that is proportional to system pressure. Piezoresistive elements in a bridge configuration are used

to provide accurate pressure readings. Solid-state pressure sensors are an excellent choice for systems with frequent rapid cycling because they are rated at a lifetime of over 10 million cycles before expected failure.

## SELF-TEST

1. Both hydraulic and pneumatic systems require control of system ________.
2. A pressure-measuring device that operates a switch at a particular setpoint is called a pressure ________________________.
3. A pressure-measuring device that reports pressure continuously is called a pressure ________________________.
4. Both the bellows and the diaphragm pressure sensors will ____________ under increasing system pressure.
5. In a piston pressure sensor, the system pressure forces the piston to move against pressure from a(n) ________________________.
6. The Bourdon tube is a curved, flat metal tube. Increasing system pressure causes the Bourdon tube to ________________________.
7. The solid-state pressure sensor outputs a(n) ________________________ caused by a change in resistors that are buried in a silicon diaphragm.
8. The resistances in solid-state pressure sensors are arranged in a(n) ________ ________________________ configuration.
9. For higher-pressure systems, the diaphragm in a solid-state pressure sensor may be made of ________________________.

## QUESTIONS/PROBLEMS

1. You are designing a pressure control system for a pneumatic robot. The system requires that the compressor must be shut down if the pressure increases to 125 PSI. Would you select a pressure switch or a pressure transducer for the job?
2. You are assembling a pressure control system and making final adjustments. The system includes a 0–50 PSIG solid-state pressure sensor with a linear current loop output. The output signal is a standard ISA 4–20 mA signal. (At 0 PSIG, the output is 4 mA and at 50 PSIG, the output is 20 mA.) A pressure gage in the system tells you that the system pressure is 25 PSIG. What output current would you expect from the pressure sensor?

# 6

# Measuring Liquid Level

Another important measurement for industry is the level of liquid in a container. This measurement is used to control inventory, mix chemicals, and even control the buoyancy of a submarine. In this chapter, we will explore some of the devices that are used for this important task.

## OBJECTIVES

You will have successfully completed this chapter when you can:

1. Explain the operating principles of the float switch level detector.
2. Explain the operating principles of the resistive level detector.
3. Explain the operating principles of the capacitive level detector.
4. Explain the operating principles of the pressure level detector.
5. Explain the operating principles of the purge bubbler level detector.

## 6.1 THE FLOAT SWITCH LEVEL DETECTOR

Probably the simplest and most often used level detector is the simple float switch. Hundreds of thousands of float switches operate sump pumps in basements of homes and commercial buildings, protecting them from flooding. In even greater numbers, these simple and reliable devices provide industry with accurate and repeatable liquid level measurement.

Float switches are another special application of the limit switch. A float, moving directly with the liquid's surface, actuates a limit switch to indicate that the liquid has reached a preset level. Early float switches operated on a pivot and operated a snap-action limit switch. More recent float switches have magnet-equipped floats that act on a magnetic reed switch. The reed switch is hermetically sealed and can operate millions of times without failure because the contacts are not exposed to atmospheric dust and corrosive fumes. Figures 6.1a and 6.1b illustrate both types of float switch.

Float switches are mounted directly in the tank from the top, the bottom, or even the sides. When the tank is sealed and the inside of the tank is inaccessible, the bottle-type switch provides an easy solution. This is shown in Figure 6.1c.

Float switches are reliable, much used standards in industry. They are available in a wide variety of materials to withstand virtually any type of liquid and extremes of temperature. They are easily installed and require virtually no maintenance, which is an important consideration. When a switch closure (or opening) will do the job, the float switch is the preferred sensor.

When a more exact reading is required, you may still find a float that will do the job. In one type of float-operated absolute level sensor, the float slides up and down a long stem containing many reed switches. The controller can determine the exact level by taking note of which switch is closed. For more exact readings, the float may be attached to an arm that operates a potentiometer. The changes in resistance now report liquid levels. Your car gas gage is probably this type.

## 6.2 THE RESISTIVE LEVEL DETECTOR

Resistive level detectors may be used when the liquid is conductive. They determine the level of liquid in the tank by measuring the resistance between the single probe and the tank as shown in Figure 6.2a, or between two probes as shown in Figure 6.2b. The detector head contains solid-state circuitry to convert the resistance into a voltage or a current output for the controller.

Some special considerations are involved when using resistive detectors. Corrosion is a major cause of failure in any liquid-metal interface. It will cause serious malfunction in resistive detectors if the probe and housing materials are not carefully selected. Also, if electrolysis is permitted to occur, the liquid may be contaminated by ions from the metal probes. Both of these problems may be allayed somewhat by using an ac excitation voltage. These are usually at some frequency between 5 kHz and 10 kHz.

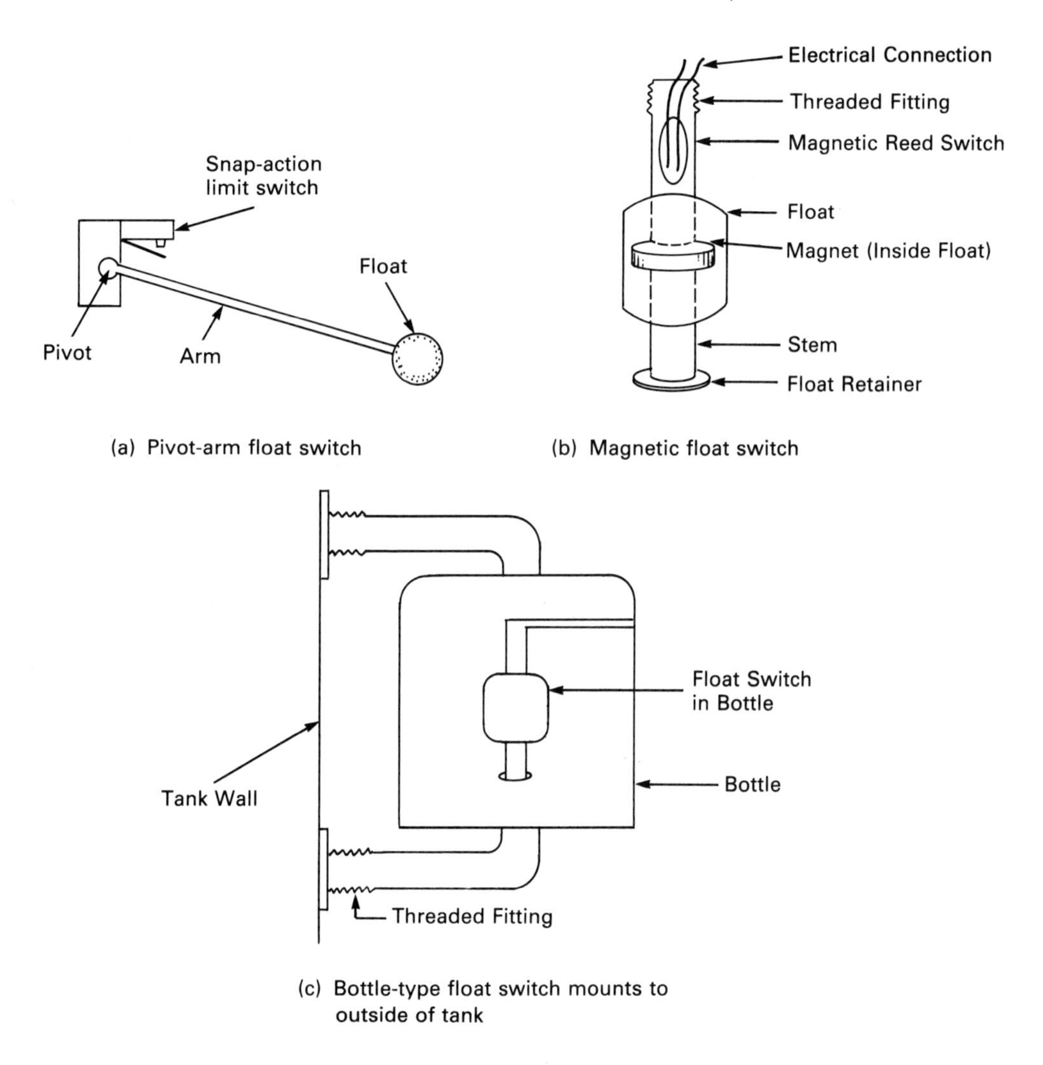


**Figure 6.1** The float switch

## 6.3 THE CAPACITIVE LEVEL DETECTOR

For nonconductive liquids, a capacitive detector may be substituted. The dielectric constant of air or vacuum is one. The dielectric constant of most liquids is much higher. By measuring the capacitance of a pair of dip rods in the liquid, we can determine the exact depth of the liquid. This approach may also be used with highly corrosive materials since the rods can be enclosed in a thin layer of protective material without seriously affecting operation. This idea is shown in Figure 6.3.

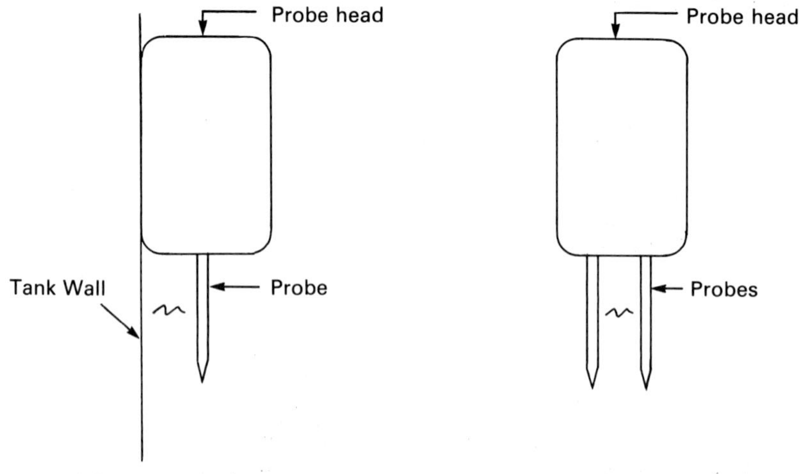


**Figure 6.2** Resistance level detection

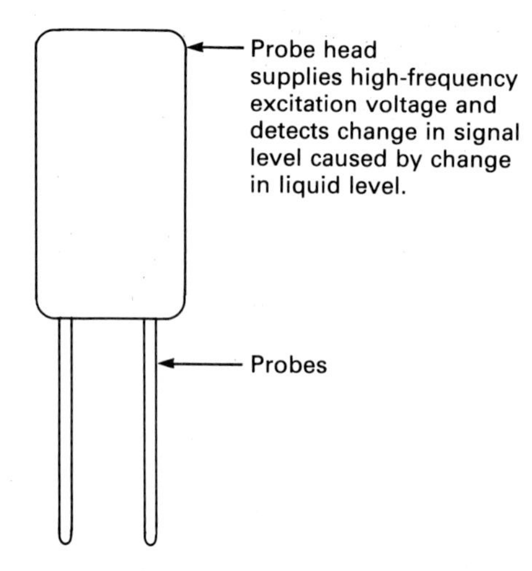


**Figure 6.3** Capacitive level sensing

## 6.4 THE PRESSURE LIQUID LEVEL DETECTOR

Pressure sensors may also be used to measure the depth of liquid in a tank. As any scuba diver knows, the pressure of the water in the ocean increases as you descend. This simple fact enables very accurate level sensing with some of the same sensors we examined in Chapter 5.

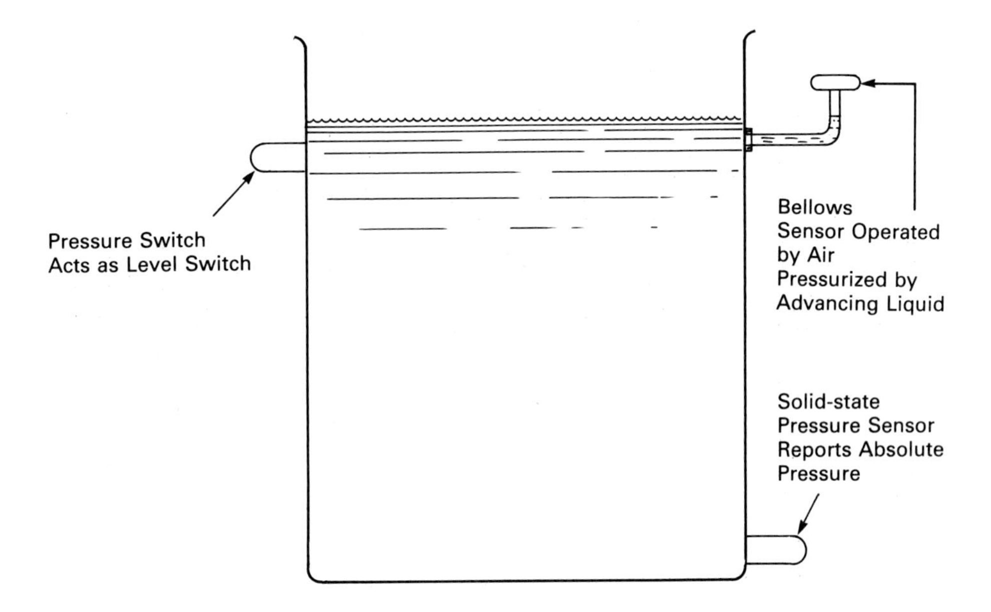


**Figure 6.4** Pressure sensors as level sensors

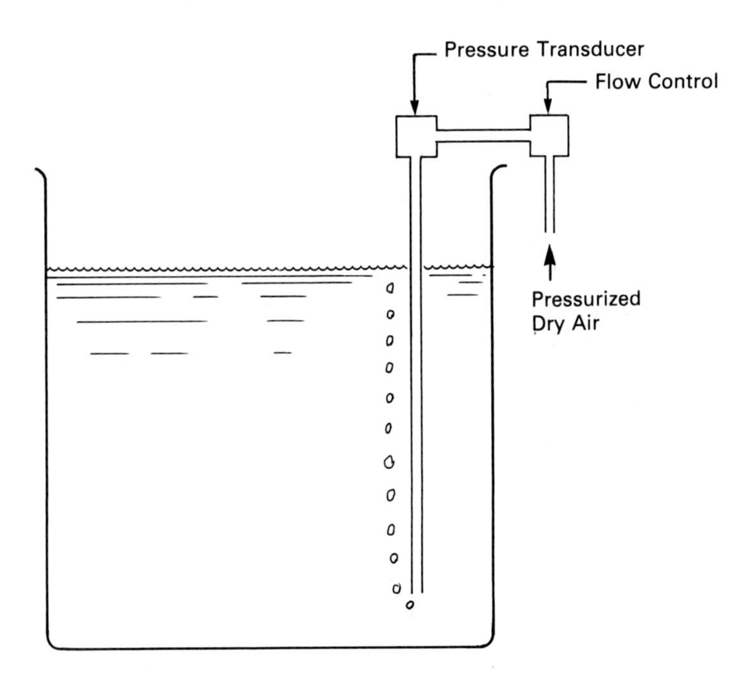


**Figure 6.5** The purge bubbler level sensor

The sensor is mounted in the tank at an appropriate place. For absolute measurement, the sensor is in the bottom of the tank. Sensors used in this capacity must be capable of withstanding the full weight of the liquid. For level switch operation, the sensor may be mounted at or near the level to be detected. The pressure switch may even be installed in such a way that it is operated by air pressure and never comes in contact with the liquid. These applications are shown in Figure 6.4.

The purge bubbler level sensor is another industry standby. This technique is useful for measuring absolute levels to an accuracy better than 0.1 foot. A dip tube of suitable material is inserted into the liquid and dry air is forced through the tube with a controlled rate of flow, about 1 PSIG, and allowed to bubble through the liquid. For viscous liquids like crude oil or molasses, higher air pressures may be used, up to about 20 PSIG. Since the pressure required to maintain the constant flow is directly related to the depth of the liquid, measuring that pressure will report the liquid's depth. This is shown in Figure 6.5.

Purge bubblers may be operated with compressed nitrogen when the liquid would be contaminated by air. Whatever the gas, they typically operate at pressures between 1 and 20 PSIG. They are used in liquids as varied as water, ink, sewer sludge, and molten metal.

## SUMMARY

In this chapter, we have looked at some techniques for measuring the level of liquids in their containers. We have seen that:

1. The float switch is probably the most used and one of the most reliable level detectors available. Its simplicity should make it a first choice when it will do the job.
2. The resistive level detector relies on the conductivity of the liquid. It can report exact depths in conductive liquid.
3. The capacitive level detector can report exact depths in either conductive or nonconductive liquids. It is also useful in corrosives.
4. The pressure level detector can report exact depths when connected in the bottom of a tank, or it can act as a level switch when mounted in the sides or top of the tank.
5. The purge bubbler uses gas under pressure to detect the pressure at the bottom of a tank of liquid. It can be very accurate and is useful in both corrosive and viscous liquids.

## SELF-TEST

1. The most commonly used level sensor is the ______________________.
2. The float switch is used to determine when the liquid in a tank has reached a(n) ______________________.
3. Modern float switches use ______________________, which are not susceptible to atmospheric dust and corrosive fumes.
4. A resistive level sensor relies on the ______________________ of the liquid.
5. The resistance between the sensor rods of a resistive level sensor will measure ______________________ when the liquid level is below the rods.
6. When the liquid reaches the rods of a resistive level sensor, the resistance will ______________________.
7. A capacitive level sensor is used for ______________________.
8. When the liquid reaches the rods of a capacitive level sensor, the capacitance will ______________________.
9. Pressure level detectors are used to determine the exact level of liquid in a tank because as the liquid gets deeper, the pressure ______________________.
10. The purge bubbler level sensor actually measures the pressure required to force a stream of air bubbles through the liquid at a constant ______________.

## QUESTIONS/PROBLEMS

1. You have been asked to select a level sensor for an application where it is only necessary to determine if the liquid storage tank is full or nearly empty. Which type of level sensor would you choose? Explain why.
2. The sensor rods of a capacitive level sensor have been damaged and must be replaced. The replacement rods are slightly shorter than the original rods, but it has been determined that the shorter length will not be a detriment. What effect will their length have on the capacitance when the tank is full?
3. A purge bubbler system has been in use measuring the level in an oil tank. The type of oil in the tank is being changed from crude oil, which is very thick, to a more refined oil, which is less viscous than the crude. Will the pressure be increased or decreased?

# 7

# Measuring Fluid Flow

Measuring the flow of fluids is an important task in many industrial plants. Whether for process control or inventory control, accurate measurement is often critical to successful operation. In this chapter, we will examine some of the devices used to measure fluid flow.

## OBJECTIVES

You will have successfully completed this chapter when you can:

1. Explain the operating principles of the differential pressure flowmeter.
2. Explain the operating principles of the positive displacement flowmeter.
3. Explain the operating principles of the velocity flowmeter.
4. Explain the operating principles of the mass flowmeter.

## 7.1 AN OVERVIEW OF FLUID FLOW

Four different classes of fluid flowmeters are in wide use in modern industry. They serve as the sensor arm of control for fluids as diverse as compressed gases, gasoline, sewage sludge, and slurries made of finely powdered solids suspended in liquid.

Flowmeters may report an exact rate of flow for these fluids or simply act as flow detection devices. In all cases, correct installation and maintenance is of major importance to the industry: failure may result in severe damage to the plant and endangerment of workers.

When installing flowmeters, provide a length of straight pipe on the input side of the meter to avoid problems caused by fluctuating pressures in the pipe as fluids negotiate curves. The length requirement varies from four to 50 times the pipe diameter, depending on the type of flowmeter. A shorter length of straight pipe on the discharge side is also recommended. Other considerations include the nature of the fluid, ambient temperature, and pressure loss caused by insertion of the meter. Correct selection is best made through close cooperation with the vendor's engineering staff.

The four classes of flowmeter are the differential pressure flowmeter, the positive displacement flowmeter, the velocity flowmeter, and the mass flowmeter. Each type has advantages and disadvantages, depending on the nature of the fluid being controlled. It is beyond the scope of this text and the duties of a technician to attempt to select the correct type for any particular job. In this chapter, we will discuss the operating principles of all four types.

## 7.2 THE DIFFERENTIAL PRESSURE FLOWMETER

The most common flowmeter is the differential pressure flowmeter, also known as a head meter. The differential pressure flowmeter works on the Bernouli effect, the same effect that keeps an airplane in the sky. As a fluid passes over a curved surface or through a smaller opening than the pipe that carried it, the velocity of the liquid increases and the pressure of the liquid decreases proportionately. By measuring the difference in pressure before and after the change, we can determine the rate of flow. From this and the cross-sectional area of the pipe, we can determine the flow in gallons per minute or in other convenient units. Applying this principle has led to many variations of the differential pressure flowmeter. Some of these variations are illustrated in Figure 7.1.

Of the flowmeter assemblies in Figure 7.1, the orifice shown in Figure 7.1a is the simplest and the most popular. It is easy to install between pipe flanges and is relatively inexpensive, but it does cause a fairly large pressure drop, which must be allowed for in system design. The venturi tube meter in Figure 7.1b causes less loss of system pressure at large volumes of flow.

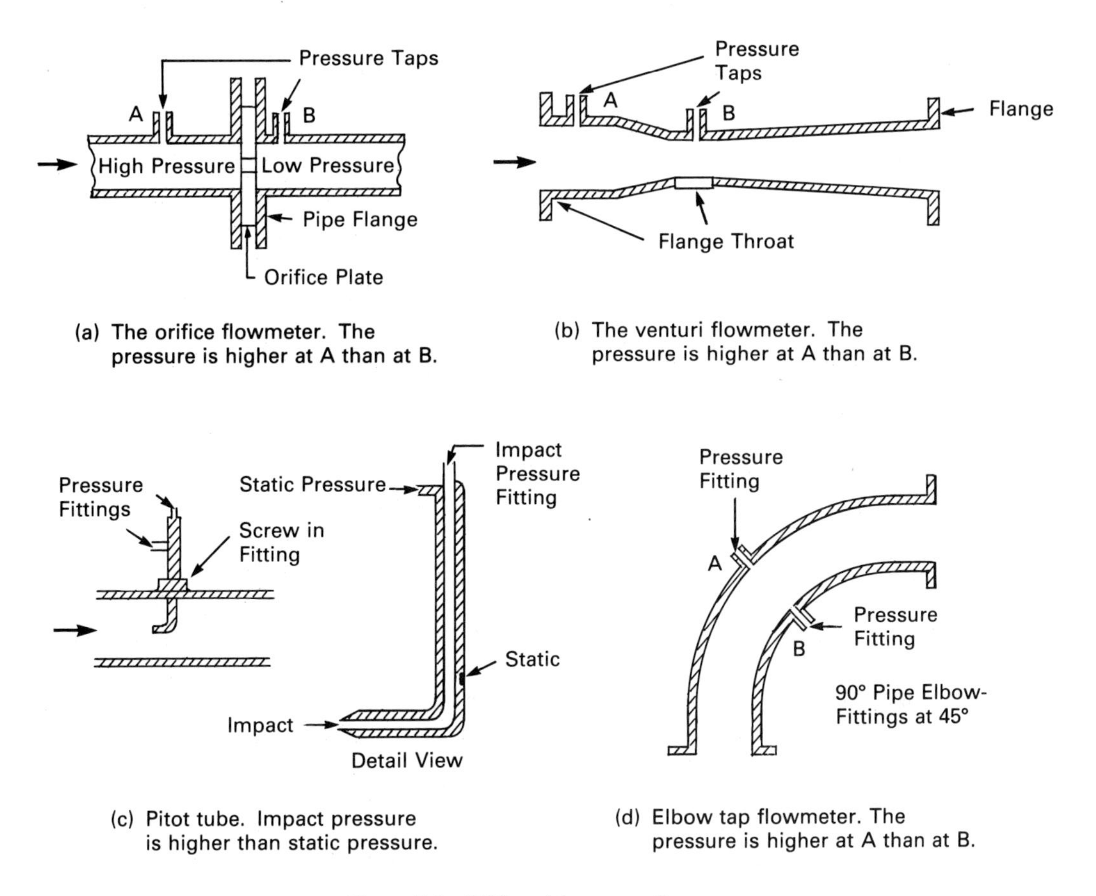


**Figure 7.1** Differential pressure flowmeters

The Pitot tube, Figure 7.1c, measures two pressures at the same time through two separate tubes. The tubes are usually mounted with one inside the other. The inside tube is open at the end and bent at right angles so that it is pointed at the source of flow. It measures the impact pressure of the fluid. The outer tube is sealed at the end and has a small slot in the side. This slot measures static pressure. The difference between the two pressures is proportional to the rate of fluid flow. Pitot tubes are limited to clean fluids because the small openings are easily clogged.

The elbow tap flowmeter is the simplest of all differential pressure meters. Any 90-degree pipe elbow can be used as a flowmeter by placing two small holes at the center of the elbow, Figure 7.1d. The fluid along the outside of the turn will have a higher pressure than the fluid along the inside of the turn. Once again, the difference in pressure is a measure of the flow rate.

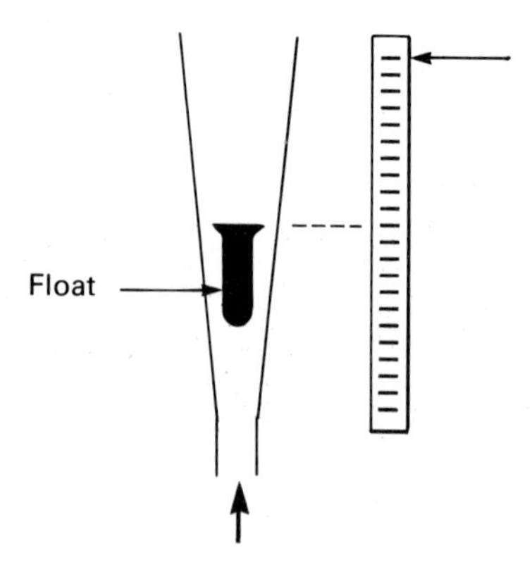


**Figure 7.2** The variable area flowmeter. As flow increases, the float moves upward.

The variable area meter shown in Figure 7.2 is a variation of the differential pressure meter. Its operation depends on the area of the tapered metering tube, the float, and the volume of moving fluid. When the fluid flow is slight, only a small area of the tube needs to be open to accommodate the flow, and the float moves only a small distance. When the flow of fluid is large, the float moves further upward into the throat of the metering tube to allow the greater volume of fluid to pass. This meter, often called a rotameter, can be read directly or fitted with electronic sensors when it is necessary to report to a controller. It is often used to meter oxygen flow to patients in the hospital because of its simplicity and reliability.

## 7.3 THE POSITIVE DISPLACEMENT FLOWMETER

The positive displacement flowmeter works by dividing the fluid into exact increments and passing them along the pipe. The number of increments is counted by an electronic counter and reported to either a display or a controller.

The most common of these very accurate meters is the rotary piston flowmeter, Figure 7.3a, which is more properly called an oscillating piston meter, since the piston does not really rotate a full 360 degrees. The principles of operation are shown in Figure 7.3b. The fluid enters the inner measuring chamber through the inlet port marked *E* in position 1. It fills the chamber and forces it to rotate clockwise on its eccentric center. Liquid now enters the outer measuring chamber and continues to force the piston to rotate, as shown in positions 2 and 3. When the rotating piston reaches position 4, the fluid in the inner chamber is discharged through the outlet port. Continued pressure from the fluid in the outer chamber now forces the piston to the original starting position. A magnetic coupling counts the number of oscillations of the rotary piston, thus directly measuring fluid volume.

**Figure 7.3a** Rotating piston flowmeters with indicator readouts *(Courtesy of Atlantex Industries, Inc.)*

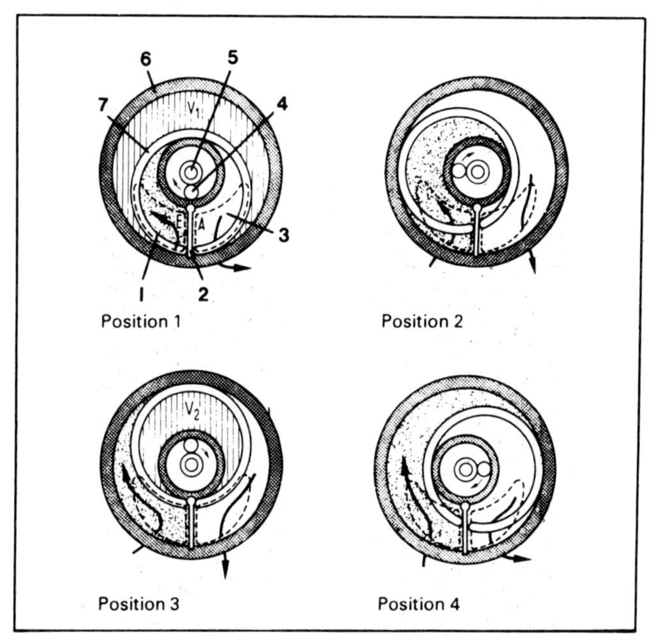


1. Inlet
2. Partition Wall
3. Outlet
4. Piston Guide Pin
5. Center Pin Of Chamber
6. Chamber Wall
7. Rotary Piston

**Figure 7.3b** Operating details of the rotating piston flowmeter *(Courtesy of Atlantex Industries, Inc.)*

The nutating disk flowmeter is also popular. Chances are that the meter that measures your water consumption from your local utility company is of this type. The disk is mounted on a sphere and contained within a spherical walled chamber. The action of flowing fluid causes the disk to rock back and forth, alternately filling and discharging the chamber. The rocking is counted by either a mechanical or an electronic counter.

The rotary vane meter, also called a fan-wheel flowmeter, contains a rotating impeller that is mounted inside the meter housing. The impeller is divided into measured compartments that fill and empty to measure volume. Other variations of the positive displacement meter use oval or helical gears or reciprocating pistons to measure fluid flow.

All of the positive displacement meters are extremely accurate measuring devices, but they have moving parts that are in direct contact with the fluid. They are susceptible to both wear problems and to clogging from dirty liquids or slurries. They should be used only with clean liquids. Periodic maintenance is vital to their continued operation.

## 7.4 THE VELOCITY FLOWMETER

Velocity flowmeters measure the rate of fluid flow directly. The most common of these is the turbine flowmeter illustrated in Figures 7.4a and b. The flow of fluid causes the rotor to spin. The rotational speed is directly proportional to the rate of the fluid flow.

Rotational speed is sensed by a magnetic proximity sensor and converted by a controller or display unit into gallons per minute or some more convenient unit of measure. The turbine flowmeter's rotor is in direct contact with the fluid and its use is therefore limited to clean fluids. Another potential problem is that the rotor bearing may wear rapidly in some liquids. Once again, periodic maintenance is essential.

The vortex meter shown in Figure 7.5 relies on the fact that when a flat object is placed in a stream of fluid, it creates a series of vortexes (swirls) downstream from the object. The frequency of this vortex shedding is proportional to the rate of flow. Sensors downstream from the blunt object can count the vortexes. The count can then be converted into rate of flow for the controller. The vortex meter has no moving parts, but it does introduce an obstruction into the pipe. Its use is normally limited to clean fluids to avoid pipe clog.

The electromagnetic meter shown in Figure 7.6 is useful for conductive liquids. It does not introduce any obstructions in the pipe. It relies on the fact that when a conductor moves through a magnetic field, a voltage is induced. The electromagnetic flowmeter consists of externally mounted magnetic coils, which generate the field, and a pair of electrodes, which are mounted in the wall of the pipe so the voltage can be read. Since it does not include any obstructions in the pipe, it is useful for slurries and dirty liquids.

**Figure 7.4a** Cutaway view of a turbine flowmeter *(Courtesy of Flow Technology, Inc.)*

**Figure 7.4b** Typical turbine flowmeters *(Courtesy of Atlantex Industries, Inc.)*

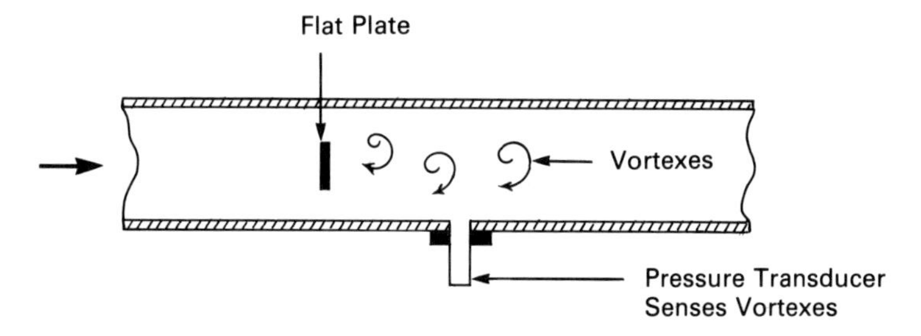


**Figure 7.5** The vortex flowmeter

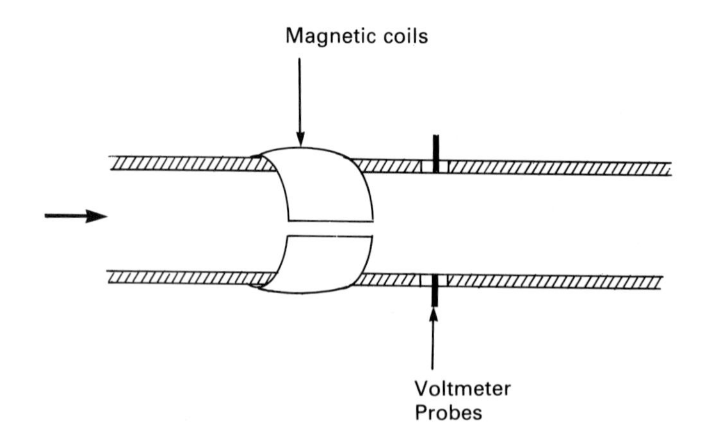


**Figure 7.6** The electromagnetic flowmeter for conductive fluids

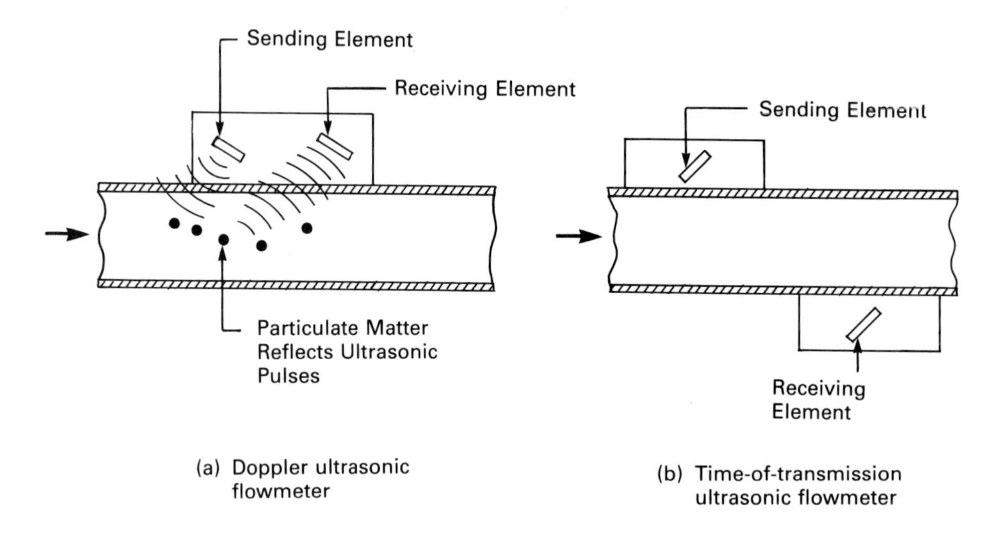


**Figure 7.7** The ultrasonic flowmeter

The ultrasonic Doppler flowmeter is mounted to the outside of the pipe as shown in Figure 7.7a. The transmitted ultrasonic pulse is reflected from particles or bubbles in the fluid to the receiver. Because the fluid is moving toward the receiving element, the frequency of the received pulse is higher than that of the transmitted pulse. The difference in frequency is a measure of fluid flow rate. This unit is useful for dirty liquids and slurries because it does not introduce an obstruction in the pipe.

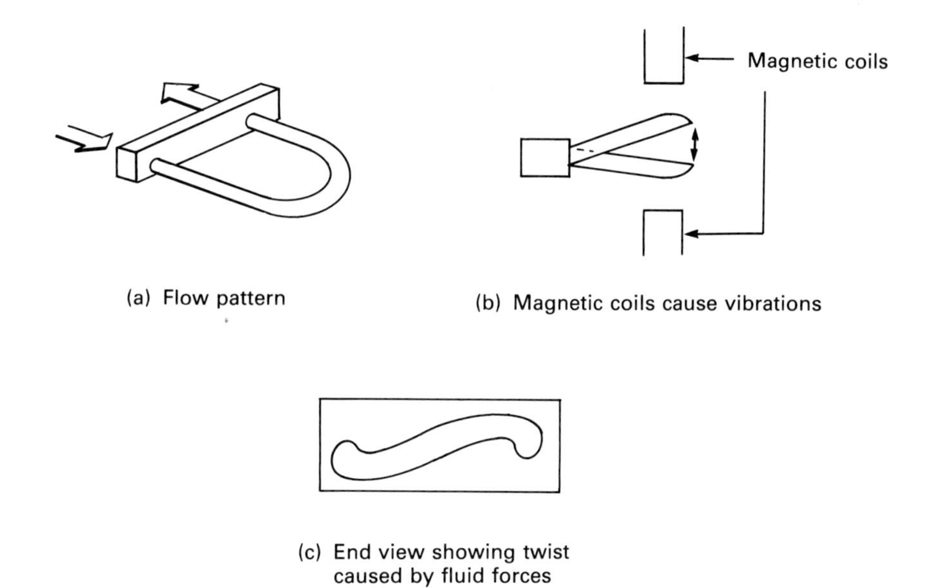


**Figure 7.8** The Coriolis mass flowmeter

The time-of-transmission ultrasonic meter shown in Figure 7.7b uses transducers mounted on opposite sides of the pipe at 45-degree angles from each other. The speed of the pulses traveling between the transmitter and receiver increases or decreases depending on the direction and rate of flow. The time-of-transmission meter requires clean fluids that are free of bubbles for proper operation.

## 7.5 MASS FLOWMETERS

Mass flowmeters perform extremely accurate measurements because they measure the mass of the fluid, rather than its rate of flow. This is very useful in controlling mass-related processes such as chemical reactions. Mass flowmeters may be divided into two types, the Coriolis meter and the thermal-type meter.

The Coriolis meter consists of a U-shaped flow tube mounted in a sensor housing, Figure 7.8. The tube is made to vibrate at its natural frequency, much like a tuning fork, by electromagnetic coils mounted near the U-bend. Liquid flowing into the U-bend resists this vibration because of the kinetic energy generated by its speed and its mass. Once forced into vibrating in a given direction, the liquid resists reversing directions as the tube's oscillation changes. The result is that the tube twists sideways. The amount of twist is directly proportional to the mass of liquid passing through the U-bend. It is measured by magnetic sensors on either end of the tube.

The thermal flowmeter consists of a heated sensing element that is not in the flow path of the fluid. The sensing element may be mounted in a Tee in the pipe so

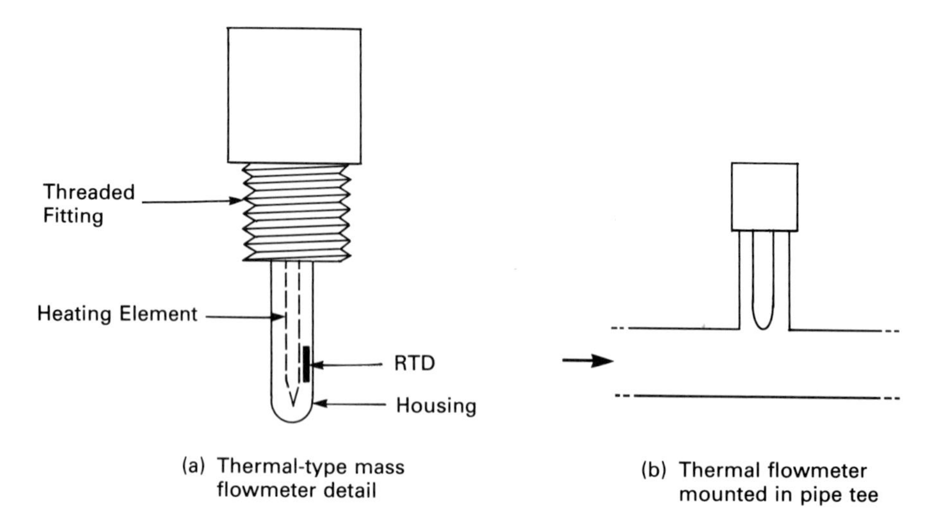


**Figure 7.9** The thermal mass flowmeter

that it is in contact with the fluid but not in the flow path itself, Figure 7.9. The passing fluid carries away some of the heat. RDTs measure the change in heat and report to the controller. The current needed to maintain the sensing element at a constant temperature is a measure of the mass of passing fluid. Thermal flowmeters are most often used to measure the flow rate of gases.

## SUMMARY

In this chapter, we have looked at some of the ways to measure the flow of fluids. We have seen that:

1. When installing flowmeters, there should be a length of straight pipe in front of the inlet side of the meter to reduce turbulence in the fluid.
2. The differential flowmeter measures the change in pressure as the fluid passes through an obstruction or around a bend. This can be converted directly into rate of flow.
3. The variable area flowmeter is a special variation of the differential pressure flowmeter.
4. The positive displacement flowmeter is very accurate, but it places an obstruction in the line. The loss of line pressure must be allowed for in system design. The positive displacement flowmeter also has moving parts that must be maintained.

5. The velocity flowmeter measures the rate of fluid flow, rather than fluid volume.
6. The turbine flowmeter has moving parts that must be maintained.
7. The electromagnetic flowmeter and the ultrasonic flowmeter do not place an obstruction in the line.
8. Mass flowmeters are extremely accurate. They are especially useful for controlling mass-related processes.

## SELF-TEST

1. An orifice flowmeter is classified as a(n) ________________ flowmeter.
2. A Doppler ultrasonic flowmeter is classified as a(n) ________________ flowmeter.
3. A rotary piston flowmeter is classified as a(n) ________________ flowmeter.
4. A Coriolis flowmeter is classified as a(n) ________________ flowmeter.
5. Positive displacement flowmeters all have ________________ and require periodic maintenance.
6. A slurry is made up of solids suspended in a liquid. A flowmeter used with slurries should not introduce any ________________ in the line.
7. A velocity flowmeter measures the ________________ in the pipe.
8. A(n) ________________ flowmeter measures the actual volume of fluid that passes through the pipe.
9. The ________________ flowmeter measures the voltage induced in a conductive fluid as it passes through a magnetic field.
10. The Doppler flowmeter measures the ________________ of ultrasonic pulses that are reflected from solids or bubbles in the fluid.

## QUESTIONS/PROBLEMS

1. You have been asked to recommend a flowmeter for dispensing a light oil product. The product is very costly, and it is sold by volume. What type of flowmeter would you choose? Why?
2. Phosphate ore is used to manufacture fertilizers. The ore is strip-mined using a high-pressure stream of water, and the resultant slurry is transferred by pipeline to processing plants nearby. It is important to control the rate of flow of this corrosive slurry. What type of flowmeter would you use? Why?

# 8

# Measuring Quantity: Counting

Whether the product is automobiles or bagged cement, aspirin tablets or beer cans, the bottom line of profitability is quantity: how many did we make? Overproduction is a costly manufacturing error and underproduction is every bit as bad — worse from a sales point of view. Certainly no consumer wants to purchase 250 vitamin tablets and receive 200. Just as certainly, no manufacturer wants to charge for 100 machine screws and put 250 in the box.

In this chapter, we will examine some techniques that industry uses for counting.

## OBJECTIVES

You will have successfully completed this chapter when you can:

1. Explain the use of LSI integrated circuits for counting in industry.
2. Describe some of the uses of counting in industry.
3. Describe the operation of an events-per-unit-of-time counter and some of the uses for this counter.

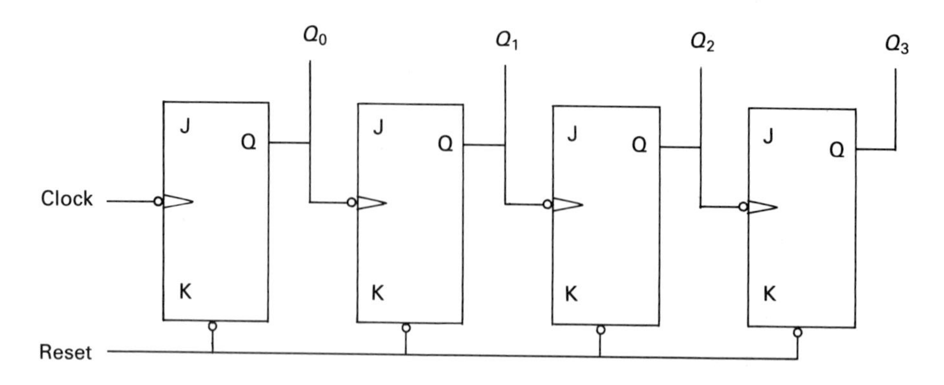


**Figure 8.1** Counting with JK flip-flops, a divide-by-16 counter.

## 8.1 COUNTING WITH INTEGRATED CIRCUITS

Any study of digital logic involves much time spent working with counter circuits developed for counting to various moduli. Flip-flops and gates can be combined to count to any desired number.

Most digital counters are based on the basic binary counter shown in Figure 8.1. These counters can count up to $2^n$, where $n$ is the number of JK flip-flops in the circuit. There is no need here to explain the operation of such a counter because of the advances in IC manufacturing technology.

Large-scale integration (LSI) has gone well beyond the simple counting scheme of Figure 8.1. Single ICs are available with three-, four-, and six-digit counters included on one chip. The chips may have multiplexed seven-segment or BCD outputs. They have carryout outputs and inputs and borrow outputs and inputs, and many include a preset feature with terminal count outputs to signal when a preset count has been reached. With careful selection, the task of developing a counter for a custom job is greatly simplified.

Figure 8.2 shows a typical four-digit counter chip. This is not the block diagram for any specific chip, but it illustrates the features that are common to most LSI counters. The BCD output is a form that is acceptable to most microprocessor-based controllers. The seven-segment output is in the standard format. The multiplex (MX) outputs are the control signals that indicate which of the four digits is being output; digit $A$ is the least significant and digit $D$ is the most significant digit. These outputs are used to turn on the correct seven-segment display at the correct time. The carry output allows cascading for longer counts. The terminal count output pulses when the

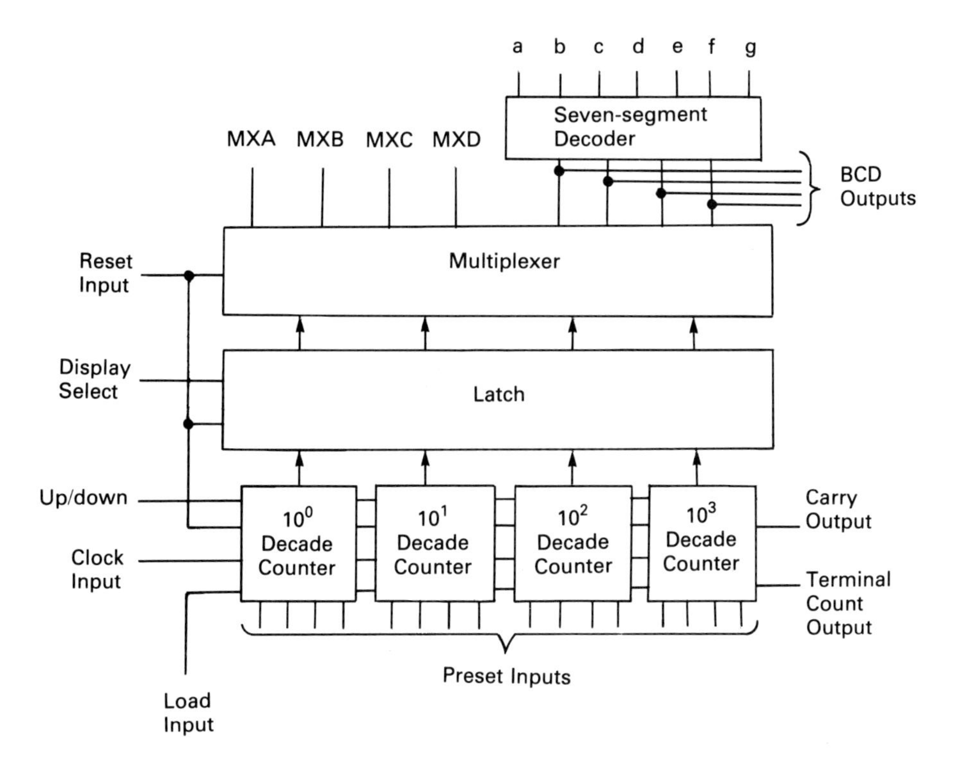


**Figure 8.2** A four-digit counter chip

count reaches zero. This is most useful when counting down from a preset count. The pulse may be used to operate a relay, a solenoid, or some other actuator.

On the input side, the clock input accepts the pulses to be counted, and the up/down control input determines the direction of the count. The preset input allows the counter to be set to any desired count prior to counting, and the load input causes the preset inputs to be clocked into the latches. The terminal count output may be used to strobe this input so that an operation may be repeated. The display select input allows the display to show either the ongoing count or the latch outputs. This is useful when using the IC as a frequency counter or tachometer, where the count is constantly being updated.

When selecting an IC for a custom counting task, you should first determine which features you need. Avoid using ICs that have much more capability than you are likely to need. Chips that have varying combinations of the features described in Figure 8.2 are available off the shelf. Careful selection will make custom counting easy.

**Figure 8.3** A typical industrial counter *(Courtesy of Automatic Timing and Controls Company, Inc.)*

## 8.2 INDUSTRIAL COUNTERS

Unless there is a need for some special kind of count sequence, there is little need to use an LSI counter chip. Many manufacturers of industrial controls sell industrial-grade counters. These units come complete with power supplies, LED, LCD, or fluorescent displays, housings or rack mounts, and bezels (front panel covers). Over the long run, these industrial units are more reliable and less expensive than units made in the factory shop. Figure 8.3 shows a typical industrial counter.

Industrial counters offer many of the same features you can get from LSI counter chips. They may be programmable from a rear-panel-mounted DIP switch for up/down counting, presetability, and terminal count output pulse. Some will offer multiplexed BCD outputs, which can be coupled into a computer or digital controller. Others may have RS232 outputs, which can be interfaced with virtually any computer. Most will have the capability of being reset electrically from a remote switch or by a computer.

Generally speaking, industrial counters also contain some signal processing components. These will accept practically any signal, from the millivolt level to 15-volt pulses. They also often have Schmitt trigger inputs to eliminate the chance of false counts from slowly changing pulse waveforms. Many even include debounce circuits for counting mechanical switch closures.

## 8.3 COUNTING EVENTS PER UNIT OF TIME

One major use of counters is counting rates: the number of events in a given unit of time. A familiar example of this is frequency, which is measured in cycles per second. The events here are either positive or negative alternations, depending on the counter. The unit of time is one second. Other events per units of time found in industry include revolutions per minute (RPM) and rate of flow, which may be in gallons per hour (GPH), gallons per minute (GPM), or some other volumetric measure.

The main difference between an ordinary counter and an events-per-unit-of-time counter is that the counting is controlled by a time base. The counter begins in the reset condition (all zeros) and the pulses to be counted are enabled by the time base. At the end of the preset time, the count is latched and displayed, the counter is reset, and the process repeats.

The flow indicator in Figure 8.4 can function as an events-per-unit-of-time indicator. It is designed to be used with a turbine flowmeter, such as the ones discussed in Chapter 7. Recall that a magnetic sensor outputs a pulse as each of the turbine blades passes its position. Signal conditioning in the unit allows it to be sensitive to signals as small as 100 mV. The *maximum* input pulse voltage is 50 volts.

With a one-inch pipe operating at a flow rate of 50 GPM, the turbine flowmeter will have a maximum output frequency of 2000 Hz and will output 2400 pulses per gallon. The flow indicator can be programmed by setting the ten-position rotary switches (visible on the indicator with the cover removed) to read directly in GPM. Other programming options allow the counter to *totalize* — provide a count of the total flow since it was last reset.

**Figure 8.4** A flow indicator — gallons per minute *(Courtesy of Atlantex Industries, Inc.)*

**Figure 8.5** An events-per-unit-of-time counter *(Courtesy of Electro Corporation)*

The unit shown in Figure 8.5 is an events-per-unit-of-time counter. It is specially designed for use with magnetic sensors and can reliably count the frequency of a signal as small as 30 mV peak to peak. It also has a logic level input that recognizes a voltage of less than 4 volts as a low and a voltage greater than 8 volts as a high. Maximum input voltage is 14 volts. This unit is accurate for frequencies ranging from a low of 2 Hz to a high of 1 MHz.

Counters such as this may be used to count RPM, to measure flow rate, or to measure the number of units that pass on a conveyer per minute or second. In fact, any quantity can be measured on a rate per minute basis if pulses per unit of quantity can be generated.

Conveyer belts are often controlled by an events-per-unit-of-time counter. Factory experience indicates the optimum speed for workers along the conveyer. Speeds that are too fast cause pileups, and speeds that are too slow increase labor costs and decrease profitability. The counter maintains the optimum speed, neither allowing too many nor too few units to arrive at work stations down the line.

Many industrial sensors overcome the electrically hostile environment of industry by outputting a frequency instead of a voltage. A frequency counter with correct prescaling can convert the frequency into a direct readout in the proper units. Using this method allows long runs of sensor cable without undue problems.

## SUMMARY

In this chapter, we have seen the main points of counting as used in industry. They include:

1. Counters made with discrete JK flip-flops have many parts and a high degree of wiring complexity when more than one digit of counting is needed.
2. LSI integrated circuits offer multiple digit counting and many other features in a single integrated circuit.
3. Industrial counters are readily available to perform routine counting tasks.
4. Industrial counters are self-contained units, complete with power supply, case, rack mount, display, and bezel.
5. Events-per-unit-of-time counters contain a time base and allow rate measurements, frequency measurements, and RPM measurements.

## SELF-TEST

1. A counter made of five JK flip-flops has a maximum count of ________.
2. When counting with an LSI counter, the terminal count output will pulse when the count reaches ____________________.
3. The ____________________ inputs of an LSI counter allow the count to begin at some predetermined number.
4. A(n) ____________________ counter is preferable to one made in the factory shop for routine counting chores.
5. In order to measure how many events occur in a given time, a(n) ________ ______________ must be added to the counter.
6. Some transducers convert their output into a frequency in order to avoid problems with ____________________.

## QUESTIONS/PROBLEMS

1. Explain why it is preferable to use ready-made industrial counters rather than to build a custom counter for an industrial application.
2. You are servicing an events-per-unit-of-time counter and discover that the time base is 10% too long. What effect will this have on the count?

# 9

# Measuring Time

Industry must often measure time. Many processes require very precise timing in order to be successful. In this chapter, we will look at some of the ways to measure time and generate time delays.

## OBJECTIVES

You will have successfully completed this chapter when you can:

1. Explain the importance of time measurement to industry.
2. Describe the use of low RPM synchronous motors for timing applications.
3. Explain why synchronous motors are not suitable for very precise timing intervals.
4. Describe the operation of the 555 timer as a monostable timer.
5. Describe the operation of the 555 timer as an astable timer.
6. Explain the basic operation of a digital timer.

## 9.1 MEASURING TIME: AN OVERVIEW

"Time is money!" the sage tells us, and nowhere is this more true than in manufacturing. The best texture of a box cake is dependent on exactly the right baking time. Flash frozen green beans are spoiled by a too long exposure to the extreme cold of the freezing room. Proper annealing of heat-treated steel tools cannot be done without an exactly correct time in the heat treatment oil bath. These and many other processes are time dependent — in order for the processes to work, we must be able to create precise time delays.

There are three approaches to measuring and controlling time. The mechanical approach uses synchronous clock motors to measure time. The analog approach uses RC networks and the RC time constant to measure time. The digital approach uses a crystal-controlled frequency source and frequency divider circuits to measure time. In this chapter, we will look at all three approaches.

## 9.2 MEASURING TIME MECHANICALLY

The speed of a synchronous motor is determined only by the frequency of the supply voltage. Changes in voltage have no effect on the RPM of such a motor. Consequently, a synchronous motor that is being powered by a constant frequency ac voltage will operate at a constant RPM. Analog clocks operate on this principle. The clock motor revolves at a constant speed that is reduced by a gear train to drive the hands of the clock.

The same principle is used in the drum controllers discussed in an earlier chapter. The low RPM, synchronous clock motor in a drum controller turns the timing cams, which actuate lever-type snap-action switches. The position of the indentations or lobes on the cams determines the timing sequence. This type of controller is common in industry, in business machines, and in household appliances.

Mechanical time control is reliable, inexpensive, and easy to modify if the timing requirements change. What it lacks is precision. While time sequences can be controlled to the second, shorter time intervals are not practical. Mechanical considerations make more exact spacing of the switch actuators very difficult.

Also, although the power line frequency in North America is fixed at 60 Hz, that is an average. In fact, the frequency may drift somewhat up or down. The power company monitors the frequency drift by counting cycles. When the count gets to be about 180 cycles above or below what it should be, it is corrected by allowing the alternators to slow down or speed up for an appropriate time. Over a period of 24 hours, the count will be exactly right, but at any given time, the frequency may be slightly above or below the expected 60 Hz. When timing requirements are very precise, this drift precludes using the power line frequency as a time standard.

## 9.3 ANALOG TIMERS

The need for creating short time delays and pulses has led to an abundance of circuit designs. Practically every type of active component has been used in time delay and pulse generation. Bipolar transistor circuits and FET transistor circuits have enjoyed great popularity in their day. Today, those discrete component circuits have given way to the integrated circuit.

Of the many integrated-circuit timers, perhaps the most common is the 555. The 555 can be operated as a monostable or an astable multivibrator. This IC has such popularity that it is available in the 555 single-timer and the 556 dual-timer versions, and in a low-current CMOS version as well. The 555 operates over a supply voltage range of 4.5 V to 16 V. When it is operated at 5 V, it has a TTL compatible output.

Figure 9.1 shows the functional block diagram of the 555 IC. The resistive voltage divider made up of $R_1$, $R_2$, and $R_3$ provides reference voltages for both the upper and lower comparators. This arrangement makes the IC relatively independent of changes

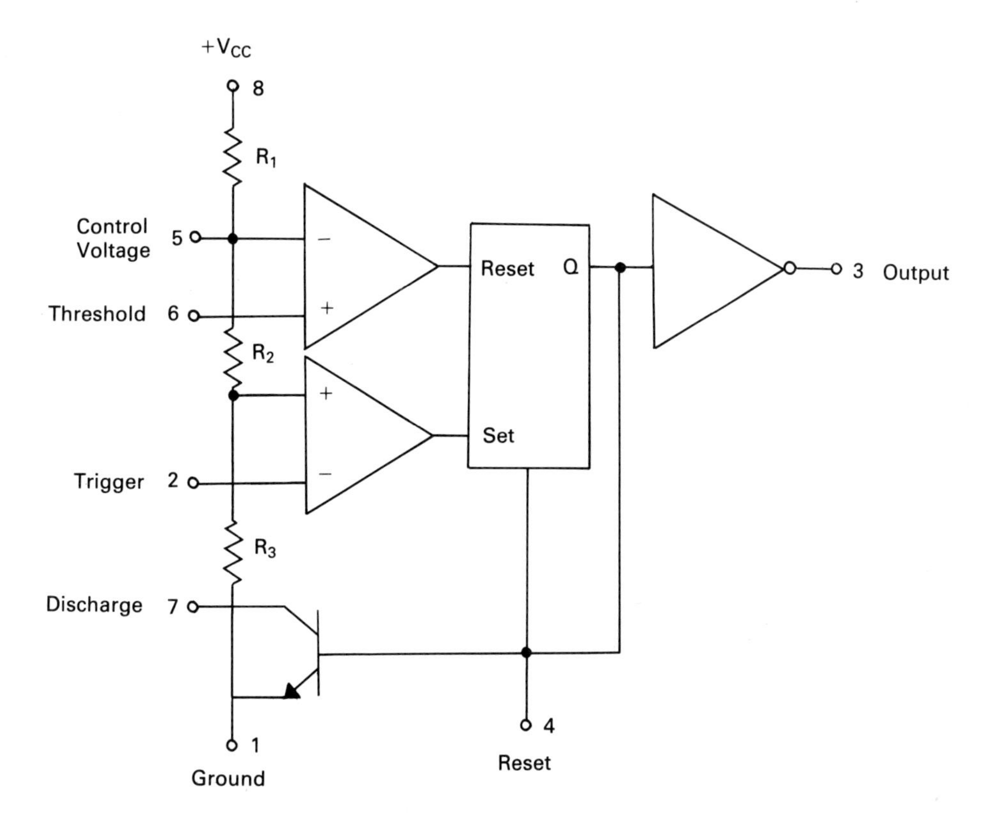


**Figure 9.1** The 555 timer

in the power supply voltage. The comparator outputs set or reset the control flip-flop. The control flip-flop output operates a high-current inverter, which provides the output. It also operates a transistor switch, which grounds the discharge pin (7) when the chip output is low.

A brief look at the pin functions of the 555 will help in understanding how the IC works:

- Pin 1 is the ground connection.
- Pin 2 is the trigger input. A low pulse on this pin sets the latch and causes the output of the chip to go high. The trigger voltage must be less than one-third of $V_{CC}$. The trigger pulse must be shorter than the timing cycle or the IC will retrigger and deliver a timing cycle that is too long. Trigger current is 500 nA, which sets the upper limit on series resistance.
- Pin 3 is the output pin, which is driven by a high-current transistor totem-pole output.
- Pin 4 is the reset input. It will reset the latch and force the output low, no matter what else is happening with the IC.
- Pin 5 is the control voltage pin. This allows the modification of the timing for some special purposes, but most circuits using the 555 connect this pin to ground through a 0.01 $\mu$F capacitor.
- Pin 6 is the threshold input, which is one of the inputs to the upper comparator. When the voltage on pin 6 is greater than two-thirds of $V_{CC}$, the latch is reset and the output from pin 3 is low. A dc current of 100 nA must flow into this pin, and it is this requirement that sets the upper limit on the external timing resistor.
- Pin 7 is the discharge pin. When the output at pin 3 is low, the open collector circuit of pin 7 is also low. This pin is used to discharge the timing capacitor in the external RC network. Internal circuits control the discharge current and prevent damage from the surge current of large capacitors. This means that there is no theoretical limit on the size of the timing capacitor.
- Pin 8 is the $V_{CC}$ connection, the positive end of the power supply.

Figure 9.2 shows the basic connections for using the 555 in the monostable mode in its simplest form. While the trigger input is high, the output pin is low and the open collector of the discharge pin is also low, shorting out $C_1$. When the trigger is pulsed low, the output pin goes high and the discharge transistor is cut off, allowing $C_1$ to charge through $R_1$. When the charge on $C_1$ reaches two-thirds $V_{CC}$, the threshold input triggers the upper comparator and resets the latch, forcing pin 3 low and turning the discharge transistor on once more. The length of the timing pulse is calculated with the formula

$$T = 1.1R_1C_1$$

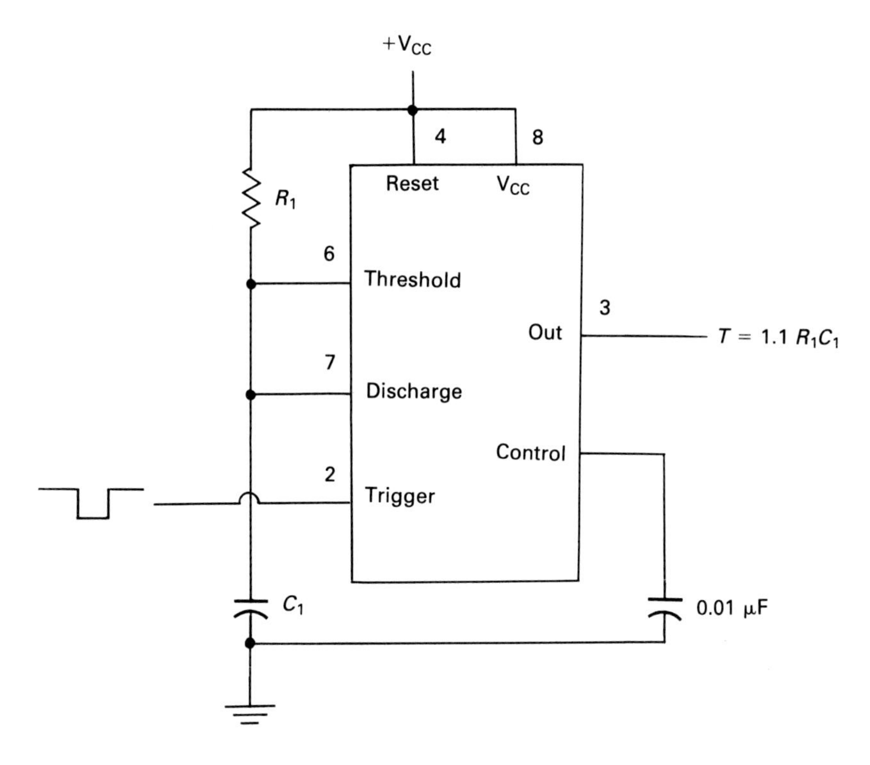


**Figure 9.2** The 555 as a monostable

Variations on the circuit are mainly connections to condition the trigger input.

The 555 monostable output may be used to trigger other 555 timers, providing a means for sequencing multiple operations. Figure 9.3 shows how this is done. The circuit in this figure operates three LEDs in sequence, each with a different "on" time. These LEDs might be optical isolators controlling machine tools or any other devices.

Figure 9.4 is the 555 used in its simplest astable connection. We will follow its operation from the instant of turn on so that we can see its total operation. When $V_{CC}$ is first turned on, there is no charge on $C_1$. Since the timing capacitor is connected to the trigger input as well as the threshold input in this circuit, the IC is automatically triggered and pin 3 goes high. This also causes the discharge transistor to be cut off, which allows $C_1$ to begin charging through $R_1$ and $R_2$. When the charge on $C_1$ reaches two-thirds of $V_{CC}$, the discharge transistor turns on and begins to discharge $C_1$ through resistor $R_2$. When the capacitor has discharged to one-third of

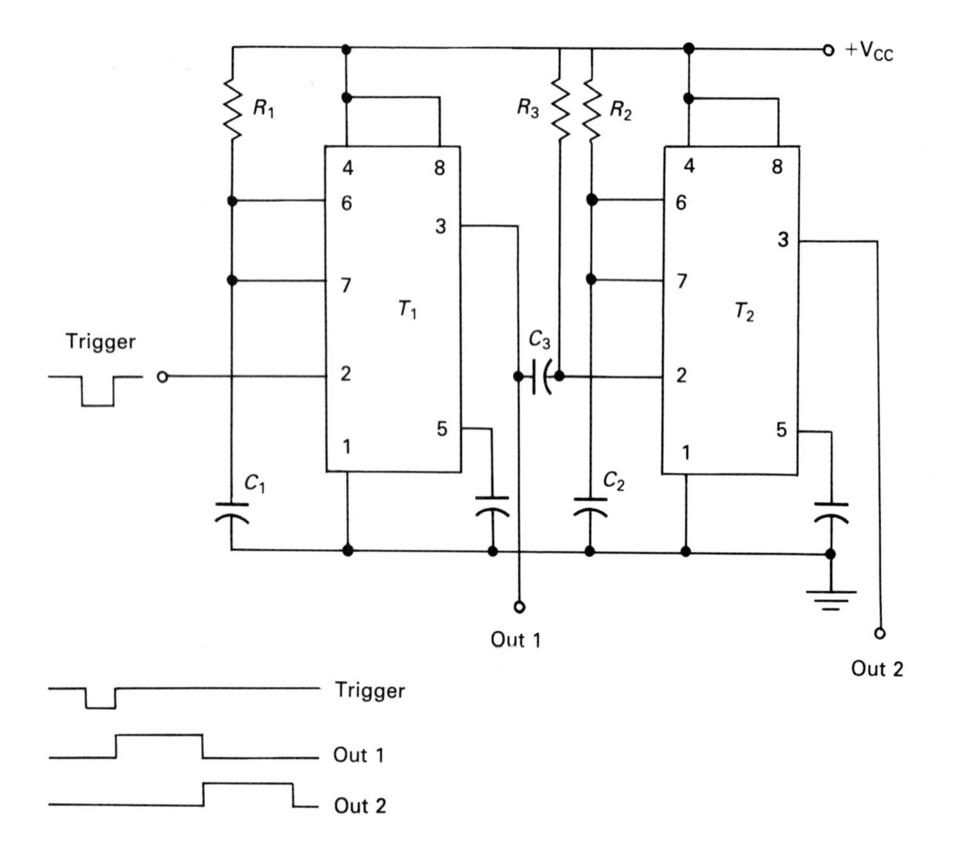


**Figure 9.3** Cascaded monostables. $R_3C_3$ forms differentiator to trigger $T_2$.

$V_{CC}$, the trigger input is activated and the process repeats. Frequency is found with the equation

$$f = \frac{1.44}{(R_1 + 2R_2)C_1}$$

Note that the capacitor must charge through both resistors and discharges only through $R_2$. This means that the output waveform of this circuit cannot be symmetrical. The time high will always be longer than the time low because of the added resistance of $R_1$. Making $R_1$ small with respect to $R_2$ can reduce the difference, but if a true square wave is needed for output, it is best to operate at twice the needed frequency and use the output to clock a JK flip-flop divide by two circuit.

The accuracy of timing with a 555 depends on the quality and tolerance of the timing capacitor. When very close tolerances must be met, tantalum capacitors are

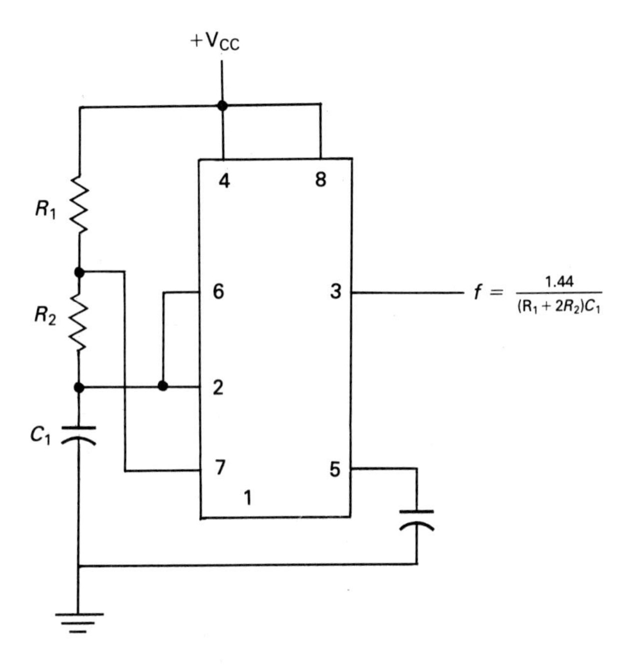


**Figure 9.4** The 555 astable

used because of their tolerance rating. As a technician, you should be aware that an ordinary aluminum electrolytic capacitor cannot be used to replace a tantalum capacitor. The resistors in 555 circuits should also be of the highest precision. One-percent metal film resistors are inexpensive enough in today's market that substitutions need not be made.

## 9.4 DIGITAL TIMERS

Digital time control relies on a crystal oscillator and a counting circuit. This arrangement can create time delays that are very precise. This kind of precision is rarely needed for controlling large machinery, but the technique is popular for its simplicity.

Figure 9.5 shows the basic circuit for a digital timer. The crystal oscillator is made up of two inverters, $U_1$ and $U_2$. The third inverter, $U_3$, acts as a buffer to isolate the oscillator from the rest of the circuit and prevent loading from affecting the frequency. The oscillator output is divided by an appropriate amount to arrive at the desired pulse width.

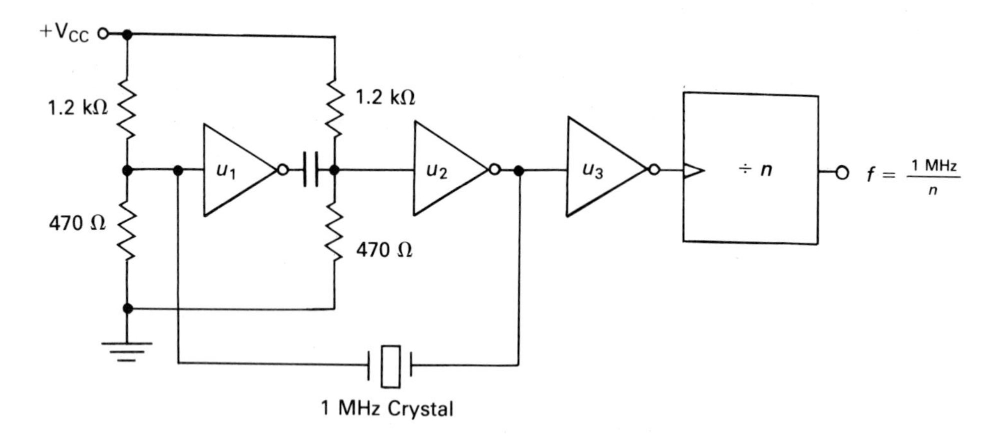


**Figure 9.5** Digital timer with oscillator

Special chips have been designed to simplify the task of frequency division. One of the most common and inexpensive crystals today is the 3.5 MHz color burst crystal used in all color TV sets. Several ICs that contain the circuitry needed for both an oscillator and a 17-stage divider have been developed. They output either 60-Hz or 100-Hz clock pulses. All that is required is the color burst crystal and IC to generate very accurate time bases.

## SUMMARY

As we have seen in this chapter, there are three basic ways to generate timed pulses.

1. Mechanical timers use synchronous clock motors, which run at a constant RPM, determined by the power line frequency.
2. Mechanical timers actuate limit switches with cam lobes or indentations.
3. Analog timers generate time delays that are controlled by the RC time constant. They can be more precise than mechanical timers.
4. Digital timers operate by dividing a crystal-controlled frequency with counter chips.
5. Digital timers are the most precise timing circuits.

## SELF-TEST

1. The mechanical timer relies on the fact that the speed of a synchronous motor is controlled by the ________________________.
2. A clock motor may drive one or more cams. Lobes or indentations on the cams operate ________________________.
3. A controller driven by a clock motor is called a(n) ____________________ controller.
4. The 555 IC is an example of a(n) ________________________ timer chip.
5. The 555 IC can be used as an astable or as a(n) ____________________ multivibrator.
6. A 555 timer chip is wired as a one shot. The timing resistor is 15 kilohms, and the timing capacitor is 0.27 μF. What is the pulse width of the output?

   ________________________
7. A 555 timer chip is wired as an astable. $R_1$ is 150 ohms, $R_2$ is 2 kilohms, and $C_1$ is 0.01 μF. What is the frequency of this circuit? ____________________
8. Several 555 timers can be wired in ________________________ to create a sequencer.
9. A digital timer has a 1 MHz crystal. The circuit divides the crystal by 256. What is the pulse width of the output pulse? ____________________
10. The most precise time delays can be generated using ____________________ timers.

## QUESTIONS/PROBLEMS

1. You are servicing a 555 monostable circuit. The output pulse is supposed to be 1 mS +/− 1%. Resistor $R_1$ is burned up, but $C_1$ is 0.001 μF. What is the correct size for $R_1$?
2. You are setting up a time base which uses a 2 MHz crystal. The output of the crystal oscillator passes through a presettable divider to establish the final time base. The divider can divide by any value in the range from 1 to 10,000. Your application calls for a 1 mS time base. What setting will you use on the presettable divider?
3. The crystal used for problem 2 is marked +/− .005%. What is the range of time base periods you might expect from the system?

# 10

# A Summary of Measurement — Some Applications

In Section II, we have been examining some of the many different measurement transducers used in industrial and control electronics. We have seen the transducers most popularly used to measure temperature, position, and force, the pressure level and flow of liquids, and quantity and time. In this chapter, we will see a few of these units in practical application. We will also study some techniques that are useful in troubleshooting sensing transducers.

## OBJECTIVES

You will have successfully completed this chapter when you can:

1. Explain the operating principles of a waterbed heater control.
2. Describe the use of inductive position sensors for measuring material thickness.
3. Describe the use of a potentiometer position sensor to control a robotic arm.

## 10.1 A WATERBED HEATER CONTROL

Waterbeds are recommended widely by doctors and chiropractors for a restful night's sleep. The water envelope conforms to the shape of the body and provides uniform support to the sleeper. An unheated waterbed, however, can rapidly draw the warmth from the body; the sleeper awakens chilled and achy.

A waterbed heater, essentially a 300-watt waterproof heating pad under the waterbed mattress, can solve the problem. But if it is left on too long, the waterbed heater can make the waterbed temperature as uncomfortably warm as an unheated mattress is uncomfortably cold. A safe way to control the heater operation is required.

The traditional temperature control for a waterbed heater is shown in Figure 10.1. The bulb temperature sensor is filled with freon, a common refrigerant fluid. The bulb is connected to a bellows through a thin copper tube. The pressure of the freon in the bulb, tube, and bellows is a direct function of the temperature of the freon. When the water is cool, the freon is mostly in its liquid state and the pressure is low. Under low pressure, the bellows pressure-sensor switch is in its nonactuated state. The NO contacts are used to turn the heater on. As the temperature gets warmer, the freon boils and the pressure increases. This expands the bellows, which actuates the limit switch, turning the heater off.

Figures 10.2a and b show an electronic approach to this same task. The sensor/controller is an LM3911 temperature-controller IC. This IC contains a temperature sensor and an op amp. The output of the temperature sensor is 10 mV per degree Kelvin, referenced to the $V^+$ supply pin. This translates to an output voltage difference of 3.1 volts between the $V^+$ pin and the output pin at the normal body temperature of 98.6°F, or 37°C.

This measurement signal is internally connected to the noninverting input of the op amp. The op amp output and the inverting input are available on the IC pins, as shown in the figure. In the working circuit, the inverting input is connected to the wiper of the "set" potentiometer, which acts as a voltage divider between the supply voltage and system ground. Thus, the op amp is being used as a comparator. When the output of the sensor is more positive than the setpoint voltage, the op amp output

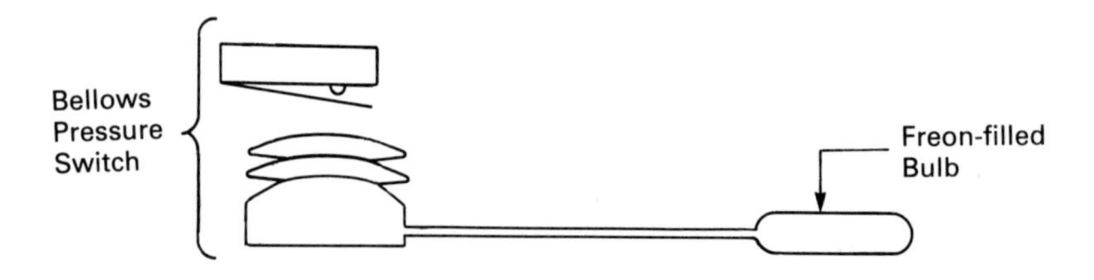


**Figure 10.1** A bulb-type temperature controller

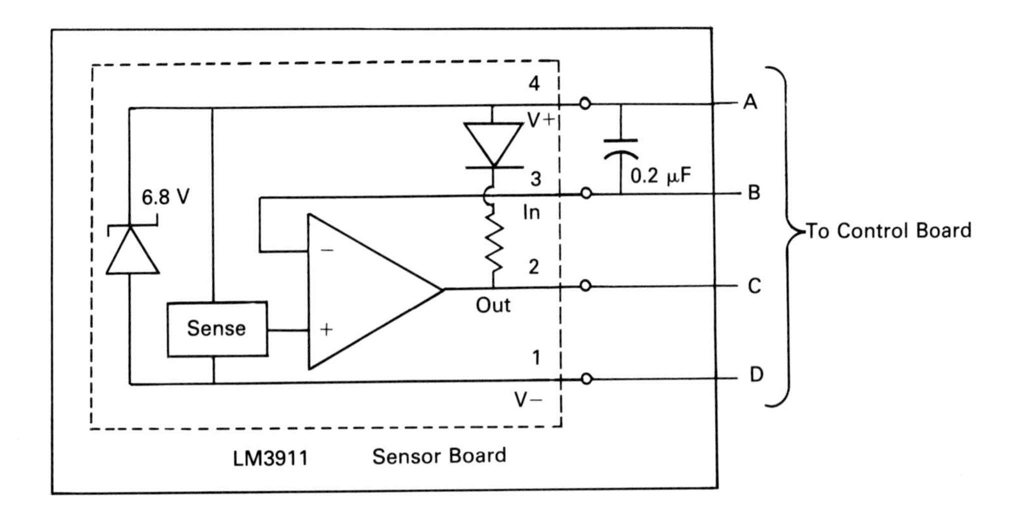


**Figure 10.2a** An electronic waterbed controller: the sensor board.

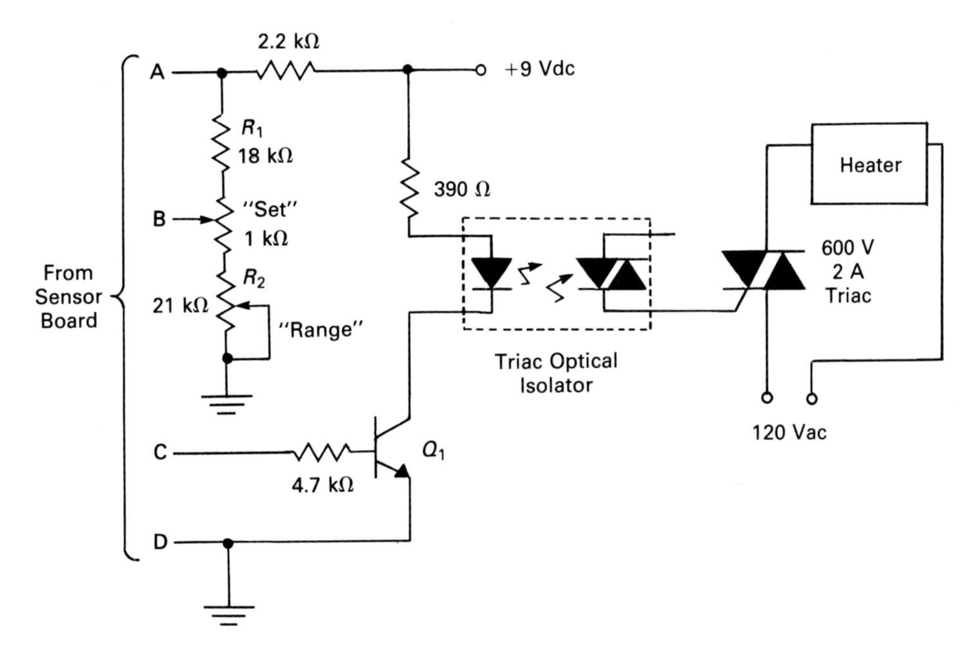


**Figure 10.2b** An electronic waterbed controller: the control board.

is high. When the output of the sensor is less positive than the setpoint voltage, the output of the op amp is low.

The desired operating range for the controller is between about 85° and 115° Fahrenheit, or 29° and 46° Celsius. By adding 273.2 to convert Celsius to Kelvin, we find that the operating range must be between 302°K and 319°K. That means that the noninverting input will sense a range of voltages between 3.02 V and 3.19 V *below the IC's* $V^+$ *pin*. Because of the 6.8 V Zener diode built into the chip, we can translate the voltage range on the noninverting input to be

$$6.8\text{ V} - 3.02\text{ V} = 3.78\text{ V at } 85°\text{F}$$
$$\text{and } 6.8\text{ V} - 3.19\text{ V} = 3.61\text{ V at } 115°\text{F}$$

This is a range of only 170 mV between the lowest and highest voltages.

An ordinary potentiometer rotates through its entire range in 270° of rotation. Connected directly across the 6.8 V reference, the potentiometer will change 680 mV in a tenth of a turn. For the 170 mV range of voltages needed in this project, we need to be able to adjust the potentiometer only a fortieth of a turn. This adjustment is far too sensitive for our needs.

The solution chosen was to use a 1 kilohm potentiometer situated between two resistors, $R_1$ and $R_2$, in the circuit. $R_1$ is an 18-kilohm resistor and $R_2$ is a ten-turn potentiometer adjusted to 21 kilohms. The voltage at the top of the potentiometer is 3.78 V. The voltage at the bottom of the potentiometer is 3.61 V, which corresponds to the required adjustment range.

Recall that when the temperature is low, the output of the comparator is high. This high output is used to saturate transistor $Q_1$, which in turn operates the LED of an optical isolator, which then operates a high-current TRIAC to operate the water-bed heater. The optical isolator ensures that the 120 Vac that operates the heater does not get into the sensor board, a hazardous situation since this board must be located between the frame of the bed and the mattress.

This control is certainly not an industrial controller, but it follows the same principles. The problems solved in its design are similar to those found in designing a similar control for an industrial process, and the sequence of operation is the same: measure, decide, act, then repeat. The same circuit can be adapted to temperature control over a wide range of applications.

## 10.2 CONTROLLING MATERIAL THICKNESS

Sheet metal is manufactured from steel ingots through a series of rolling operations. The ingots are first rolled into 6-inch-thick slabs. The slabs are then reheated and rolled into 3/4-inch plates. These plates are further rolled and reduced to thinner gages, cleaned of scale, and rolled into coils for easy handling. The coils are then delivered to the cold-reduction mill for the last manufacturing step.

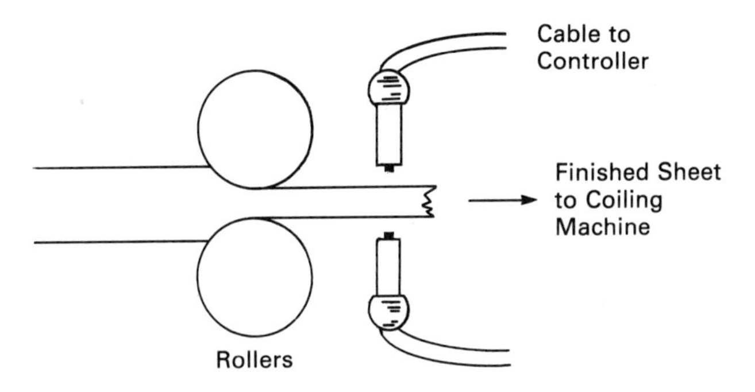


**Figure 10.3** Controlling sheet metal thickness

The last step in manufacturing sheet steel is another rolling operation, but in this one the sheet metal is under considerable tension between the rolling mills and the last set of rollers and the coiling machine. This tension provides a fine finish texture and also improves the accuracy of the gage (thickness) and flatness of the finished steel sheets.

The pressure on the rollers and the tension on the steel, which contribute to the final gage of the steel, must be carefully controlled. The inductive proximity detector described in Chapter 3 is used in this important sensing task.

Figure 10.3 shows the placement of two inductive proximity sensors to measure the thickness of steel. Recall that the proximity detector radiates a high-frequency field in front of its sensor and detects the presence of eddy currents in nearby metal. The amount of eddy current is proportional to the distance between the detector and the metal. Any variation in the thickness of the metal sheet will cause a change in the output of one or both of the sensors. The two outputs are compared in a differential-type controller and appropriate control signals are sent to actuators, which adjust the pressure of the rollers or the tension on the moving sheet of steel.

The controller and actuators used in this application are more complex than the controller for the waterbed heater. They are not discussed in this chapter.

## 10.3 A ROBOTIC ARM POSITION CONTROLLER

Electronic control of robotics requires precise knowledge of the position of each joint or axis of movement of the robotic manipulator. As the arm moves through its range of motion, the load it is carrying and the forces of gravity and inertia combine to pull the motion of the arm off course. In order for the arm's motion to be repeatable and accurate, some means must be provided to detect and correct changes in the arm's position. The circuit in Figure 10.4 illustrates one approach to this task. The circuit offers full servo feedback, and "knows" and corrects the arm position 100% of the time.

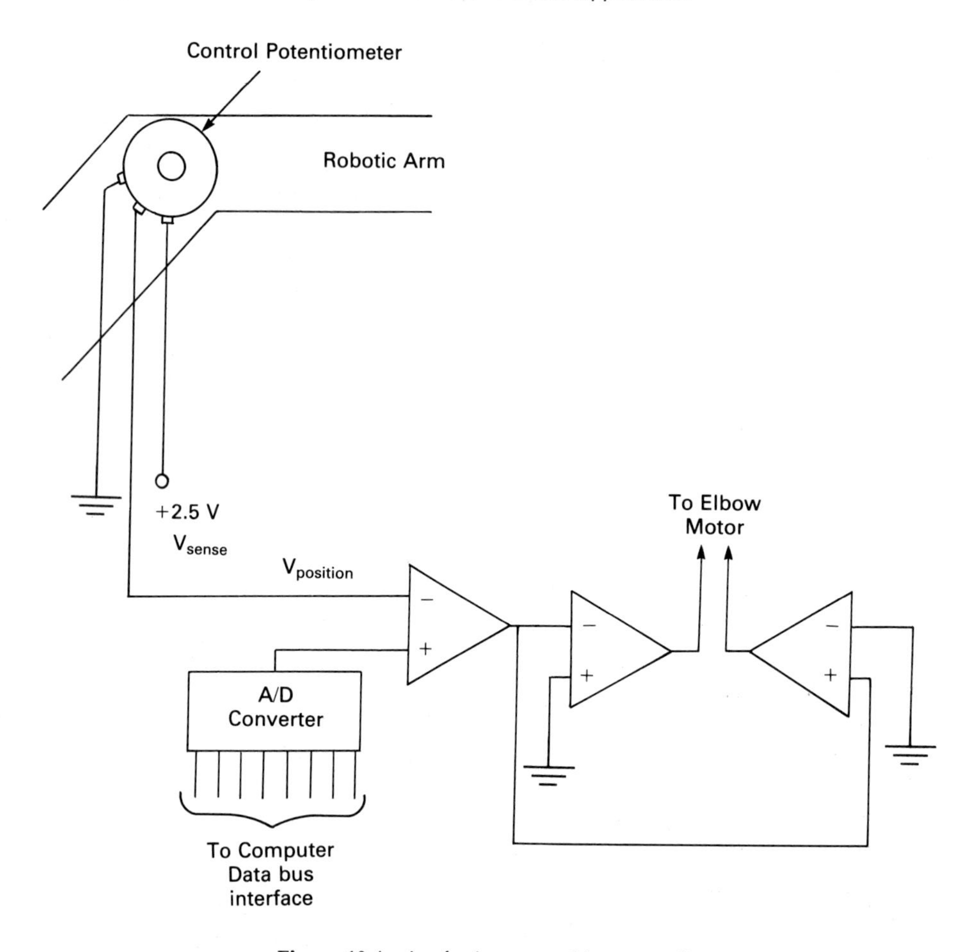


**Figure 10.4** A robotic arm position controller

The circuit operation begins with a well-regulated reference voltage, $V_{sense}$, of 2.5 Vdc. This voltage is carried over a pair of control lines to the position-sensing rotary potentiometer on the robotic joint — in this case, the elbow of the arm. The position of the potentiometer wiper is determined by the position of the elbow. Thus, each position of the elbow through its entire range of movement is represented by a voltage output, $V_{position}$.

The controller program supplies a second voltage, $V_{control}$, in the range from 0 to 2 Vdc. In this example, the controller is a small personal computer. $V_{control}$ is generated by a digital-to-analog-converter IC on the control interface board. $V_{control}$ and $V_{position}$ are compared by a comparator that outputs an error signal, $V_{error}$.

The error signal drives, in turn, a bridge amplifier whose output drives the dc elbow motor. The magnitude of the motor voltage determines the speed of the motor, while the polarity determines the direction. Since the level of the motor voltage is large when the error is large and small when the error is small, this type of control is called *proportional control.*

## SELF-TEST

1. The 10 mV per degree K output of the LM3911 temperature controller is referenced to the ______________________________ pin of the IC.
2. Because of the built-in voltage regulator, the $V^+$ pin of the LM3911 is at ______________________________ referenced to the $V^-$ pin.
3. Why was an optical isolator used in the waterbed heater controller?
4. The inductive proximity detectors in the sheet steel measuring circuit detect the ______________________________ in nearby metal.
5. The A/D converter in the robot control circuit accepts binary numbers from the computer and converts them into ______________________________.
6. The elbow motor in the robot controller is driven by a(n) ______________ amplifier.
7. When the elbow of the robot is at its mid position, the sensor potentiometer will return ______________________________ volts to the controller board.

## QUESTIONS/PROBLEMS

1. At 25°C, the LM3911 sensor will apply ______________________________ volts to the noninverting input of the op amp.
2. Explain how you would modify the controller to operate over a range of 20°C to 30°C.

# Troubleshooting Sensor Circuits

The sensing transducers in an industrial system are the components that are the most exposed to the unfriendly manufacturing environment. It should come as no surprise, then, to realize that the sensors and their wiring fail more often than any other single component in a control system. The ability to troubleshoot these circuits is an important skill for the industrial technician.

Sensory systems may be conveniently divided into two parts — the wiring and the sensor itself. We will begin with the techniques for troubleshooting the wiring, the easier of the two techniques and the one best suited to field repair.

## TROUBLESHOOTING SYSTEM WIRING

The first step in troubleshooting the wiring in a system is to inspect the connections at both ends of the wire. Most such connections are made on insulated screw terminal strips that have barriers between adjacent connections. Look for signs of corrosion, for wires that have broken at the screw terminal, and for tiny strands of wire that may have created short circuits between terminals. The wire will often break *inside the insulation* near the terminations. Test for this by pulling firmly on the conductor. One frequent sign of a broken conductor is a sharply angled bend in the wire.

If the wiring passes the visual inspection, *disconnect the wire at both ends.* When you make the disconnections, be sure that the wires are not contacting each other or the conduit box or equipment housing. Be careful to see that not even a single tiny strand of the wire is in contact with anything.

Once the wire is disconnected, use an ohmmeter to test it for shorts. Set the ohmmeter on a high resistance range and test each conductor against every other conductor, against conduit ground, and against the shielding, if it exists. You should read an open circuit in all cases. If you do not, you have probably found the problem.

If the wire passes the short circuit test, short all of the wires together at one end of the cable and conduct a continuity test from the other end. Once again, test every conductor in the cable with an ohmmeter, this time set on the lowest resistance range. At this point, you should read a low resistance between the conductors. There will be some resistance, depending on the wire's length and gage, but it should not be more than a few ohms. On many wire runs, the readings will be in the tenths of ohms. Any substantial variation indicates poor conductivity, which is caused by a corrosive break in the conductor. If any of the conductors prove to be open, or if there is an abnormal resistance reading, the problem has been found.

Thermocouples are often connected to the temperature-measuring electronics with *thermocouple wire*, which is specially made to match the characteristics of the thermocouple itself. This is done to avoid creating extra thermocouple junctions where the thermocouple and the system wiring join. Since this wire is composed of such things as iron and various metal alloys, the resistance readings from it will be somewhat higher than those from copper wire. Shop records should include a short-circuit resistance continuity reading for each thermocouple run in a system. Compare your readings against this standard.

If the wiring proves defective, it must be replaced. Most such wiring is run inside of electrical metal conduits (pipe). The old wire must be pulled out and new wire inserted in its place. When replacing thermocouple wire, be sure to use the same type of wire to prevent calibration problems. Seek help from the plant electrician for this task.

## TROUBLESHOOTING TRANSDUCERS

If the system wiring has passed its tests, the problem is either the sensor or the controller. Reconnect the wiring to the sensor, being careful to avoid single-strand wire shorts to other terminals or to ground. When the sensor requires a voltage source or a current source from the controller, be sure that this is present before reconnecting the wiring to the controller. Check voltage sources with a voltmeter and check current sources with a milliammeter with a series resistance about equal to the resistance of the transducer added.

If the controller is supplying the proper levels of voltage or current, reconnect the system wiring and verify that the voltage or current source is still functioning. Check the voltage source after all of the wires are connected. Check the current source by placing your milliammeter in series with the current supply wire after all of the other wiring has been connected. Large variations may indicate problems with the system power supply, as well as with the transducer being tested.

If the system passes these tests, check it for normal operation before proceeding to any further checks. Your work with the wiring may have corrected a problem that your visual inspection missed. While such miraculous cures leave the technician with a sense of uncertainty, they are far more common than we would like to admit.

If the system still does not function, it is necessary to continue with tests of the sense transducer. No matter what the type, the simplest, fastest, and most reliable test for a failed sensor is *substitution*. Most well-equipped plants have replacement transducers on hand. Replace the suspect transducer with a new one and test for system operation. When safety allows, this can be done without physically removing the old transducer. If the replacement transducer does not fix the problem, the problem is in the system controller.

When such replacement fixes the problem, your work is done. Most sense transducers should not be repaired; the cost saving of such repair does not compensate for the lack of reliability. In the event that a particular sensor is expensive, have it repaired by the manufacturer. Remember that *reliability* is the operative word in manufacturing.

When circumstances dictate that a repair be attempted, the technique depends on the type of sensor. Many thermocouples have their junction connected to ground. Check here for a poor connection, which is usually caused by dirt and corrosion under the mounting bolt. Scrape the area clean and replace the thermocouple bolt.

Both slide and rotary potentiometers are subject to the infusion of dirt. Occasionally, a temporary repair can be made with a few squirts of contact cleaner. Squirt the cleaner into the potentiometer and run the potentiometer back and forth several times. Even if this works, order a replacement and install it as soon as it arrives.

Some electromechanical sensors develop problems where the mechanical and electrical elements join. Check and tighten the setscrews. Look for dirt and particles of waste material between the mechanical and electrical components interfering with proper action.

Inspect the actuating levers of snap-action switches for accidental bending. These levers are made of a soft, easily bent metal. Use a voltmeter to verify that these and any other switches are operating. The voltage across an open set of contacts should be the applied voltage. The voltage across a closed set of contacts should be very nearly zero. Testing these contacts with an ohmmeter may not bring a set of arced contacts to light.

Troubleshooting is as much an art as it is a science. An expert troubleshooter relies on years of experience and practice to acquire an "instinct" for what is wrong. The preceding steps can guide a new technician in the proper direction. Short of living through the same number of years, with the same amount of experience, shop talk can add immeasurably to your troubleshooting skills.

# Section III

# Signal Conditioning

The transducers in Section II of this book measure the controlled variable in a process. Usually, these devices output information as small changes in resistance, current, or voltage. These small signals will need to be amplified and conditioned before they can be used.

In this section, we will examine the techniques that are used to amplify these signals, and to filter out the effects of electromagnetic and radio frequency interference. Much of the information in this section demonstrates practical applications for information you learned earlier in your study of the operational amplifier. It includes a brief review of op-amps for this reason.

# 11

# The Bridge Network

Measurement transducers often report to the system controller with a change in resistance. Some devices, such as the thermistor, have large resistance swings, which are easily measured. Other devices, such as the RTD or the bonded wire strain gage, have very small resistance swings, which are more difficult to measure. Other measurement transducers report to the controller with a change in voltage, which may also be very small and difficult to measure or transmit.

Strangely enough, both small changes in resistance and small changes in voltage are measurable with variations of the same circuit — the bridge network. In this chapter, we will discuss the basic Wheatstone bridge circuit, then expand the basic bridge into even more sensitive measurement bridges.

## OBJECTIVES

You will have successfully completed this chapter when you can:

1. Analyze the current and voltage in the galvanometer leg of a Wheatstone bridge.
2. Explain the difficulties of measuring accurately with a Wheatstone bridge.
3. Describe and explain the operation of a three-wire bridge network.
4. Describe and explain the operation of a four-wire bridge network.
5. Explain the use of a bridge network for measuring voltage.

## 11.1 MEASURING RESISTANCE

The conventional ohmmeter approach to measuring resistance is shown in Figure 11.1. The meter battery, $V_M$, provides current for the circuit. The adjust resistor is used for calibration, and the resistor to be measured is labeled $R_X$. Resistances $R_1$ and $R_2$ represent the resistance of the ohmmeter leads, which is negligible for bench instruments and readily compensated for when the ohmmeter is calibrated.

When $R_X$ is an RTD or a strain gage, however, it may be located a considerable distance from the meter. In this case, lead resistance may well become a significant part of the measurement. Lead resistance will also vary with temperature, depending on the temperature coefficient of the lead wiring. Another problem is that such circuits require relatively large currents, contributing self-heating errors to the system. The measurement accuracy also becomes subject to variations in $V_M$, since operation is most likely continuous. All of these problems combine to require frequent calibration — not a very desirable requirement for automatic industrial control systems.

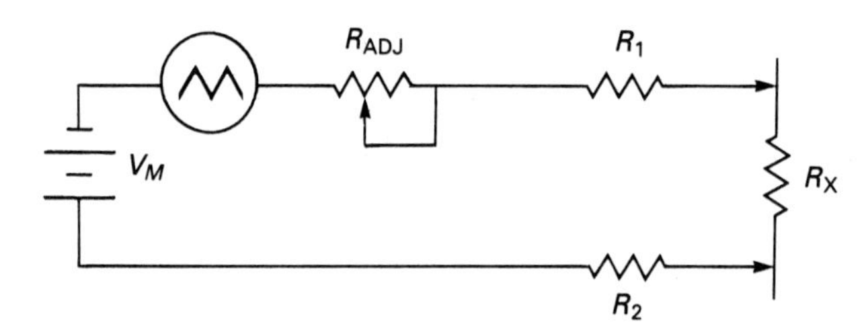


**Figure 11.1** Measuring resistance

## 11.2 THE WHEATSTONE BRIDGE

The Wheatstone bridge has long been a standard measuring device in the field of electronics and electricity. Capable of measuring values of resistance with a high degree of accuracy, it is little wonder that power companies, telephone companies, and industrial electricians have long considered the Wheatstone bridge as a standard measurement tool in their inventory of instruments. Automation and process control have adapted this versatile tool to suit their needs as well.

Figure 11.2 illustrates the basic Wheatstone bridge configuration used for measuring resistance. The bridge has two pairs of terminals. Terminals $C$ and $D$ are connected to a source of excitation voltage, while terminals $A$ and $B$ are connected to a sensitive galvanometer. Resistances $R_A$ and $R_B$ are usually equal and quite large to limit and control the overall circuit current. $R_C$ is a precision potentiometer with a calibrated dial, while $R_X$ is the resistance to be measured.

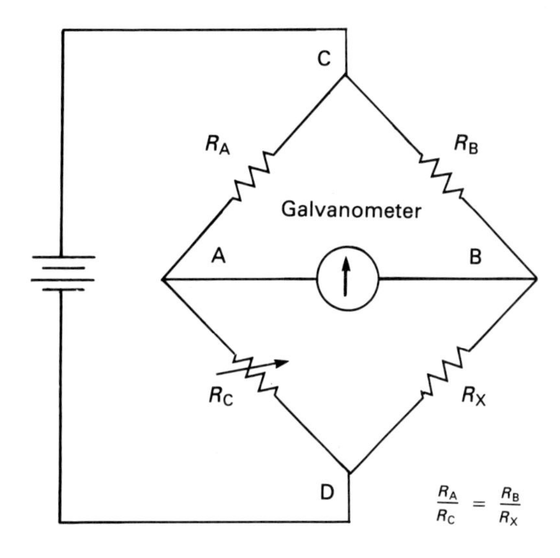


**Figure 11.2** The Wheatstone bridge

In use, $R_C$ is adjusted for zero current in the galvanometer leg of the bridge. This indicates that the bridge is *balanced*. The value of $R_X$ can now be calculated from the equation shown in Figure 11.2. In the usual case, when $R_A = R_B$, then $R_X = R_C$, and the resistance can be read directly from the calibrated dial of the potentiometer. This, of course, does not lend itself well to automatic control. The need to adjust the arm of the potentiometer to take a reading would require mechanical linkage and cause frequent recalibration and maintenance due to wear.

Fortunately, we can also determine the value of $R_X$ by measuring the current in the galvanometer leg. This does not lend itself well to simple Ohm's law calculations. It is best solved using Thevenin's theorem. Thevenin's theorem states that any network can be replaced with a constant voltage source in series with a single resistance. To Thevenize the Wheatstone bridge, we must calculate both the voltage, $V_{TH}$, and the series resistance, $R_{TH}$.

The first step in Thevenizing a circuit is to remove the section in question. In this case, we want to know the current through the galvanometer leg, so that is the portion of the circuit removed. The Thevenin voltage is the voltage between the Thevenin terminals, points $A$ and $B$ in the bridge. The Thevenin resistance is the resistance between these same two points when the excitation voltage has been replaced by a short circuit. This is more easily seen by example.

Figure 11.3 illustrates the necessary steps to Thevenize a Wheatstone bridge. We will use the resistance values shown in the drawing for our example. Figure 11.3a shows the Wheatstone bridge redrawn for clarity. $R_G$ represents the resistance in the galvanometer leg. In Figure 11.3b, we can see that $R_G$ has been removed, the first step in Thevenizing. The Thevenin voltage, $V_{TH}$, is the difference between the voltage across $R_C$ and the voltage across $R_X$, both now easily calculated with Ohm's law.

$$I_1 = \frac{12\ V}{100\ k\Omega + 20\ k\Omega}$$

$$I_1 = 100\ \mu A$$

$$V_{RC} = 20\ k\Omega \times 100\ \mu A$$

$$V_{RC} = 2\ V$$

$$I_2 = \frac{12\ V}{100\ k\Omega + 10\ k\Omega}$$

$$I_2 = 109\ \mu A$$

$$V_{R_X} = 10\ k\Omega \times 109\ \mu A$$

$$V_{R_X} = 1.09\ V$$

$$V_{TH} = 2\ V - 1.09\ V$$

$$V_{TH} = 0.91\ V$$

Figure 11.3c shows the circuit with the power leads shorted. The circuit is redrawn for clarity in Figures 11.3d and 11.3e. From Figure 11.3e, we can see that it is a simple resistance calculation to the final simplification, which is shown in Figure 11.3f.

$$R_{EQ_1} = 100\ k\Omega \parallel 100\ k\Omega$$

$$R_{EQ_1} = 50\ k\Omega$$

$$R_{EQ_2} = 20\ k\Omega \parallel 10\ k\Omega$$

$$R_{EQ_2} = 6.67\ k\Omega$$

$$R_{TH} = 50\ k\Omega + 6.67\ k\Omega$$

$$R_{TH} = 56.7\ k\Omega$$

In the final circuit, $V_{TH}$ and $R_{TH}$ are connected in series with $R_G$. Now we can solve for the actual current through $R_G$.

$$I_G = \frac{0.91\ V}{56.7\ k\Omega + 1\ k\Omega}$$

$$I_G = 15.8\ \mu A$$

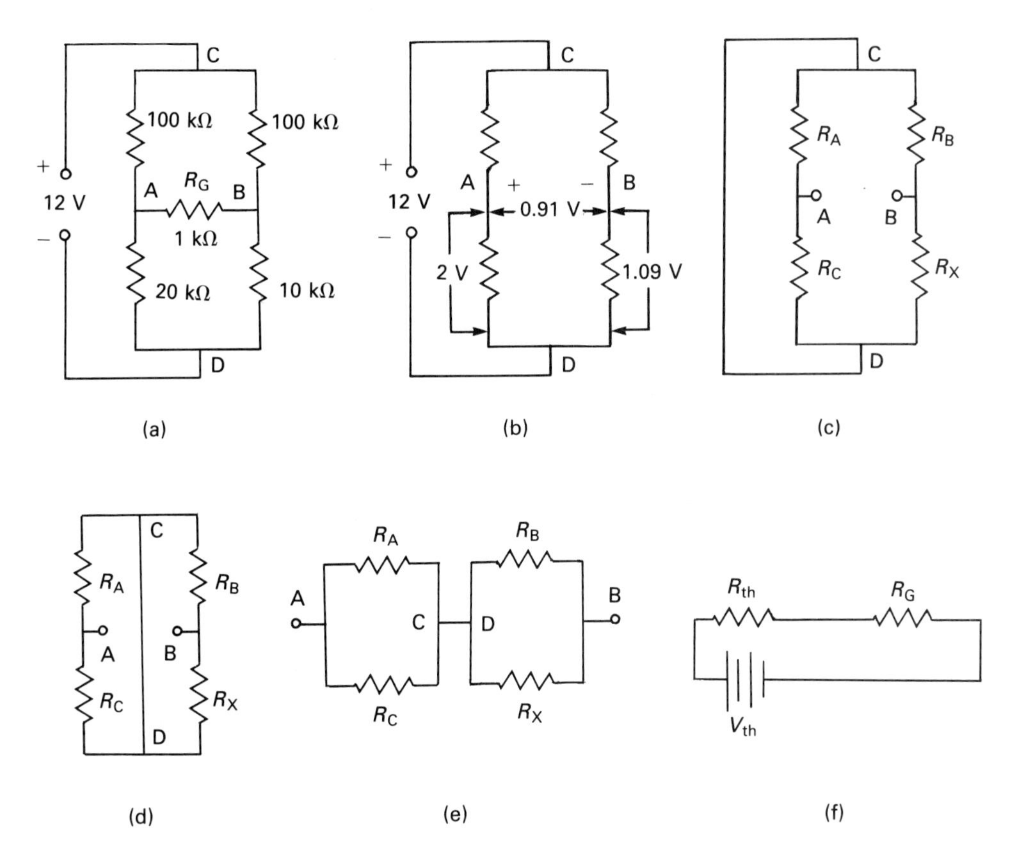


**Figure 11.3** Thevenizing the bridge step by step

This calculation is easily programmed in a computer in BASIC. Furthermore, the current in the galvanometer will vary in a linear manner with changes in $R_X$. That means that a current-to-voltage converter can feed the "value" of $R_X$ to a controller directly from the bridge.

Even better, Thevenin's theorem problems can be solved directly in a real world circuit. First, remove the galvanometer leg, as we did on paper with our example. Now measure the voltage from $A$ to $B$. This is $V_{TH}$. Now replace the excitation voltage with a short circuit and measure the resistance between $A$ and $B$. This is $R_{TH}$. Now you can draw the Thevenin equivalent circuit and calculate $I_G$.

A variation of the Wheatstone bridge replaces the galvanometer with a high-impedance digital voltmeter (DVM). The current in the galvanometer leg is negligible. The meter measures $V_{TH}$, which is directly proportional to $R_X$.

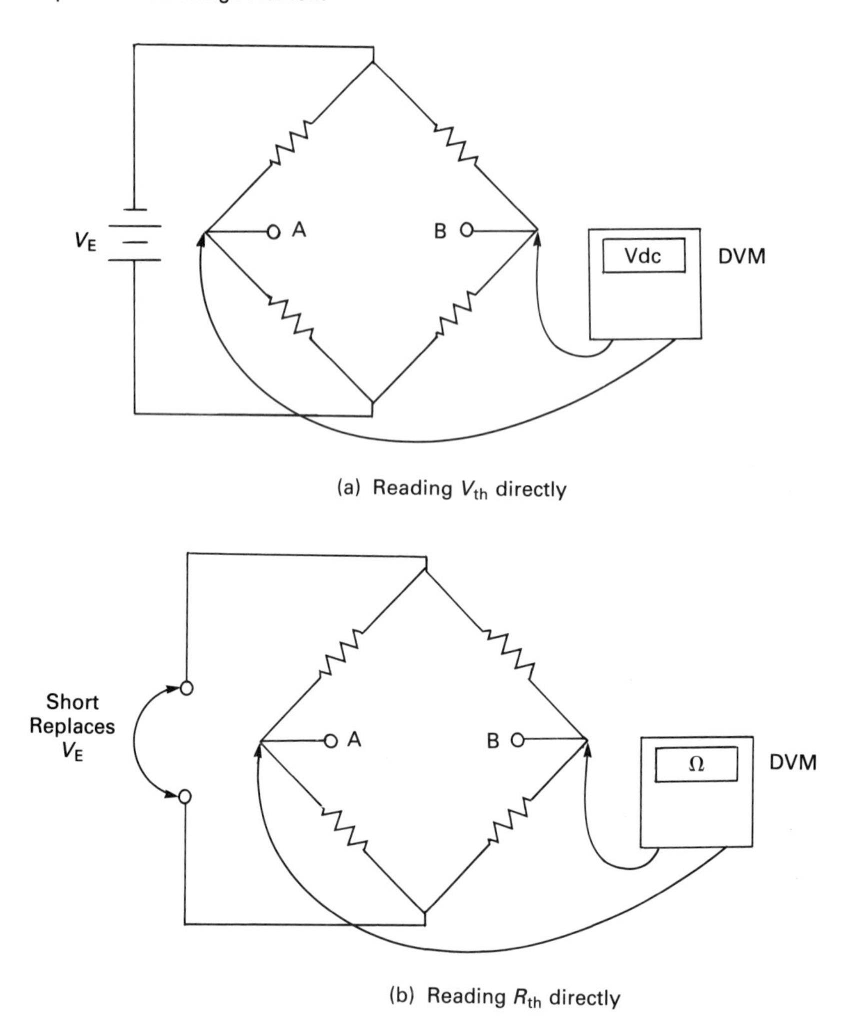


**Figure 11.4** The easy way to Thevenize

We can demonstrate the relationship between $V_{TH}$ and $R_X$ with an example using the circuit in Figure 11.5. Note that the resistors in the bridge network have been rearranged. This configuration is used to simplify the mathematics. $R_A$ and $R_B$ are equal. The bridge is designed around a particular value of $R_X$, and $R_C$ is selected or adjusted to be equal to $R_X$ at the setpoint.

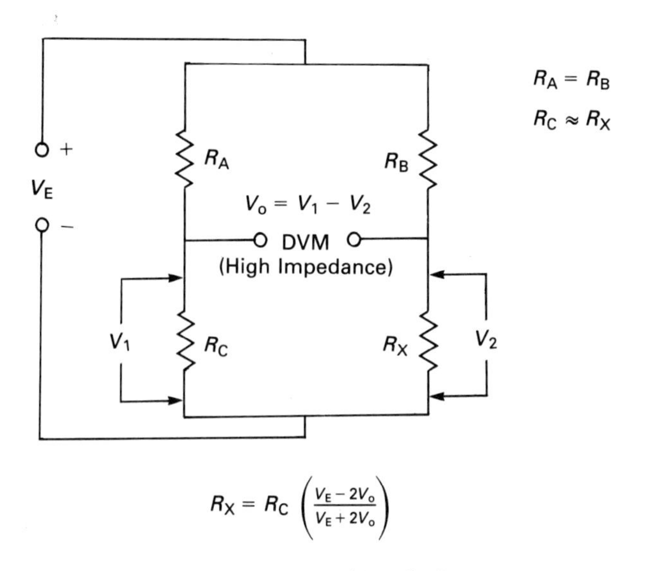


**Figure 11.5** Solving for $R_X$

Now $V_1$ will be equal to one-half of the excitation voltage, $V_E$, and $V_2$ can be calculated using the formula for a two-resistor voltage divider.

$$V_2 = \frac{R_C \times V_E}{R_C + R_X}$$

$$V_1 = \frac{V_E}{2}$$

$$V_o = V_1 - V_2$$

When $R_X = R_C$, then $V_o = 0$. When $R_X$ does not equal $R_C$, we can use $V_o$ to calculate $R_X$ by transposing the formula.

$$R_X = \frac{R_C\,(V_E - 2V_o)}{V_E + 2V_o}$$

## 11.3 THE THREE-WIRE BRIDGE

When it comes to applying the bridge to remote sensing, as in RTD applications, problems pop up once again. In order to avoid putting the bridge itself into the high-temperature atmosphere, we must once again use wire leads with their inherent resistance.

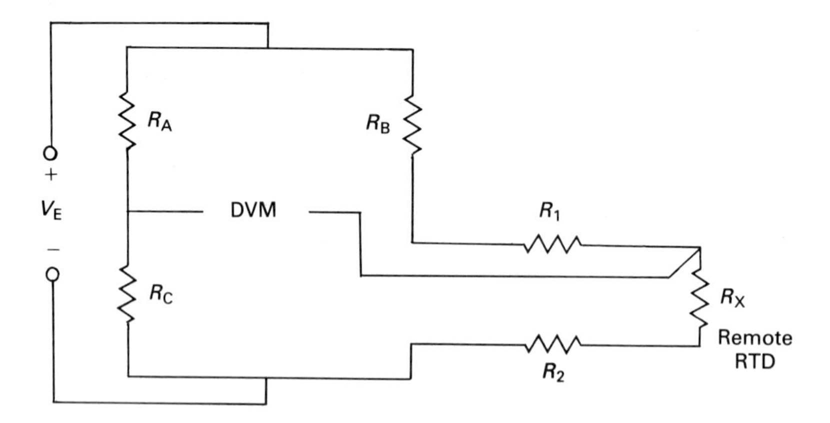


**Figure 11.6** The three-wire bridge network

The three-wire bridge, Figure 11.6, solves some of the problems of remote sensing, making the bridge measurements more accurate.

This approach adds a third wire to the leads to the remote resistance, $R_X$. Leads one and two must be kept at exactly the same length, so that their lead resistances are equal. Since the location of the sense lead connection at the RTD places $R_1$ in one leg of the bridge and $R_2$ in the other, the lead resistances cancel out. No current flows in the sense lead, so its resistance is not important since it has no effect on the reading.

The three-wire bridge is quite accurate when the bridge is near balance and $V_o$ is very small. Strain gages, which have only very small changes in resistance over their range, are well suited to measurement with the three-wire bridge. RTDs, however, may vary considerably over their sensing range. As $V_o$ becomes larger, so does the error introduced by lead resistance. If you can measure the resistance of the leads, the error can be corrected. If not, it may be necessary to calibrate the instrument very carefully against a standard and use a software table for corrections.

## 11.4 FOUR-WIRE RESISTANCE READINGS

If a constant current source is connected to a resistance, the voltage drop across the resistance is directly proportional to the resistance. This simple statement of Ohm's law provides the clue to a very accurate method of measuring the resistance of RTDs, strain gages, and other resistive sensors. The system requires four lead wires to the sensor, but the wire lengths and lead resistances are completely eliminated from consideration. The extra wire seems a small price to pay.

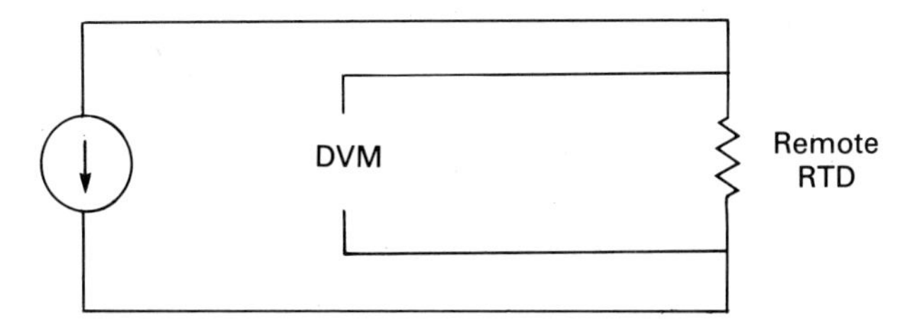


**Figure 11.7** Four-wire resistance measurement

The circuit is shown in Figure 11.7. Current sources are rather easily constructed using op amps or transistors. They may also be found in IC form at low cost. The current source is capable of supplying a constant current to any size load. The current will remain constant whether the load increases or decreases in size. Thus, the voltage across the load must change.

The sense leads, of course, carry no current, so their lead resistance doesn't matter. The voltage measured is that which is directly across the resistive sensor, so the voltage drops in the lead resistance of the current loop are also not important. This simple expedient provides accuracy at any distance.

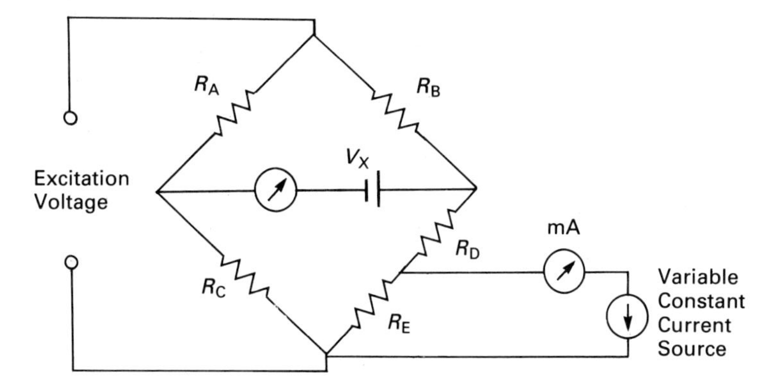


**Figure 11.8** Measuring voltage with a bridge

## 11.5 MEASURING VOLTAGE WITH A BRIDGE

Many of our transducers output tiny voltages which must be measured. Most such sensors are sensitive to the loading effect caused by measuring devices that draw current. If you recall that the current in the galvanometer leg of a balanced bridge is zero, you may begin to see how the bridge can be used to measure these millivolt-level voltages without drawing current.

Figure 11.8 shows one circuit used for bridge voltage measurement. In this bridge, the resistive portion of the bridge is balanced to begin with. Only the unknown voltage, $V_X$, in the galvanometer leg can put the bridge out of balance. When this happens, a controlled current source adds an external current source, $I_3$. The path for this extra current in the bridge is through $R_E$ only. The amount of current needed to restore balance to the bridge is directly proportional to $V_X$. No current flows in the galvanometer leg when the bridge is balanced. The circuit offers infinite impedance to the sensor.

This circuit, which was developed before the very high-impedance FET voltmeter and DVM were available, is found only in older systems. Its use should not be considered for new installations. It is included here for reference.

## SUMMARY

In this chapter, we have examined the use of the Wheatstone bridge for measuring resistance and voltage. The Wheatstone bridge is a very sensitive device that is capable of measuring small changes in resistance with high accuracy. We have seen:

1. The current in the galvanometer leg of the Wheatstone bridge can be calculated using Thevenin's theorem.
2. When the galvanometer is replaced with a DVM, the unknown resistance may be easily calculated using the output voltage.
3. Lead resistance can add significant error to these voltage readings when the sensors are remote from the bridge.
4. The three-wire bridge can overcome some of the effects of lead resistance when the bridge is near balance.
5. When the system must deal with large changes in sensor resistance, the four-wire resistance measuring technique must be used.

## SELF-TEST

1. Standard ohmmeter measurement of resistive sensors is inconvenient because the meters require frequent ______________________________.

2. Ohmmeter measurements are also inaccurate for remote sensing because of the effects of ____________________________.
3. When a Wheatstone bridge is in balance, the current in the galvanometer leg is ____________________________.
4. In the circuit shown in Figure 11.2, when $R_X$ is larger than $R_C$, the voltage at point *A* will be ____________________________ with respect to point *B*.
5. The bridge in Figure 11.2 is brought into balance by adjusting __________ ____________________________.
6. The ____________________________ bridge circuit can reduce the effects of lead resistance when remote sensing must be used.
7. The three-wire bridge is accurate only when the bridge is ______________.
8. When a resistive sensor is expected to undergo large resistance changes, the ____________________________ resistance measurement is used.
9. The four-wire resistance measurement technique requires a(n) __________ source.
10. The Wheatstone bridge may also be used to measure millivolt-level sensor voltages without ____________________________ effect.

## QUESTIONS/PROBLEMS

1. A 100-ohm RTD is connected in a Wheatstone bridge. The galvanometer leg of the bridge contains a high impedance DVM, so there is no appreciable current in this leg. Resistors $R_A$ and $R_B$ are equal to 100 kilohms. Resistor $R_C$ is adjusted so that the bridge is in balance when the RTD measures exactly 100 ohms. A change of 10 °C will cause a change in the RTD's resistance of four ohms. How much change will there be in the DVM reading for a 10 °C change? The excitation voltage is 12 VDC.
2. A one-kilohm thermistor is connected in a four-wire resistance measurement circuit. The constant current source supplies one mA of current. The thermistor changes 40 ohms per degree of temperature change. What voltage change would you read for a five degree change in temperature?

# 12

# The Operational Amplifier — A Refresher

The integrated-circuit operational amplifier is one of the most important building blocks of electronics. From its early and somewhat troubled beginnings when the output would suddenly lock up and refuse to change, the op amp has evolved into a reliable, simple-to-use component. It is hard to find a branch of electronics where the op amp does not find some use.

In process control work, the op amp serves to buffer and isolate the sense transducers. It amplifies tiny transducer signals and filters out undesirable noise. Op amp circuits perform summing and comparison functions, add damping to the system, and even make the decisions that control output signals.

In this chapter, we will refresh your understanding of this valuable tool.

## OBJECTIVES

You will have successfully completed this chapter when you can:

1. Explain the operating principles of the inverting op amp circuit.
2. Explain the operating principles of the noninverting op amp circuit.
3. Explain the operating principles of the voltage follower op amp circuit.

## 12.1 THE IDEAL AMPLIFIER

Engineers often find it convenient to analyze a device from the *black-box* perspective. The black box equivalent of an amplifier has two input terminals and two output terminals. From the way these terminals interact with each other, we can measure meaningful parameters for describing the amplifier.

The amplifier shown in Figure 12.1 shows the general idea of the black-box amplifier. Some of its important parameters are shown in the diagram:

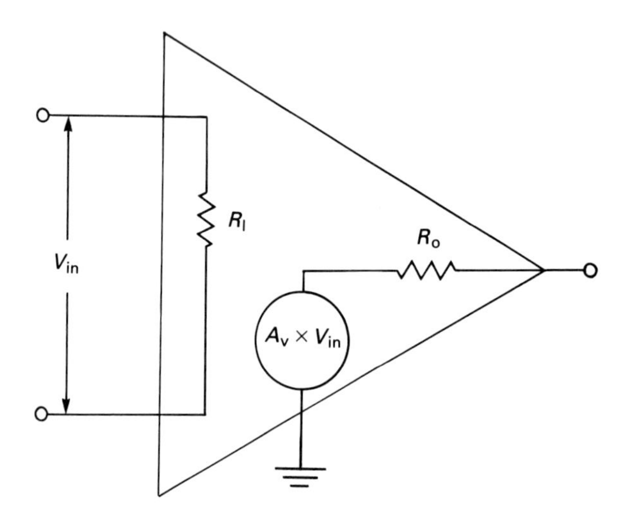


**Figure 12.1** The amplifier as a black box

$R_I$ = the input resistance. This is the resistance "seen" by the signal source looking into the input terminals. This need not be an actual resistor. The input resistance acts like a resistor in series with the signal source and its internal source resistance.

$A_v$ = the amplifier gain. This is illustrated as a voltage source in the black-box diagram. It acts like a generator whose output is equal to $A_v \times V_{in}$.

$R_o$ = the output resistance. This is the resistance "seen" by the load looking back into the amplifier's output terminals. This resistance acts like a resistance in series with the load. It need not be a real resistor.

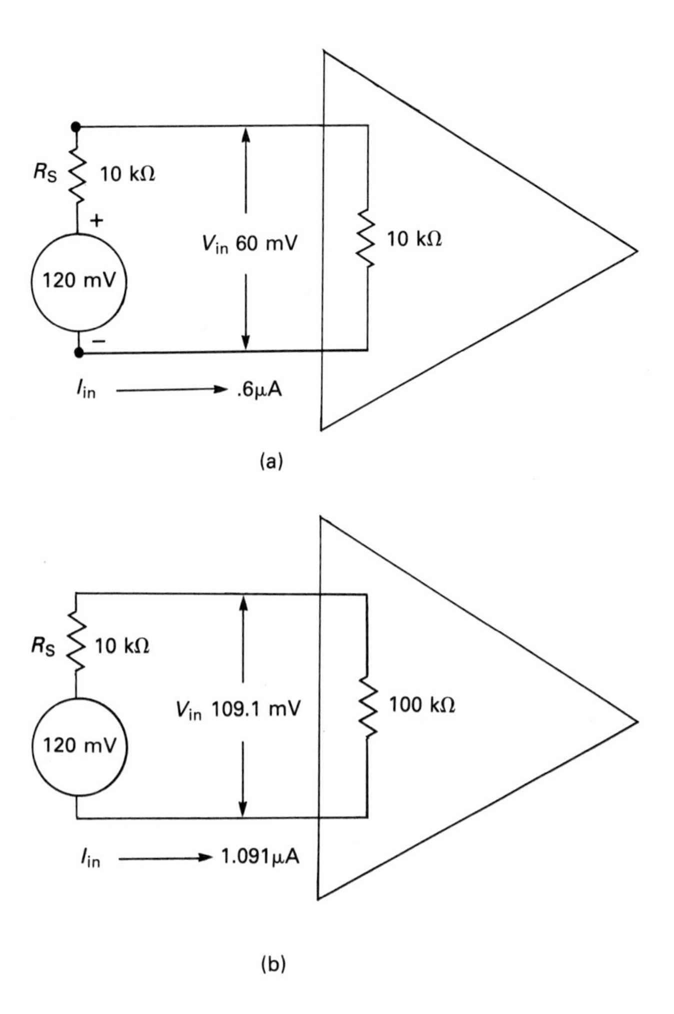


**Figure 12.2** The effect of $R_{in}$

In Figure 12.2, we can see the effect of $R_{in}$ on the performance of an amplifier. Every source of signal, whether it is a microphone or an RTD, has its own characteristic source impedance. This impedance makes the signal source appear to be a generator with a resistance in series. For our example, we will use the same figures throughout.

$$V_{sig} = 120 \text{ mV}$$
$$R_S = 10 \text{ k}\Omega$$

In Figure 12.2a, $R_I$ of the amplifier is 10 kilohms. As we can see, the input voltage at the amplifier terminals is only *half* of the signal voltage, or 60 mV.

In Figure 12.2b, $R_I$ of the amplifier is 100 kilohms. In this circuit, most of the signal voltage is available at the input terminals to be amplified. The higher input resistance means that less of the signal voltage is lost to $R_S$.

From this example, we can see that one of the characteristics of an ideal amplifier is a high input resistance. The ideal amplifier's input resistance is infinite. If no current can flow from the source to the amplifier input, none of the input signal will be lost to the source impedance.

Figure 12.3 shows the effects of $R_o$ on the output voltage. In our examples, we will use the same figures throughout:

$$V = 1\ \text{V}$$

$$R_L = 1\ \text{k}\Omega$$

In Figure 12.3a, we see the effect of an $R_o$ of 1 kilohm. Because of the voltage drop across $R_o$ in series with the load, only half of our signal is available across the load.

In Figure 12.3b, $R_o$ has been reduced to 1 ohm. Virtually all of the signal is now impressed across the load and very little is lost to $R_o$.

Since signal lost to $R_o$ is felt mainly as heat by the amplifier, the reduction in $R_o$ provides two benefits. There is a reduction in temperature rise in the amplifier, and there is more output signal available to perform work. From this, we can see that one characteristic of an ideal amplifier is an output resistance of zero ohms.

Other characteristics of an ideal amplifier are less easily seen in diagrams. These include the following:

- Bandwidth = infinity. The amplifier should amplify signals of all frequencies equally.
- Gain = infinity. This will allow us to use negative feedback with predictable and accurate results.
- Distortion = zero. The output signal should be a perfect reproduction of the input signal.

## 12.2 THE INTEGRATED CIRCUIT OP AMP

The operational amplifier began its life in the early days of analog computers and process control. Hardly recognizable by today's standards, these amplifiers were made of discrete components: vacuum tubes, resistors, and capacitors. Later variations of the operational amplifier substituted transistors for the vacuum tubes, but they were otherwise the same. Both of these operational-amplifier designs were highly specialized and found little use outside of their special niche in control and calculation.

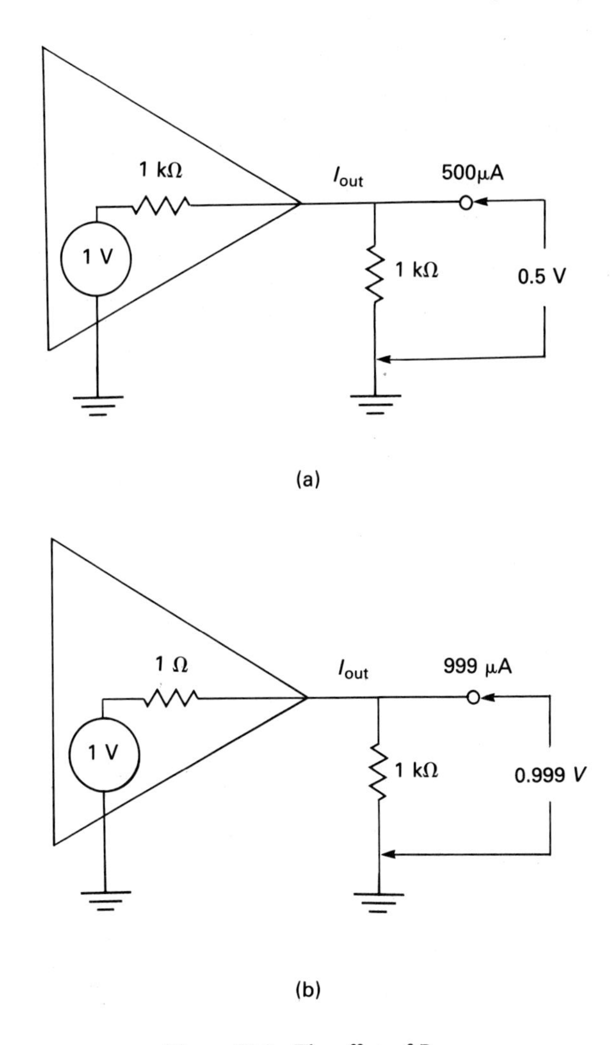


**Figure 12.3** The effect of $R_o$

The integrated-circuit operational amplifier (IC) launched a new boom in op amp use. Suddenly, large amounts of gain were available in a very small and reliable package. As understanding of op amp circuitry grew, so did the uses for the IC. Today, there is hardly a branch of electronics that has not felt the influence of the operational amplifier.

From among the many designs for op amp circuits, a few have emerged as industry standards. These are ICs that are known for dependable, predictable operation.

Other op amps are special variations of these few standard units, offering higher gain, improved performance, or meeting some special need. For our purposes in this chapter, we will concentrate on the UA741 op amp, which is probably the standard of standards.

The specifications for the 741 are modest compared to what is available in a premium op amp. They are:

$$A_{OL} = 200{,}000$$
$$R_I = 2 \text{ megohms}$$
$$R_o = 75 \text{ ohms}$$

These figures are all *open loop*, without feedback. The addition of negative feedback brings drastic improvement to the characteristics of any amplifier. Because $A_{OL}$ of the op amp is so high, the improvement brought to the op amp by negative feedback is excellent.

Returning some portion of the output signal back to the input is called *feedback*. If the feedback signal is in phase with the input voltage, it *adds* to the input signal, increasing the gain. It also decreases stability, reduces the bandwidth, increases $R_o$, and decreases $R_I$. Positive feedback, which is used mainly in oscillators and filter circuits, is beyond the scope of this chapter.

When the feedback signal is 180° out of phase with the input signal, it *subtracts* from the input signal. This is *negative* feedback. It reduces gain and increases stability. Its effect on $R_I$ and $R_o$ depends on the way the circuit is built. We will look at three basic circuits and see the effect of negative feedback on the 741 specifications for each of them.

In the circuit shown in Figure 12.4, we can see that the input signal is connected to the noninverting input of the amplifier, while the feedback is applied through a

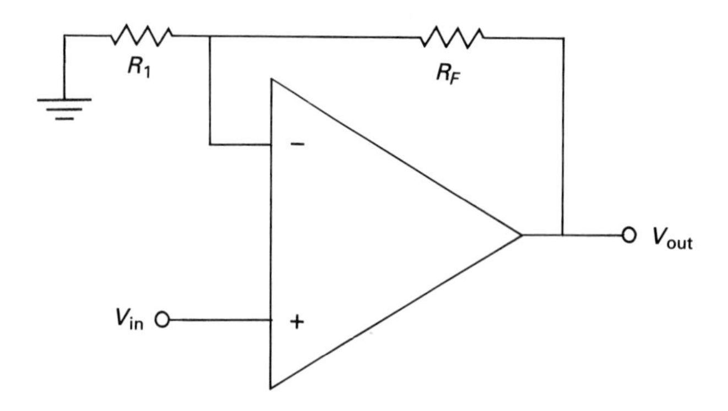


**Figure 12.4** The noninverting amplifier

voltage divider to the inverting input. The amount of feedback in the circuit is determined by the two resistors in the feedback network.

The voltage gain is found with the equation

$$A_v = A_{CL} = 1 + \frac{R_F}{R_1}$$

The *loop gain* of an amplifier circuit is a "figure of merit" for the amplifier. Loop gain is found with the equation

$$LG = \frac{A_{OL}}{A_{CL}}$$

This gives us an indication of why a very high $A_{OL}$ is important in an amplifier. The input resistance of the amplifier is given by the equation

$$R_I' = (R_I)\,(LG)$$

The output resistance is modified to

$$R_o' = \frac{R_o}{LG}$$

As we can see, the noninverting amplifier has a very high $R_I'$ and a very low $R_o'$, both characteristics that we have seen are important in an amplifier. The almost unusable, high $A_V$ has been tamed as well. Best of all, we can control all of these characteristics by selecting the ratio of two resistors.

In the circuit shown in Figure 12.5, we see the inverting amplifier. Once again, the gain is easily calculated:

$$A_v = A_{CL} = -\frac{R_F}{R_1}$$

The effect on input resistance is less desirable:

$$R_I' = R_I + \frac{R_I}{LG}$$

The effect on output resistance is the same as for the noninverting amplifier:

$$R_o' = \frac{R_o}{LG}$$

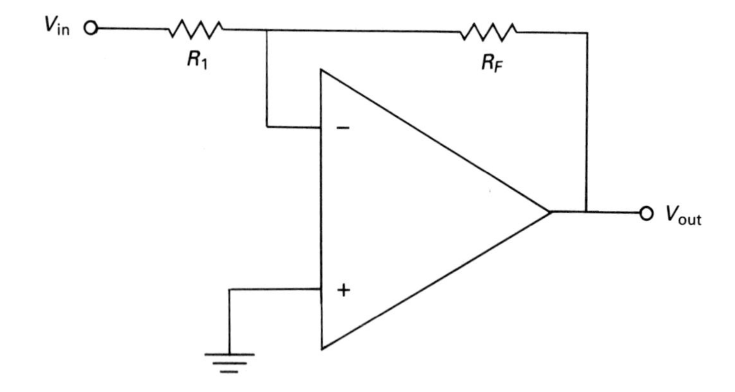


**Figure 12.5** The inverting amplifier

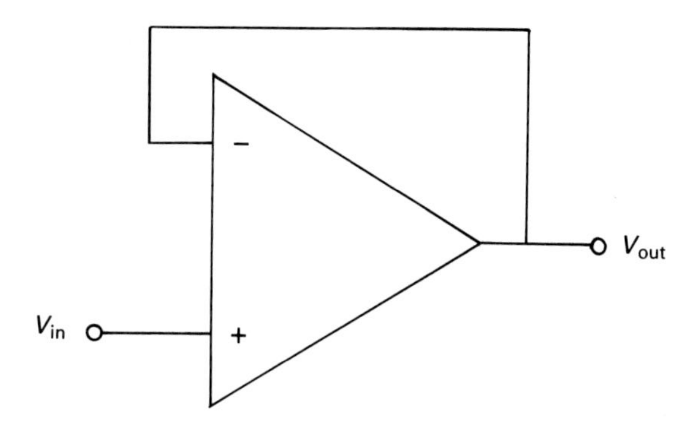


**Figure 12.6** The voltage follower amplifier

Once again, negative feedback has changed the characteristics of the amplifier circuit. While the effect on $R_I'$ is not in the desired direction, it can be allowed for in circuit design or buffered with another circuit.

The circuit in Figure 12.6 is the perfect circuit to use for buffering the input of the inverting amplifier. It is the *voltage follower.* Note that *all* of $V_{out}$ is used as feedback voltage. The gain of the voltage follower is

$$A_v = 1$$

The major effect of the voltage follower circuit is on $R_I'$ and $R_o'$:

$$R_I' = (R_I)(A_{OL})$$

$$R_o' = \frac{R_o}{A_{OL}}$$

This configuration provides the highest input resistance and the lowest output resistance of any of the amplifier circuits. That makes it the perfect buffer amplifier to separate the signal source from low-input-impedance amplifiers.

## SUMMARY

In this chapter, we have reviewed the basics of amplifiers and operational amplifiers. We have seen the following points:

1. The ideal amplifier has
   a. Infinite gain
   b. Infinite bandwidth
   c. Infinite input resistance
   d. Zero output resistance
2. Adding negative feedback can make significant changes in the performance of an amplifier. It can
   a. Reduce gain
   b. Increase bandwidth
   c. Increase $R_I$
   d. Decrease $R_o$
   e. Increase stability
3. There are three basic amplifier circuits that use negative feedback. All other op amp circuits are variations on the following circuits:
   a. Noninverting amplifier
   b. Inverting amplifier
   c. Voltage follower amplifier

## SELF-TEST

1. Using an amplifier with a high $R_I$ reduces losses due to ______________.
2. The large gain of operational amplifiers is reduced to useful levels by ______________________________.

3. Low output resistance in an amplifier reduces signal losses when the amplifier is connected to a(n) ______________________.
4. An amplifier with a wide bandwidth will amplify ______________ equally.
5. The input resistance of an inverting amplifier is approximately equal to ______________________.
6. The input resistance of a noninverting amplifier is ______________.
7. The highest input resistance can be had using the ______________ amplifier circuit.
8. The voltage follower amplifier has a gain of ______________.

## QUESTIONS/PROBLEMS

1. An inverting amplifier using a 741 op amp has the following components:

$$R_F = 100\ \text{k}\Omega$$
$$R_1 = 10\ \text{k}\Omega$$

Find $A_v$, $LG$, $R_I'$, and $R_o'$.

2. A signal source with an internal resistance of 1.5 kilohms is connected to the amplifier in question 1. The signal level is 100 mV. Find $V_{in}$, $V_{out}$.
3. You are repairing an inverting amplifier circuit that amplifies a transducer signal of 25 mV. The required output voltage is $-5$ V. $R_1$ in the circuit is 2.5 kilohms, but $R_F$ has burned in half and you cannot read its color code. What size resistor must you use to replace $R_F$?
4. You want to use a noninverting amplifier to amplify the output of a thermocouple. The thermocouple output is 1.25 mV at 25°C. You need a control signal of 250 mV. If the feedback resistor is to be 100 kilohms, what size resistor must you use for $R_1$?
5. You need an inverting amplifier circuit to provide the correct polarity of control signal, but the signal transducer has a high internal impedance that varies widely with changing temperatures. Since this would cause losses that would vary with temperature as well, the low $R_I'$ of the inverting amplifier is not acceptable. What is the correct solution to the problem?

# 13

# Instrumentation Amplifiers

During our study of the Wheatstone bridge in Chapter 11, we saw that the Thevenin, or difference, voltage across the bridge network was a measure of the imbalance in the bridge. In most bridge networks, the imbalance is very small, and the difference voltage is also very small.

In this chapter, we will see how the operational amplifier can be used to amplify this tiny voltage.

## OBJECTIVES

You will have successfully completed this chapter when you can:

1. Describe the operation of a basic voltage difference amplifier.
2. Explain the reasons that this basic voltage difference amplifier is not suitable as a bridge amplifier.
3. Describe the operation of a basic instrumentation amplifier.
4. Describe the operation of a gain instrumentation amplifier.

## 13.1 THE VOLTAGE DIFFERENCE AMPLIFIER

In the basic operational amplifiers described in Chapter 12, we saw that the op amp can be used as an inverting or a noninverting amplifier. If we connect different signals to both inverting and noninverting inputs at the same time, we can also *subtract* with an op amp.

Figure 13.1 shows the basic voltage difference amplifier circuit. Notice that there are two inputs, $V_{in}^+$ and $V_{in}^-$. When the voltage at both of these inputs is equal, the output voltage will be zero. Resistors $R_2$ and $R_3$ are used to balance the "+1" in the noninverting amplifier gain formula. Resistor $R_2$ should be equal to resistor $R_1$, and $R_3$ should be equal to $R_F$. When this is so, the output voltage is easily determined:

$$A_v = \frac{R_F}{R_1}$$

$$V_{out} = (V_{in}^+ - V_{in}^-) \times A_v$$

In Figure 13.2a, we see the basic voltage differencing amplifier *without* $R_2$ and $R_3$. Note that $V_{in}^+$ and $V_{in}^-$ are equal. The output, then, *should* be zero. But when we solve the circuit, we get a totally different answer. Solving for $A_v$ from $V_{in}^+$, we get:

$$A_{v_1} = \frac{10\ \text{k}\Omega}{10\ \text{k}\Omega} + 1 = 2$$

Solving for $V_{out_1}$, we get:

$$V_{out_1} = +2\ \text{V} \times 2 = +4\ \text{V}$$

Solving for $A_v$ from $V_{in}^-$, we get:

$$A_{v_2} = -\frac{10\ \text{k}\Omega}{10\ \text{k}\Omega} = -1$$

Solving for $V_{out_2}$, we get:

$$V_{out_2} = +2\ \text{V} \times -1 = -2\ \text{V}$$

The resultant output voltage is

$$V_{out} = V_{out_1} + V_{out_2}$$

$$V_{out} = +4\ \text{V} + (-1\ \text{V})$$

$$V_{out} = +3\ \text{V}$$

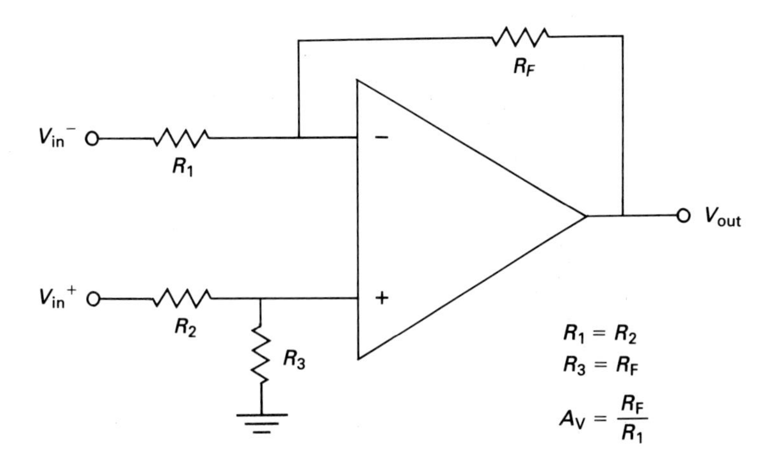


**Figure 13.1** The voltage difference amplifier

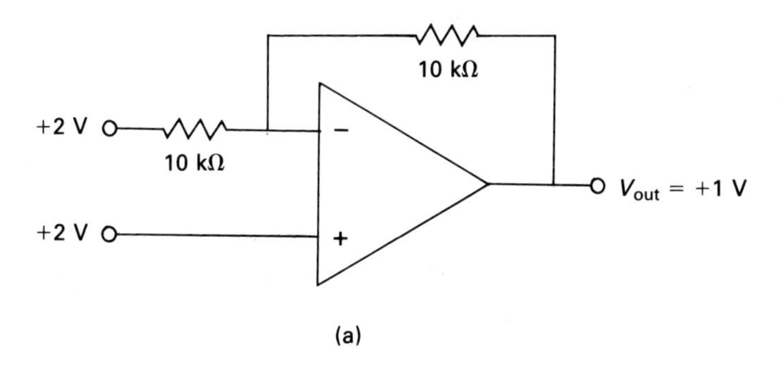


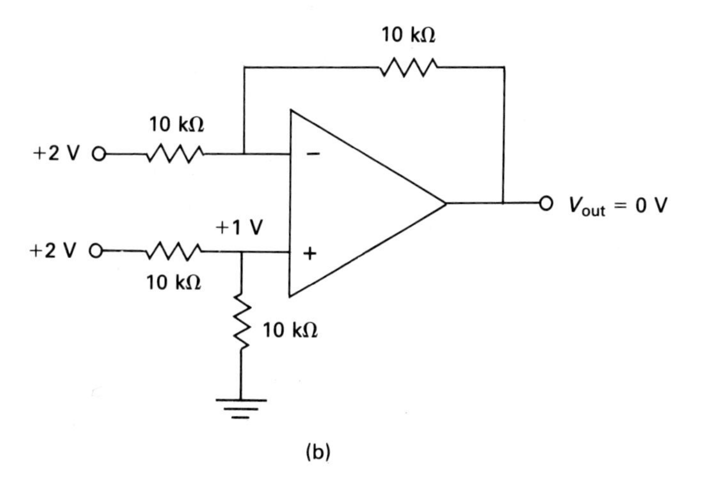


**Figure 13.2** The reason for $R_2$ and $R_3$

This incorrect output is brought on, as you can see, by the "+1" in the noninverting gain formula. In Figure 13.2b, we can see the effect of adding $R_2$ and $R_3$. These two resistors form a voltage divider, which reduces the signal level that actually reaches the noninverting input. Let's calculate again:

$$V_{in}^{+} = \frac{V_{sig} \times R_3}{R_2 + R_3}$$

$$V_{in}^{+} = \frac{2\ \text{V} \times 10\ \text{k}\Omega}{10\ \text{k}\Omega + 10\ \text{k}\Omega}$$

$$V_{in}^{+} = 0.5\ \text{V}$$

And solving for $V_{out_1}$:

$$V_{out_1} = 0.5\ \text{V} \times 2 = +1\ \text{V}$$

Since the inverting input has not changed, the output has also not changed:

$$V_{out_2} = -1\ \text{V}$$

Solving for $V_{out}$:

$$V_{out} = V_{out_1} + V_{out_2}$$

$$V_{out} = +1\ \text{V} + (-1\ \text{V})$$

$$V_{out} = 0\ \text{V}$$

This relationship holds true regardless of the values of $R_1$ and $R_F$. So long as $R_2 = R_1$ and $R_3 = R_F$, the "+1" in the noninverting gain formula will be compensated.

The problem is that now *both inputs are low impedance.* If we represent the two input impedances with $R_{in}^{+}$ and $R_{in}^{-}$, we can add them to a Wheatstone bridge circuit and visualize their effects.

From Figure 13.3, we can see that these two low impedances add extra current paths to the bridge circuit and introduce error. If the difference voltage source is a low-impedance device, the error will be very small, but that is rarely the case with instrumentation bridge circuits. The bridge resistances $R_A$ and $R_B$ are normally large to reduce the current in the bridge and its self-heating effect. In this case, the basic voltage differencing amplifier is not acceptable because the error it introduces will be quite large.

## 13.2 THE INSTRUMENTATION AMPLIFIER

The basic instrumentation amplifier is simply a voltage differencing amplifier with voltage-follower buffered inputs. Recall that the input resistance of a voltage follower

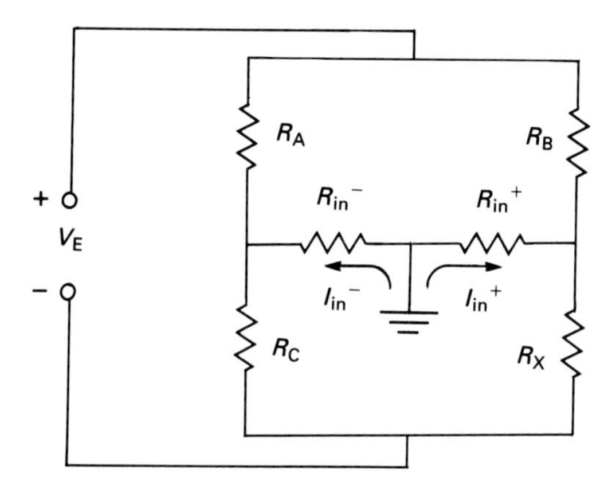


**Figure 13.3** The effect of low input impedance. Low $R_{in}^-$ and $R_{in}^+$ allow extra current paths. The inequality of the two unbalances the bridge.

is so high that it may be considered almost infinite. This high impedance draws almost no current from the bridge network. This removes the galvanometer leg from the bridge calculations — all we need to calculate is the difference voltage.

Figure 13.4 illustrates the basic instrumentation amplifier. Note that it has two sections, an input section, consisting of two voltage followers with a gain of one, and an output section, consisting of a basic voltage differencing amplifier, which may have gain.

We can also add gain to the input stage by converting the voltage followers into noninverting amplifiers. This is shown in Figure 13.5.

Notice that the noninverting amplifier input stage is designed a little differently. Resistor $R_4$ is a potentiometer, which connects the two inverting inputs together. We might imagine resistor $R_4$ as being made up of *two equal resistors in series* with their center node at ground. This view will allow us to calculate the input stage gain as follows:

$$A_{v_1} = \frac{R_5}{R_4 \div 2} + 1$$

$$A_{v_1} = \frac{2R_5}{R_4} + 1$$

If $R_4$ is set to 500 ohms

$$A_{v_1} = \frac{2 \times 5\text{ K}}{500} + 1$$

$$A_{v_1} = 11$$

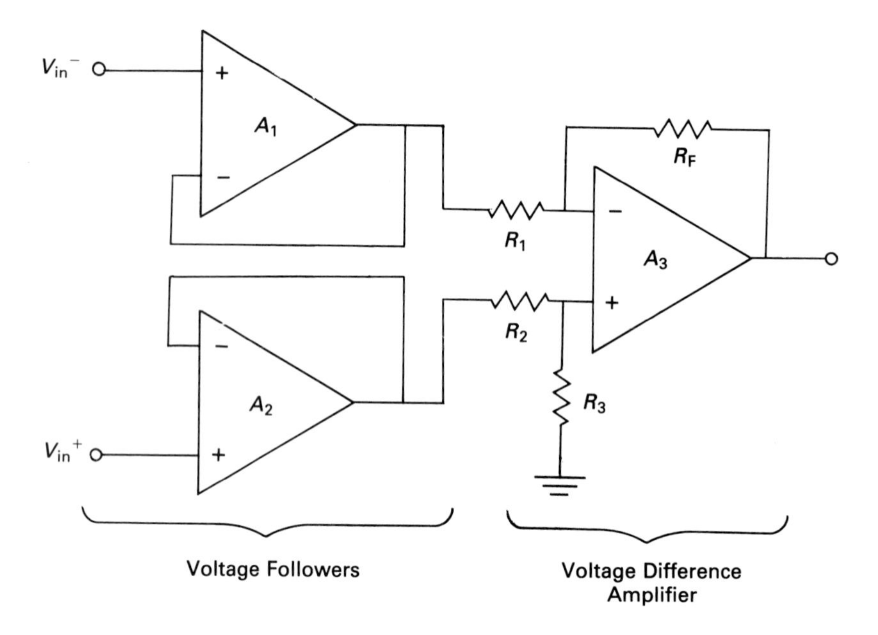


**Figure 13.4** The basic instrumentation amplifier

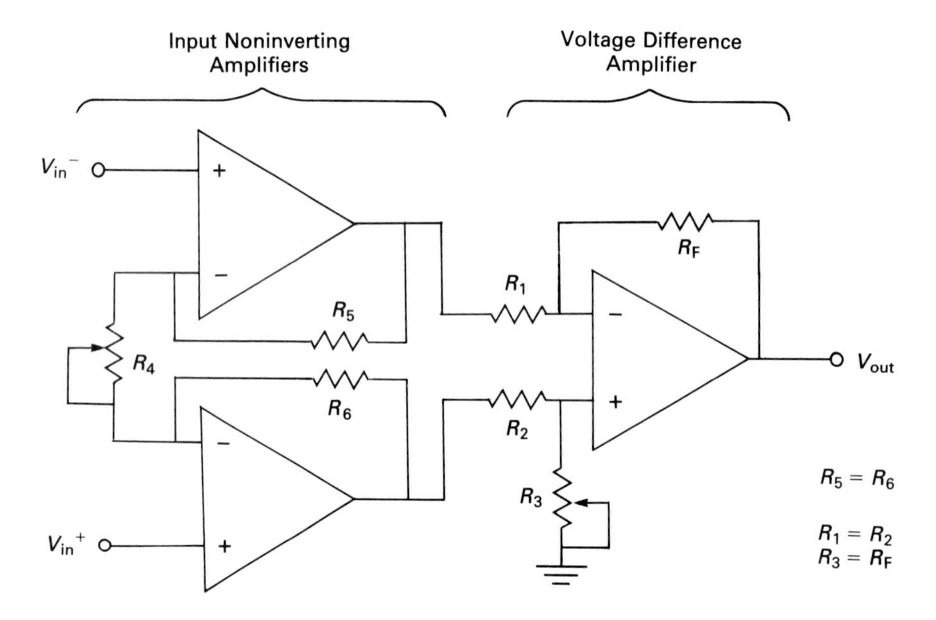


**Figure 13.5** Adding gain to the instrumentation amplifier

We can also calculate the difference amplifier gain:

$$A_{v_2} = \frac{10\text{ K}}{1\text{ K}}$$

$$A_{v_2} = 10$$

And now, we can calculate the total circuit gain:

$$A_v = A_{v_1} \times A_{v_2}$$

$$A_v = 110$$

Notice that $R_3$ in the circuit has been replaced by a potentiometer. This addition to the circuit allows us to adjust for an optimum CMRR. Small changes in this resistance help balance the signal level at the two inputs, which is present when there is a common mode voltage input to the noninverting input stage. The common mode signal is *noise*, which we want to reject.

Adding gain to the input stages improves the ratio of signal to noise, but it adds two extra op amps to the circuit. That addition can contribute to another op amp problem — drift. *Drift* is the tendency of an op amp's dc output voltage to vary with temperature. This variation is caused by imbalances within the op amp itself. If the circuit is to be exposed to extremes of temperature, low-drift op amps should be selected.

The 741 op amp, for example, has a drift of up to +/− 15 mV/degree C. Low-drift op amps have a drift of less than 10 $\mu$V/degree C. In precision work, this can be a very significant difference.

## 13.3 THE MONOLITHIC INSTRUMENTATION AMPLIFIER

In order to reduce the errors that can creep into a circuit designed with external components, manufacturers of industrial op amps produce instrumentation amplifiers on a *chip* — single ICs that contain all of the circuitry needed to implement a precision IA. These monolithic op amps simplify design and improve reliability and accuracy.

The LH0038 IA, for example, is a totally self-contained instrumentation amplifier. Gain is programmable in steps from 100 to 2000 by jumping pins on the IC. Because the gain-controlling resistors are formed on the chip itself, temperature tracking is virtually perfect. The LH0038 has an output drift of only 0.25 $\mu$V/degree C, and its CMRR is 126 dB.

The LH0036 IA requires a single external resistor to program gain between 1 and 1000. The input resistance of the amplifier is 300 megohms. The LH0036 has an output drift of 15 $\mu$V/degree C, and its CMRR is 100 dB.

These and other precision instrumentation amplifiers are recommended for new designs. We must always bear in mind that industry is more interested in reliability and accuracy than they are in low cost. It is far less expensive to provide reliable designs than it is to shut down a production line because of a component failure or the need to recalibrate an instrument.

## SELF-TEST

1. The voltage differencing amplifier performs the mathematical operation of __________________________.
2. In the design of a voltage differencing amplifier, the resistive voltage divider at the noninverting input is used to __________________________.
3. The basic voltage differencing amplifier is not suitable as a bridge amplifier because of its __________________________.
4. Adding voltage follower buffers to the inputs of a voltage differencing amplifier converts it into a basic __________________________.
5. One problem with constructing instrumentation amplifiers from ordinary op amps is the relatively large number of __________________________.

## QUESTIONS/PROBLEMS

1. You are calibrating an instrumentation amplifier for a thermistor bridge. You must have an overall gain of 150. The fixed gain of the difference amplifier is 10. The feedback resistors in the input stage are both 15 kilohms. What should be the value of the gain-setting resistor?
2. Design a complete IA around the 741 op amp, with the gain variable from 1 to 1000. Construct and test the circuit.

# 14

# Filters — Passive and Active

The industrial environment is an electrically hostile one. Large motors switch on and off at irregular times, sending electromagnetic interference, or EMI, throughout the plant's wiring system. Motor speed controls and light dimmers add their own EMI to the hash. Even fluorescent lights and hand-held variable-speed drills contribute noise.

Solid-state electronic controls are particularly subject to interference from the flood of EMI in the factory. With their high-speed operation, nanosecond pulses of interference are easily sensed and amplified, adding errors to an otherwise soundly designed system.

In this chapter, we will look at the techniques used to filter this interference out of the controllers.

## OBJECTIVES

You will have successfully completed this chapter when you can:

1. Describe the operation of passive LC filters.
2. Describe the operation of passive RC filters.
3. Describe the operation of active RC filters.

## 14.1 FILTER BASICS

A *filter* is a circuit that passes signals of a desired frequency or frequency band and attenuates all other signal frequencies. Filters are made with resistive, capacitive, and inductive components in series and parallel combination.

There are several different filter types, as shown in Figure 14.1. Figure 14.1a is the response curve of a *low-pass filter.* Note that it passes frequencies *below* $f_{co}$ easily, but that frequencies above $f_{co}$ are attenuated. Low-pass filters smooth the ripple from our dc power supplies and send only the low frequencies to the bass speakers in our stereo systems.

The frequency $f_{co}$ is the frequency where the signal level is 0.707 times the signal level at the center of the pass band. The *co* stands for *cutoff frequency*, and the 0.707 level is the *half power point*, the point where the signal power is one-half of what it is at the middle of the pass band.

The Q of a filter is a measure of its *sharpness*, how steep the slope is above $f_{co}$ in a low-pass filter or below $f_{co}$ in a high-pass filter. Since this is a measure of how well the filter will reject unwanted signals, it is often called the *quality* of the filter.

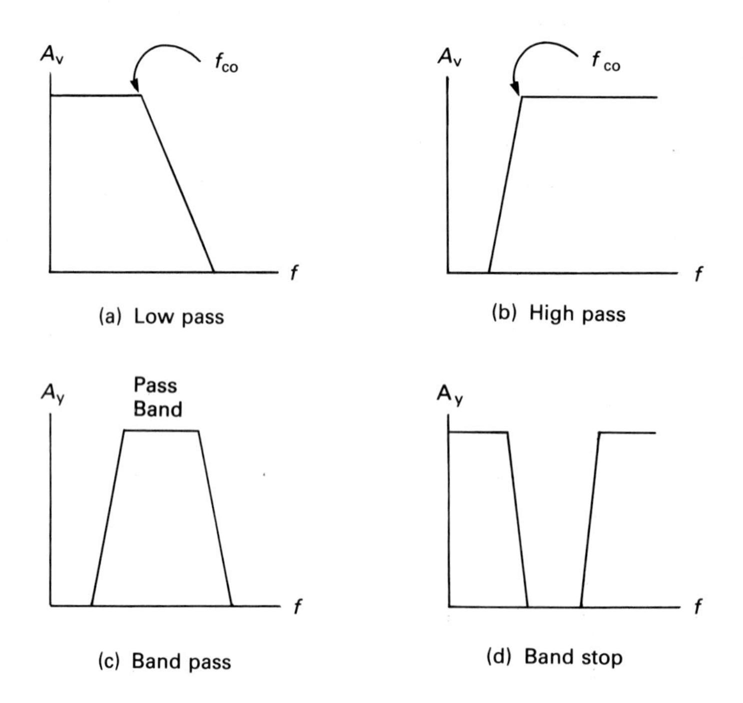


**Figure 14.1** Response curves of different filters

Note that the response curve shown in Figure 14.1b is "sharper" than the response curve in Figure 14.1a. It has a higher Q.

Figure 14.1b is the response curve of a *high-pass filter.* This filter passes frequencies *above* $f_{co}$ and attenuates those below $f_{co}$. High-pass filters send only the treble signals to our tweeters, and keep low-frequency interference out of our TV pictures.

The response curve in Figure 14.1c is that of a *band pass filter.* Band pass filters allow a certain continuous band of frequencies to pass and attenuate frequencies *above and below* the pass band. We use band pass filters when we tune in radio or TV signals.

The response curve shown in Figure 14.1d is a *band stop filter.* Band stop filters (or band reject filters) allow all frequencies *below* the stop band and all frequencies *above* the stop band to pass through, but they attenuate those frequencies *within* the stop band. Band stop filters are used to eliminate high-frequency interference from receivers. If the stop band of a band stop filter is very narrow, it may be called a *notch filter.*

## 14.2 LC FILTERS

The LC filter is made up primarily of inductance and capacitance. LC filters are used in all radio and TV receivers because of their excellent attenuation characteristics. The response curve of an LC filter may be made very sharp — with the roll-off slope steep enough so that frequencies outside of the pass band are almost totally attenuated.

The cutoff frequency is easily calculated for LC filters:

$$f_{co} = \frac{1}{2\pi\sqrt{LC}}$$

In Figure 14.2a, we see the simplest form of LC low-pass filter. The series inductance acts as a low-impedance-to-low-frequency signal and a high-impedance-to-high-frequency signal. Meanwhile, the capacitor offers a low-impedance path to ground for the high-frequency signals.

In Figure 14.2b, we see an LC high-pass filter. In this circuit, the capacitance offers a low impedance to the high frequencies and a high impedance to the low frequencies. The inductor shunts the lower frequencies to ground, but acts as a high impedance to the high frequencies.

By connecting low-pass and high-pass filters in series, we can derive a band pass filter. This simple configuration is shown in Figure 14.2c. The $f_{co}$ of the low-pass filter is set to the *upper frequency of the passband* and the $f_{co}$ of the high-pass filter is set to the *lower frequency of the passband*. Thus, no signal at a frequency above the upper $f_{co}$ is allowed to reach the high-pass section, and no frequency below the lower $f_{co}$ is allowed to pass out of the high-pass section.

There are many different types of LC filters, and their mathematical analysis fills many volumes. The interested student is referred to texts on communications for more detail about these filters and their applications.

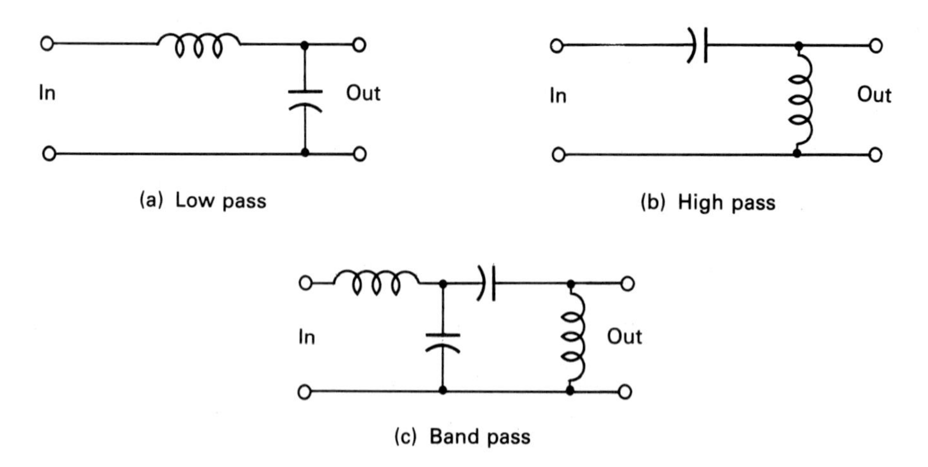


**Figure 14.2** Simple LC filters

## 14.3 PASSIVE RC FILTERS

While the characteristics of LC filters make them superior to the RC type of filter, the physical size and bulk of the inductors needed for low frequencies make them impractical for most industrial applications. Instead, industry has settled on the resistive-capacitive (RC) filter. In these filters, the inductor has been replaced by a resistor. The cutoff frequency is reached when $X_c = R$, and it is easily calculated:

$$f_{co} = \frac{1}{2\pi RC}$$

The first RC filters we will look at are called *passive filters* because they contain no active elements — amplifiers, transistors, etc. They are similar to the simple LC filters we saw in Section 14.2.

Figure 14.3a is an RC low-pass filter, Figure 14.3b is an RC high-pass filter, and Figure 14.3c is an RC band pass filter. The general operation of these filters has been described in the previous section. The major difference is that the resistance does not change with frequency. The result of this is that the roll-off or roll-in characteristic of the filter is essentially linear.

The filter illustrated in Figure 14.4a is a band stop filter. Note that it is made up of a low-pass filter and a high-pass filter in parallel. The $f_{co}$ of the low-pass filter

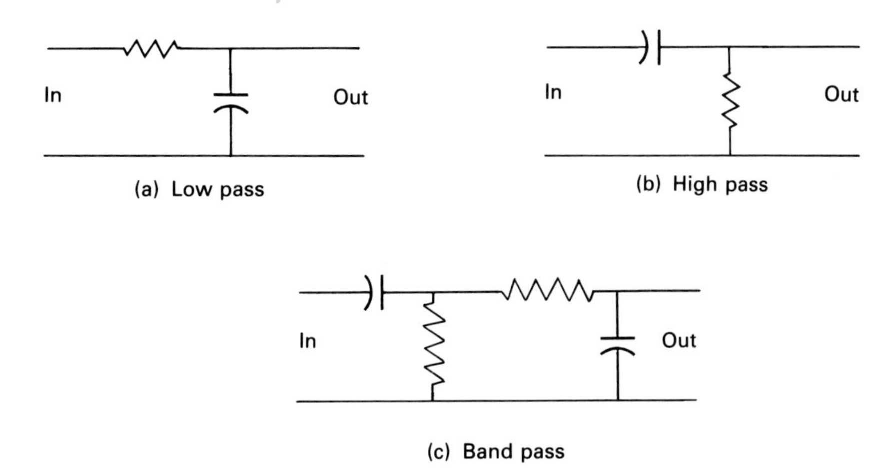


**Figure 14.3** Passive RC filters

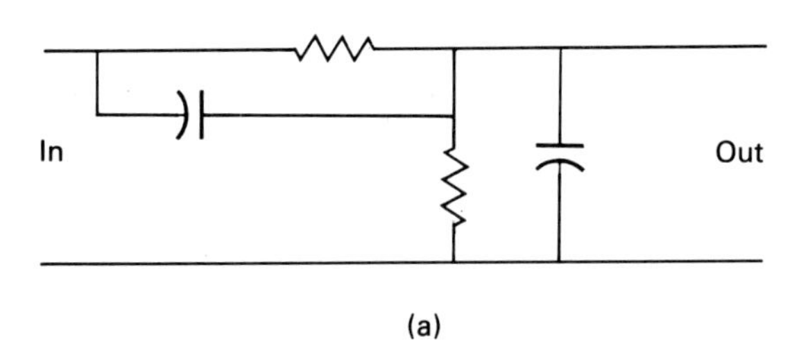


**Figure 14.4a** RC band stop filter

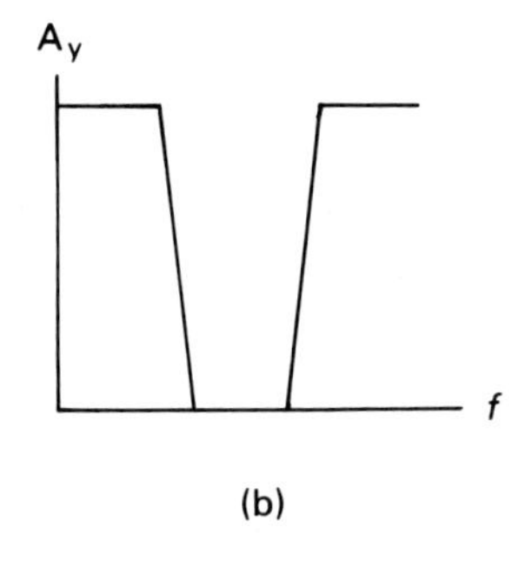


**Figure 14.4b** Bode plot of a band pass filter

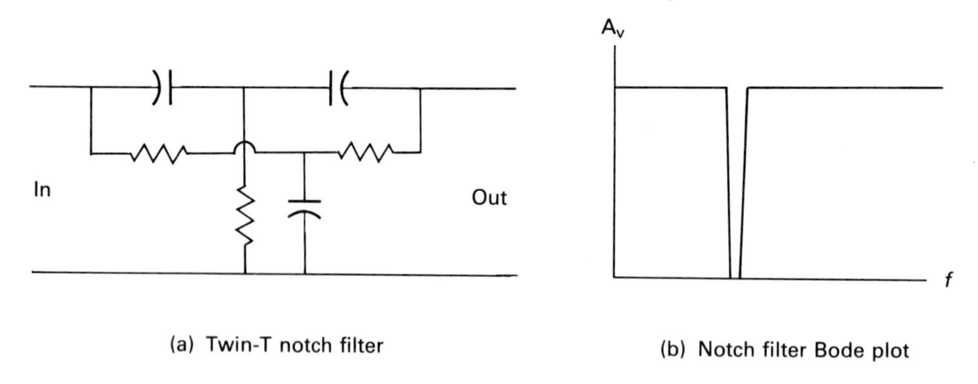


(a) Twin-T notch filter

(b) Notch filter Bode plot

**Figure 14.5** Twin-T notch filter

is set to the *lower cutoff frequency* and the $f_{co}$ of the high-pass filter is set to the *upper cutoff frequency*. Figure 14.4b shows the Bode plot of the band stop filter.

The filter shown in Figure 14.5a is a Twin-T notch filter. Both T-section filters are tuned to the same frequency. The high-pass section passes all frequencies above the selected frequency, while the low-pass section passes all frequencies below the selected frequency. If the components are carefully matched, the selected frequency is almost totally dropped out. This is shown in the Bode plot in Figure 14.5b.

Passive RC filters can be used to good effect, but care must be used in terminating them. As long as the impedance of the load on the filter is large, the general calculations are effective. When the load impedance approaches that of the resistance in the filter, however, it begins to affect the filter's characteristics. In order to avoid this, the load impedance must be on the order of ten times greater than the resistance in the filter.

Another consideration in using RC filters is the roll-off characteristic, 20 dB/decade. In some cases, it may be necessary to have a greater roll-off to eliminate a strong signal within the stop band of a filter. RC filters may be cascaded to increase the roll-off rate. Two low-pass RC filters in series create a *second-order filter* with a roll-off of 40 dB/decade. Once again, care must be exercised so that the second filter does not load down the first, affecting $f_{co}$.

## 14.4 FIRST-ORDER ACTIVE RC FILTERS

A filter is called *active* when it includes an active element such as a transistor or an op amp. RC active filters take advantage of the characteristics of an op amp to improve the operation of the filter.

The filters in Figure 14.6 are first-order active filters. The filter in Figure 14.6a is simply an RC low-pass filter with a voltage follower op amp circuit for a load.

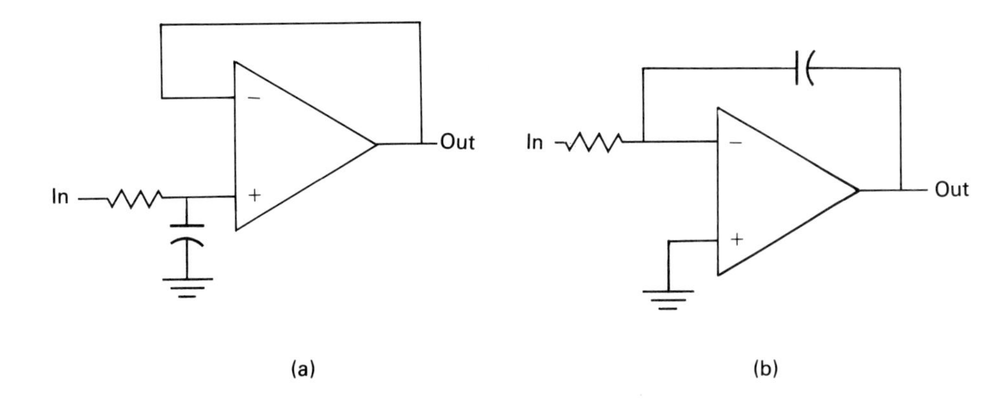


**Figure 14.6** First-order active low pass filters

This simple expedient removes all concern of loading from the filter design. As in all first-order filters, the roll-off is 20 dB/decade.

The filter in Figure 14.6b puts the reactive element (the capacitor) in the feedback loop of the op amp. Because $X_c$ is higher at low frequencies, the amplifier gain will also be higher at these frequencies. The cutoff frequency is calculated with the feedback resistor.

The high-pass filter shown in Figure 14.7a is a first-order active high-pass filter. It, too, is simply an RC filter with a voltage follower. Figure 14.7b shows that the filter may be isolated *between* voltage followers so that its characteristics will not be affected by input or output impedances.

The filter in Figure 14.8 is an active band pass filter. Remember that the high-pass section of the filter, $R_1$ and $C_1$, must be tuned to the *low-frequency cutoff*, and the low-pass section, $R_2$ and $C_2$, must be tuned to the *high-frequency cutoff*.

## 14.5 SECOND-ORDER ACTIVE RC FILTERS

While the first-order filters are useful to some degree, the 40 dB/decade roll-off of second-order filters allows a greater degree of signal rejection at frequencies near $f_{co}$. When an active element has been included in the filter design, it is easier to accomplish second-order filters with only $R$ and $C$ elements.

The filters shown in Figure 14.9 are Salen-Key second-order filters. In order for the response curve to be smooth, certain design rules must be followed.

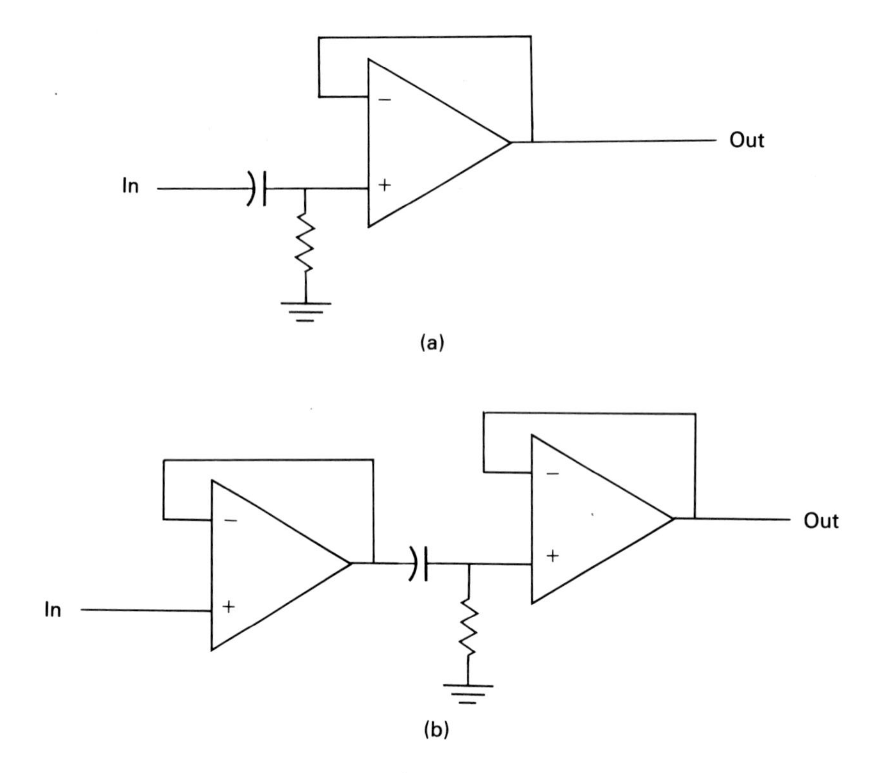


**Figure 14.7** First-order active high pass filters

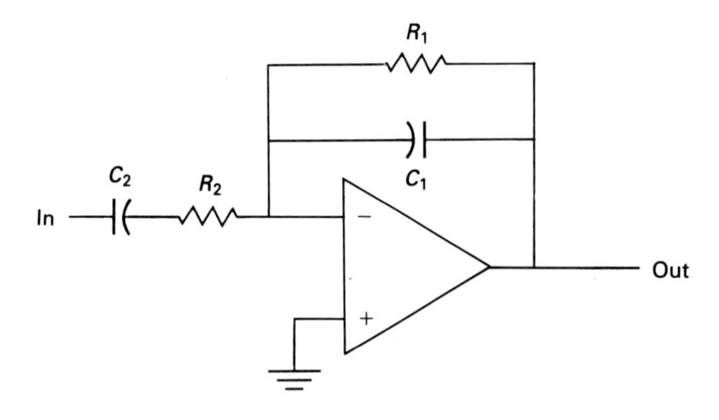


**Figure 14.8** First-order active band pass filters

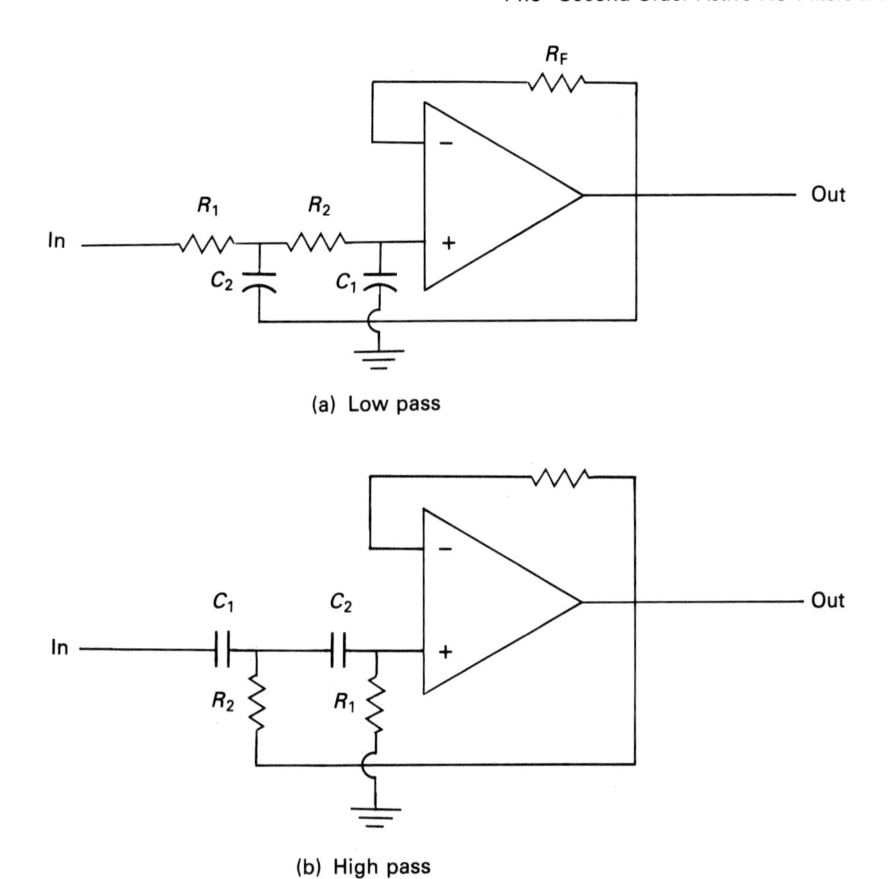


**Figure 14.9** Salen-Key second-order filter

Normally, in the design of the Salen-Key filter

$$R_1 = R_2$$
$$C_2 = 2C_1$$
$$R_F = R_1 + R_2$$

The cutoff frequency can be calculated as follows:

$$f_{co} = \frac{1}{2\pi(R_1C_1R_2C_2)^{1/2}}$$

As you can see, $f_{co}$ depends on the values of all of the components except $R_F$. The significance of this for the service technician is *not to substitute component values in any filter circuit*. The designers had a specific result in mind when the circuit was designed. Any change in component values will affect the result. The only part that is good enough is the part that is exactly right.

The filters shown in Figure 14.10 are also Salen-Key filters, but with a difference. Here, the input resistors have the same value, as do the capacitors. The cutoff frequency calculation is also much simpler:

$$f_{co} = \frac{1}{2\pi R_1 C_1}$$

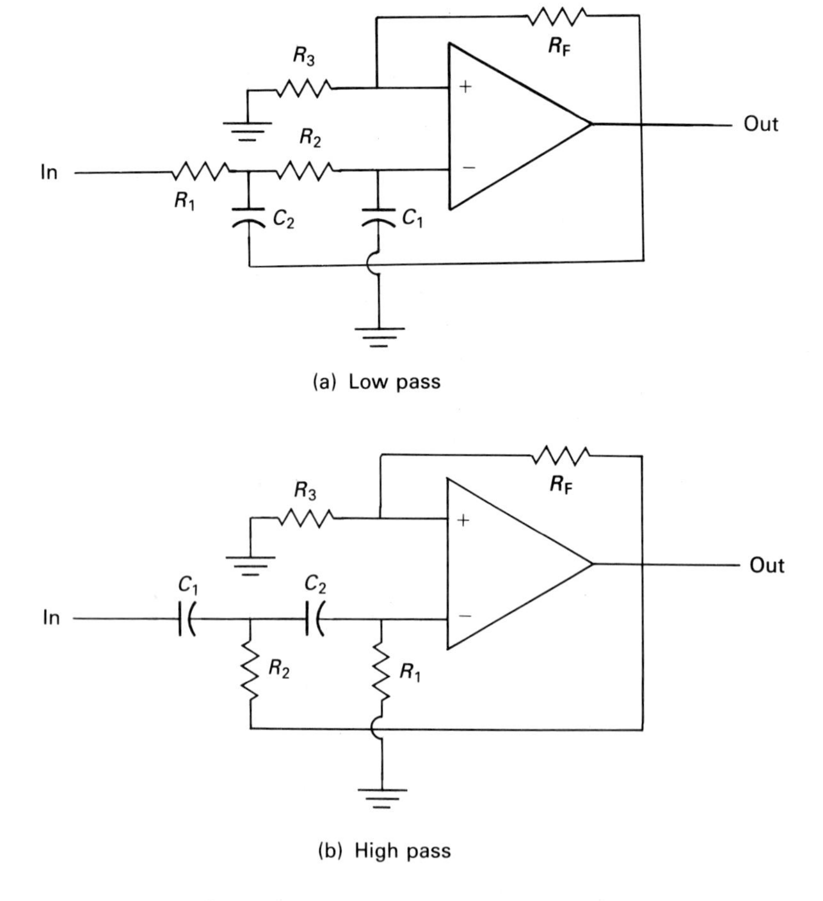


**Figure 14.10** Salen-Key equal component filters

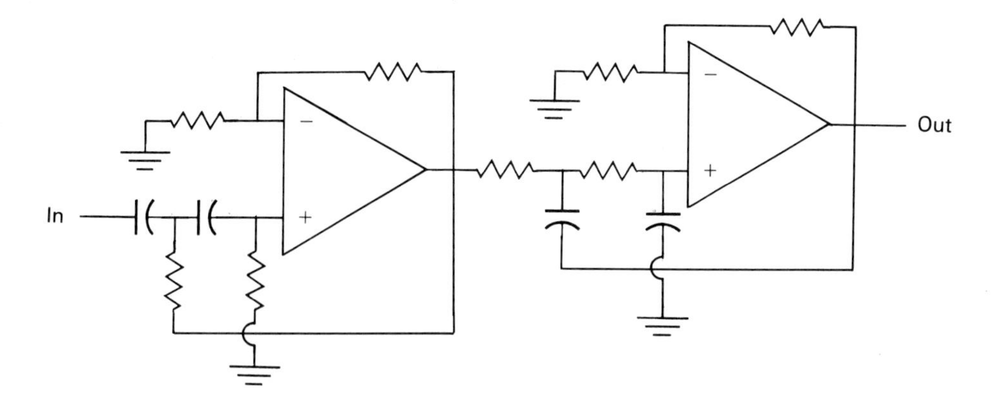


**Figure 14.11** Salen-Key band pass filter

What makes this filter easier to design is the gain control network of $R_F$ and $R_1$. The values of these resistors should be selected so that $R_F = 0.586R_1$. Any variation in this selection will result in peaks and valleys in the response curve of the filter.

The band pass filter in Figure 14.11 is the Salen-Key second-order band pass filter. Once more, the input high-pass filter is tuned to the lower $f_{co}$ and the output low-pass filter is tuned to the upper $f_{co}$.

The filter in Figure 14.12 uses another design technique — multiple feedback. This rather complicated looking filter also has some rather complicated math:

$$f_o = \left(\frac{1}{2\pi C}\right) \left(\frac{R_1 + R_2}{R_1 R_2 R_3}\right)^{1/2}$$

$$A_v = \frac{R_3}{2R_1}$$

$$Q = f_o C R_3$$

$$BW = \frac{f_o}{Q}$$

where $f_o$ is the center frequency of the pass band and $BW$ is the band width. From this, you can see that the operation of this filter is highly interactive. In order to get an idea of *how* interdependent these qualities are, we will work an example.

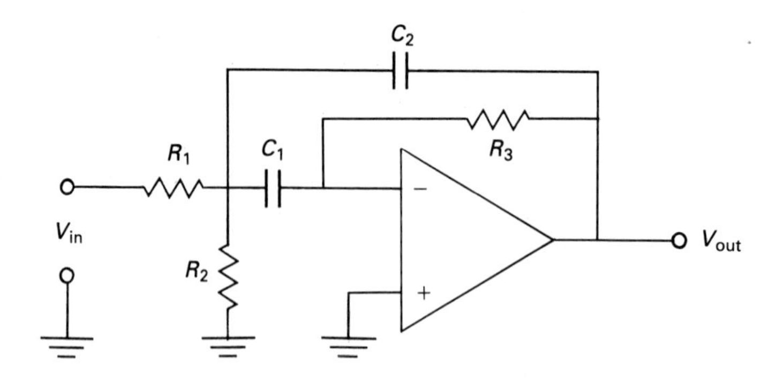


**Figure 14.12** Multiple feedback band pass filter

## Example 14.1

$$R_1 = 22\ \text{k}\Omega \qquad R_2 = 680\ \text{ohms}$$

$$R_3 = 68\ \text{k}\Omega \qquad C = 0.01\ \mu\text{F}$$

$$f_o = \left(\frac{1}{2 \times \pi \times 0.01\ \mu\text{F}}\right) \left(\frac{22\ \text{k}\Omega + 680\ \Omega}{22\ \text{k}\Omega \times 680\ \Omega \times 68\ \text{k}\Omega}\right)^{1/2}$$

$$f_o = 2376.41\ \text{Hz}$$

$$A_v = \frac{68\ \text{k}\Omega}{2 \times 22\ \text{k}\Omega}$$

$$A_v = 1.55$$

$$Q = 2376.41 \times 2376.41 \times 0.01\ \mu\text{F} \times 68\ \text{k}\Omega$$

$$Q = 1.62$$

$$BW = \frac{2376.41}{1.62}$$

$$BW = 1470.59\ \text{Hz}$$

From this, we see a band pass filter with a passband that begins at the lower cutoff frequency

$$f_{co_1} = 1640.7\ \text{Hz}$$

and ends at the upper cutoff frequency

$$f_{co_u} = 3111.7 \text{ Hz}$$

In the next example, we will change $R_3$ in an effort to narrow the passband.

**Example 14.2**

$R_1 = 22 \text{ k}\Omega$ $R_2 = 680 \text{ ohms}$

$R_3 = 120 \text{ k}\Omega$ $C = 0.01 \mu\text{F}$

$$f_o = \left(\frac{1}{2 \times \pi \times 0.01 \mu\text{F}}\right) \left(\frac{22 \text{ k}\Omega + 680 \Omega}{22 \text{ k}\Omega \times 680 \Omega \times 120 \text{ k}\Omega}\right)^{1/2}$$

$$f_o = 1788.90 \text{ Hz}$$

$$A_v = \frac{120 \text{ k}\Omega}{2 \times 22 \text{ k}\Omega}$$

$$A_v = 2.73$$

$$Q = 1788.90 \times 1788.90 \times 0.01 \mu\text{F} \times 120 \text{ k}\Omega$$

$$Q = 2.147$$

$$BW = \frac{1788.90}{6.74}$$

$$BW = 833.3 \text{ Hz}$$

The new filter has a more narrow passband, but we have also drastically affected the upper and lower cutoff frequencies. Now

$$f_{co_l} = 1372.2 \text{ Hz}$$

$$\text{while } f_{co_u} = 2205.6 \text{ Hz}$$

From these two examples, we can see that a change in only one component, $R_3$, has significantly altered the center frequency, the Q, and the bandwidth of the filter.

The design of multiple-feedback active filters is handled best by computers, which can be programmed to deal with the multiple variables that are a natural part of the solution. In attempting to design a multiple-feedback filter, the value of Q should be limited because when Q is greater than ten, the filter tends to become unstable and may cause oscillations.

## SUMMARY

In this chapter, we have looked at many filters, both passive and active. We have seen:

1. A filter is used to reject unwanted frequency signals and pass the desired frequency signals.
2. Passive filters are made up solely of reactive and resistive elements — they have no gain elements.
3. A filter that passes only signals below a certain frequency is a low-pass filter.
4. A filter that passes only signals above a certain frequency is a high-pass filter.
5. A filter that passes only signals within a certain frequency band, rejecting those above and below that band, is a band pass filter.
6. A filter that passes all frequencies except those within a certain band is a band stop or notch filter.
7. First-order passive filters have a roll-off rate of 20 dB/decade.
8. Second-order passive filters have a roll-off rate of 40 dB/decade, but the second filter section may have a bad influence on the operation of the first filter section.
9. First-order active filters use op amps to isolate the RC filter from external impedance influences.
10. Second-order active filters use positive feedback or multiple feedback to improve filter operation.

## SELF-TEST

1. The cutoff frequency for a filter circuit is also defined as the ___________.
2. The cutoff frequency is that frequency where the signal level is _________ times the signal level in the passband.
3. The Q of a filter is a measure of the ____________________________ of the slope of its response curve.
4. LC filters are rarely used at low frequencies because of ________________.
5. RC filters are effective when the ______________________________________ impedance is high.
6. First-order RC filters can be improved by ____________________________ them with op amp circuits.
7. Filters that include amplification as part of the filter circuit are called ____________________________ filters.

8. Second-order filters have a(n) ______________________ roll-off rate.
9. When servicing filter circuits, parts substitutions are ______________________ acceptable.
10. A narrow band of frequencies can be removed from a signal by using a(n) ______________________ filter.

## QUESTIONS/PROBLEMS

1. An equal component value Salen-Key low pass filter has the following component values: $R_1 = R_2 = 15$ kilohms. $C_1 = C_2 = .001\ \mu F$. What is the cutoff frequency?
2. The resistors in problem #1 are replaced with 1.5-kilohm resistors by mistake. What effect will this have on the cutoff frequency?
3. The resistors in problem #1 are replaced with resistors that are only 20% tolerance. If one of them is 20% too large and the other 20% too small, how will they affect the cutoff frequency?

# 15

# The Op Amp Analog Computer

The original application for the op amp was as an analog computer element. In fact, the name *operational amplifier* comes from the fact that the op amp can perform *any mathematical operation*. In this chapter, we will explore how the op amp can be used to perform many mathematical functions, and how this can be applied to the problem of control.

## OBJECTIVES

You will have successfully completed this chapter when you can explain how the op amp may be used to perform:

1. Addition.
2. Subtraction.
3. Multiplication.
4. Division.
5. Integration.
6. Differentiation.

## 15.1 THE OP AMP ADDER

One of the easiest and most important mathematical operations that the op amp is called upon to perform is simple addition — the summing of two or more numbers. In control applications, we will see addition used over and over again to sum the signals from different sensors in a system.

Figure 15.1 shows the circuit of an op amp adder or *summer.* In an inverting amplifier, the inverting input is called the *summing junction.* Voltage signals from different sources can be input to this simple circuit. The circuit will add or sum the input voltages, but the input voltages will have no effect on each other.

The reason for this remarkable fact is quite simple. The inverting input of the inverting amplifier is *at ground potential.* Recall that the op amp actually amplifies the *difference* or *error voltage,* the voltage difference between the inverting and noninverting inputs. Since the open loop gain of the op amp is so high that only a few microvolts of input would cause the output to swing to saturation, when the output is not saturated it must be because there is no difference voltage. When we add negative feedback to the circuit, we actually provide the op amp with a means of assuring that the difference voltage will be zero. The op amp output can now swing in such a way that it adjusts the voltage at the inverting input to equal the voltage at the noninverting input.

Once you understand that idea, you can follow the idea that the voltage potential at the inverting input is 0 V. The noninverting input in our circuit is connected to ground: it is at ground potential. If there is no voltage difference between the two

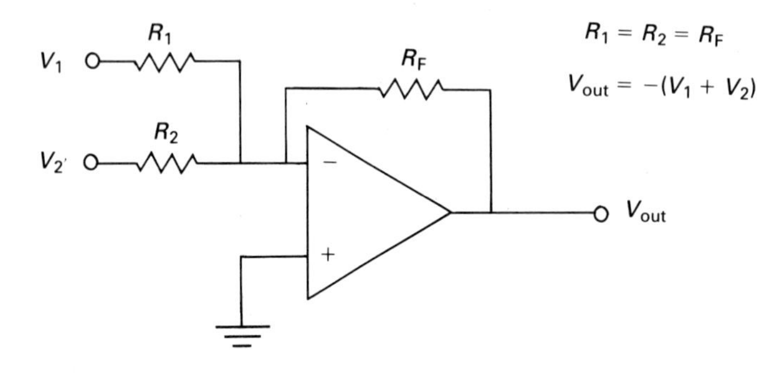


**Figure 15.1** The op amp adder

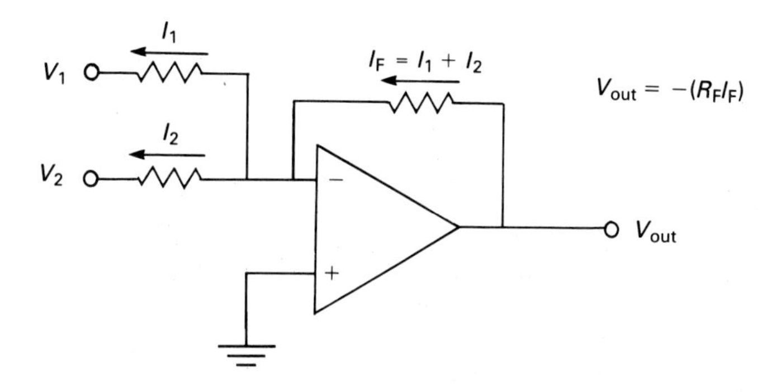


**Figure 15.2** The currents in a summing amplifier

inputs, then the inverting input is *also at ground potential.* This is known as *virtual ground.* It is not a true ground, but it is at the same voltage potential as ground.

Then, input voltage $V_1$ in our diagram "sees" only resistor $R_1$ between itself and ground, and input voltage $V_2$ "sees" only resistor $R_2$ between itself and ground. Neither input voltage can influence the other.

Voltage $V_1$ sets up a current through $R_1$, which we can find through Ohm's law. Voltage $V_2$ sets up a current through $R_2$, which we can find in the same way. And since the input of the op amp can neither sink nor source this current, it must come from the output of the op amp through resistor $R_F$. The current in $R_F$, then, is the sum of $I_1$ and $I_2$. $V_{out}$ must equal the product of $I_F$ and $R_F$.

Figure 15.2 summarizes these points. It is not usually necessary, however, to calculate the currents in an adder circuit, beyond determining whether the voltage sources can meet the current demands of the adder. The output voltage can be readily calculated as follows:

$$V_{out} = -(V_1 + V_2 + \ldots\ldots + V_n) \text{ when}$$
$$R_1 = R_2 \ldots\ldots = R_n$$

Note that this formula allows for more than two inputs. There is no theoretical limit to the number of input voltages that can be added with the op amp adder.

**Example 15.1**

In Figure 15.3, prove that $I_F R_F = V_{out}$

**Solution:**

$$I_1 = \frac{V_1}{R_1}$$

$$I_1 = \frac{0.5 \text{ V}}{10 \text{ k}\Omega}$$

$$I_1 = 50\ \mu\text{A}$$

$$I_2 = \frac{V_2}{R_2}$$

$$I_2 = \frac{1 \text{ V}}{10 \text{ k}\Omega}$$

$$I_2 = 100\ \mu\text{A}$$

$$I_3 = \frac{V_3}{R_3}$$

$$I_3 = \frac{2 \text{ V}}{10 \text{ k}\Omega}$$

$$I_3 = 200\ \mu\text{A}$$

$$I_F = I_1 + I_2 + I_3$$

$$I_F = 50\ \mu\text{A} + 100\ \mu\text{A} + 200\ \mu\text{A}$$

$$I_F = 350\ \mu\text{A}$$

$$V_{out} = -(I_F \times R_F)$$

$$V_{out} = -(350\ \mu\text{A} \times 10 \text{ k}\Omega)$$

$$V_{out} = -3.5 \text{ V}$$

Solving for $V_{out}$ by addition:

$$V_{out} = -(V_1 + V_2 + V_3)$$

$$V_{out} = -(0.5 \text{ V} + 1 \text{ V} + 2 \text{ V})$$

$$V_{out} = -3.5 \text{ V}$$

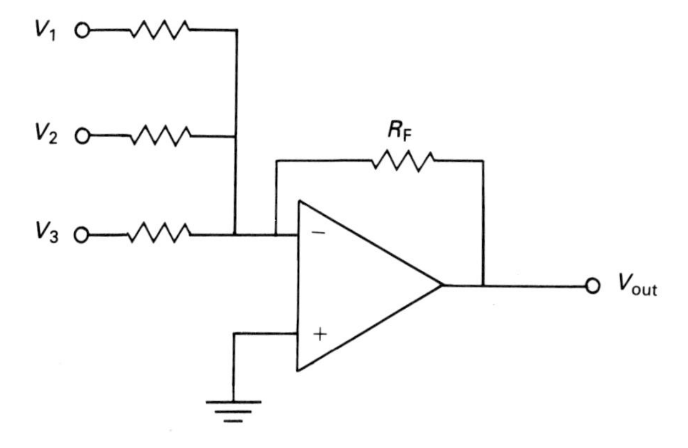


**Figure 15.3** A three-input adder

The minus sign is required because of the fact that the summing amplifier is also an inverting amplifier. The sign can be changed by adding an inverting voltage follower.

## 15.2 SUBTRACTION WITH THE OP AMP

The voltage differencing amplifier of Chapter 13 offers us one effective way to perform subtraction. Rather than reviewing that method here, we will see another technique that uses a summing amplifier.

In Figure 15.4, we see a summing amplifier once more, but now input $V_2$ is negated by an inverting voltage follower. The output voltage calculation is the same, but voltage $V_2$ is now negative.

### Example 15.2

In Figure 15.4, $V_1 = 3$ V and $V_2 = 1$ V. Find $V_{out}$.

**Solution:**

$$V_{out} = -(V_1 + (-V_2))$$
$$V_{out} = -(3\text{ V} + (-1\text{ V}))$$
$$V_{out} = -(3\text{ V} - 1\text{ V})$$
$$V_{out} = -2\text{ V}$$

Thus, we can also perform subtraction with our op amp circuit.

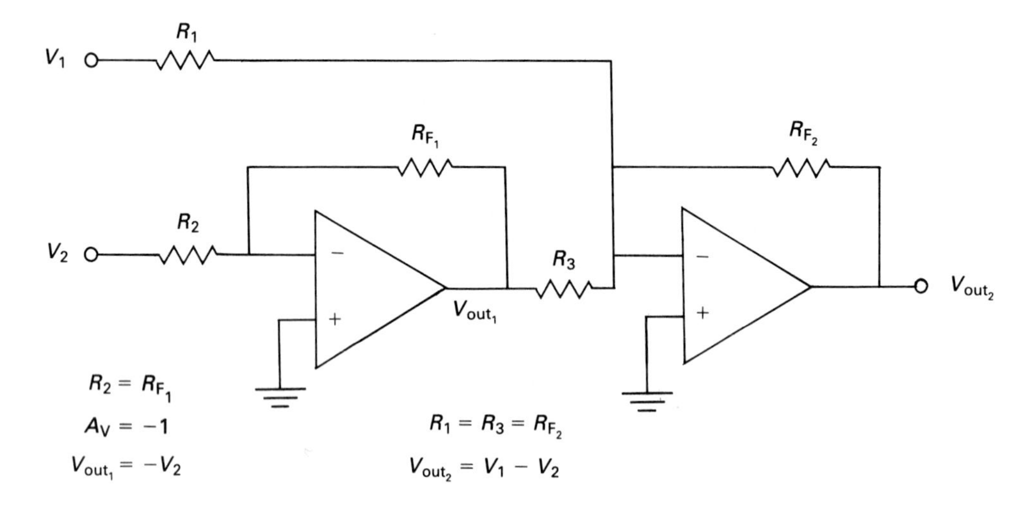


**Figure 15.4** Subtraction with a summing amplifier

## 15.3 MULTIPLICATION WITH THE OP AMP

Perhaps multiplication is the easiest of the math functions to understand with op amps. After all, amplification of a voltage *is* multiplication!

In the circuit of Figure 15.5, any input voltage will be multiplied by $-5$, the gain of the op amp. We can also combine addition, subtraction, and multiplication into one circuit.

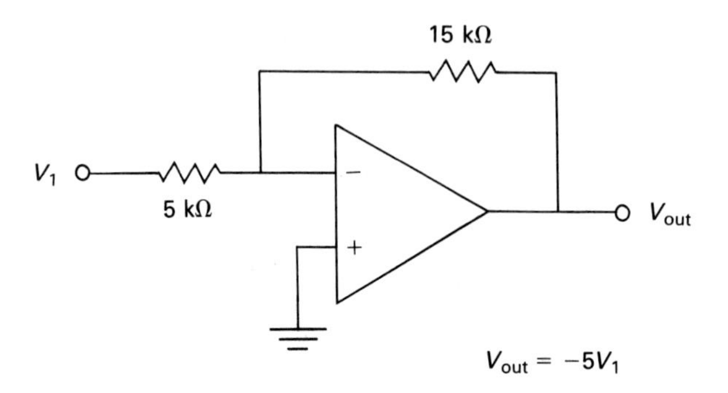


**Figure 15.5** The op amp multiplier

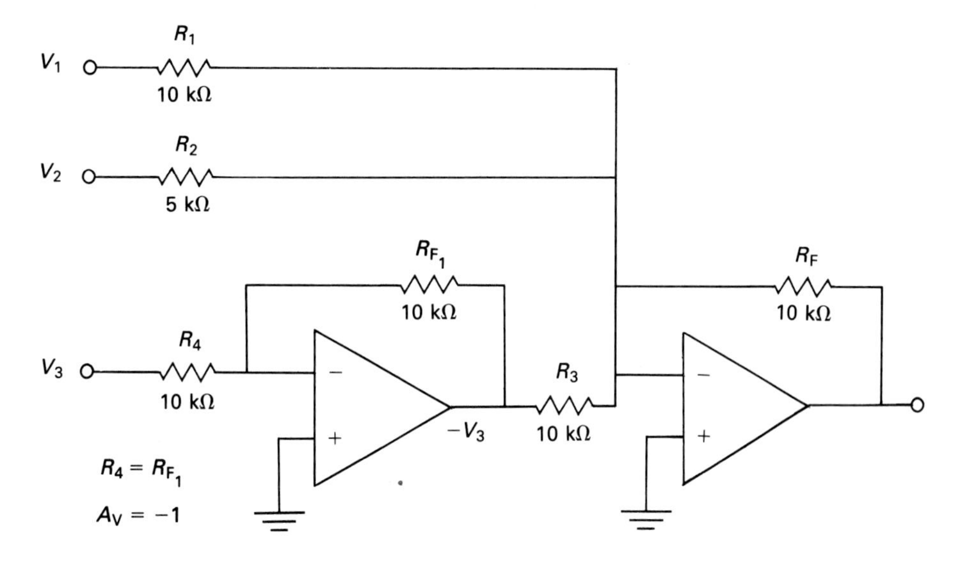


**Figure 15.6** Combined functions

The circuit in Figure 15.6 shows how this might be done. There are several ways to solve this circuit, but we will concentrate on the simplest technique by treating each input separately, as though the others were not there. We can then solve for three different answers, $V_{out_1}$, $V_{out_2}$, and $V_{out_3}$. The final answer will be the *algebraic sum* of the three. Let's solve the circuit for some specific input voltages.

### Example 15.3

In Figure 15.6, $V_1 = 1$ V, $V_2 = 0.6$ V, and $V_3 = 0.8$ V. Find $V_{out}$.

**Solution:**

$$V_{out_1} = V_1 \times \frac{-R_F}{R_1}$$

$$V_{out_1} = 1 \text{ V} \times \frac{-10 \text{ k}\Omega}{10 \text{ k}\Omega}$$

$$V_{out_1} = -1 \text{ V}$$

$$V_{out_2} = V_2 \times \frac{-R_F}{R_2}$$

$$V_{out_2} = 0.6\ \text{V} \times \frac{-10\ \text{k}\Omega}{5\ \text{k}\Omega}$$

$$V_{out_2} = 0.6\ \text{V} \times (-2)$$

$$V_{out_2} = -1.2\ \text{V}$$

$$V_{out_3} = -\quad 0.8\ \text{V} \times \frac{-R_F}{R_3}$$

$$V_{out_3} = -\quad 0.8\ \text{V} \times \frac{-10\ \text{k}\Omega}{10\ \text{k}\Omega}$$

$$V_{out_3} = -(0.8\ \text{V} \times (-1))$$

$$V_{out_3} = -(-0.8\ \text{V})$$

$$V_{out_3} = +0.8\ \text{V}$$

$$V_{out} = V_{out_1} + V_{out_2} + V_{out_3}$$

$$V_{out} = (-1\ \text{V}) + (-1.2\ \text{V}) + (+0.8\ \text{V})$$

$$V_{out} = -1.4\ \text{V}$$

This rather simple circuit has solved an equation in three unknowns. The algebraic formula for the circuit is

$$Y = -(V1_1 + 2V_2 - V_3)$$

## 15.4 DIVISION WITH THE OP AMP

When $R_F$ is *smaller* than $R_I$, our op amp multiplier functions as a divider. For example

$$\text{if } R_F = 10\ \text{K}$$
$$\text{and } R_1 = 20\ \text{K}$$

then when we solve for gain we get:

$$A_v = \frac{-R_F}{R_1}$$

$$A_v = \frac{-10\ \text{K}}{20\ \text{K}}$$

$$A_v = -1/2$$

and the circuit will divide by two.

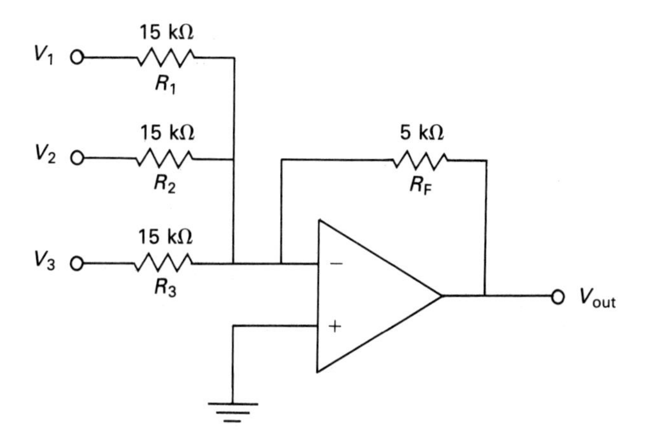


**Figure 15.7** Averaging input voltages

A special case of division is *averaging*. The op amp multiplier circuit is well-suited to this chore.

The circuit in Figure 15.7 will average the three input voltages. Note that

$$R_1 = R_2 = R_3$$

$$\text{and } R_F = \frac{R_1}{3} \quad \text{(the number of inputs)}$$

Let's solve the circuit:

### Example 15.4

In Figure 15.7

$$V_1 = 3 \text{ V}$$
$$V_2 = 1.2 \text{ V}$$
$$\text{and } V_3 = 2.7 \text{ V}$$

Solve for $V_{out}$ and prove that $V_{out}$ is the average of the input voltages.

$$V_T = V_1 + V_2 + V_3$$
$$V_T = 3\text{ V} + 1.2\text{ V} + 2.7\text{ V}$$
$$V_T = 6.9\text{ V}$$

Averaging the three:

$$V_{ave} = \frac{6.9\text{ V}}{3}$$
$$V_{ave} = 2.3\text{ V}$$

The only difference is the sign of the result, which is a consequence of the inverting amplifier. Thus, the circuit did indeed average the three input voltages.

## 15.5 INTEGRATION

*Integration* is the calculus technique for finding the area under a curve. Mathematicians tell us that they can simulate almost any process through integration. It is a technique that is often used in control work. Fortunately, our op amp can perform the integration for us, without our having to resort to the sometimes torturous paths of mathematical integrals.

Figure 15.8 is a basic op amp integrator circuit. Note that the feedback path is now through capacitor $C_F$ instead of through a resistance. Let's consider what happens in the circuit under different input conditions.

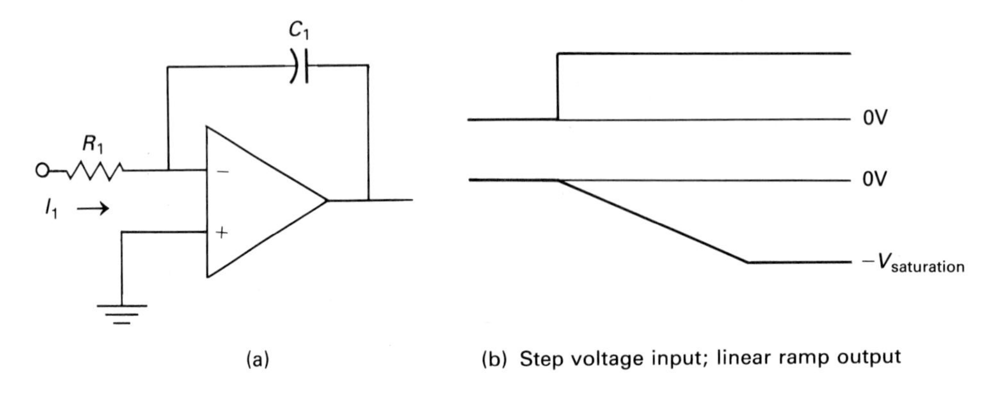


**Figure 15.8** An op amp integrator

First let's consider the circuit when $V_{in} = 0$ V. Since there is no difference voltage between the inverting and noninverting inputs, there is also no output voltage. More correctly

$$V_{out} = 0 \text{ V}$$

There is no voltage difference between input and output; therefore, $C_F$ is discharged.

Now let's input a *step voltage* as seen in Figure 15.8b. As soon as $V_{in}$ is felt on the input, current will flow in $R_1$ as shown in the diagram. But since the op amp input can neither source nor sink current, the current *must come from the capacitor.* At the same time, the output of our op amp has become negative because of the positive input voltage. Thus, the output of the op amp supplies current to the output side of the capacitor.

Since the output of an op amp is a *constant current source,* $C_F$ is charged with a constant current. When a capacitor is charged with a constant current, the voltage across the capacitor *increases linearly.* Thus, the output is a linear ramp, as seen in Figure 15.8b.

Of course, there is a practical limit to how negative the output voltage can become. After a certain length of time, the op amp output will become saturated. We cannot expect an output voltage greater than the supply voltage, after all.

Notice that the output becomes *more negative over a period of time.* In practical control work, the integrator is used to correct for the small errors that are inherent in any real-world circuit, called *steady-state errors.* The longer the error exists, the greater will be the output voltage and the greater, therefore, will be the corrective action. In other words, the output voltage is a *measure of how long the input voltage has been present.*

We can calculate the output voltage quite readily:

$$V_{out} = t \times V_{in} \times \frac{-1}{RC}$$

where $t$ = time in seconds

### Example 15.5

In Figure 15.8

$$R_1 = 500 \text{ k}\Omega$$
$$C_1 = 1\ \mu\text{F}$$
$$\text{and } V_{in} = 1.5 \text{ V}$$

Find $V_{out}$ after 0.5 seconds, and after 2.5 seconds.

**Solution:**

Solving for $RC$:

$$RC = 500\ \text{k}\Omega \times 1\ \mu\text{F}$$
$$RC = 0.5\ \text{sec}$$

Solving for $V_{out}$ (0.5 sec):

$$V_{out} = 0.5 \times 1.5\ \text{V} \times (-0.5)$$
$$V_{out} = -0.375\ \text{V}$$

Solving for $V_{out}$ (2.5 sec):

$$V_{out} = 2.5 \times 1.5\ \text{V} \times (-0.5)$$
$$V_{out} = -1.875\ \text{V}$$

In practical circuits, several other components are added to compensate for the offset voltage of the op amp and to allow the capacitor to be discharged.

The circuit in Figure 15.9 illustrates the additional components. Resistors $R_1$ and $R_F$ reduce the effects of offset voltage in the integrator circuit, while the FET can be turned on to discharge $C_F$ completely. Without these components, the circuit would eventually saturate from even the tiniest of offset voltages.

From the viewpoint of a technician, you must remember that component values in this circuit have been selected after a great deal of calculation and much field "cut and try" experience. It is not a good idea to substitute component values in an integrator control circuit.

## 15.6 DIFFERENTIATORS

Differentiation is a measurement of *rate of change.* We use differentiation to determine the acceleration of spacecraft, for example, or to determine how fast a falling body is going. Differential calculus is an important branch of mathematics, but the op amp can be used as a differentiator, saving us from having to perform the calculations.

Figure 15.10 is an ideal op amp differentiator. Notice that the resistance and capacitance have traded places from the integrator circuit. As you might guess from this, differentiation is the opposite of integration.

You may recognize this circuit as a high-pass filter from Chapter 14. The difference here is that the expected input signal is not ac signals. Control circuits use the differentiator to correct for *sudden changes in the process variable.* The input voltage is expected to be a step voltage or, more accurately, a rapidly rising ramp.

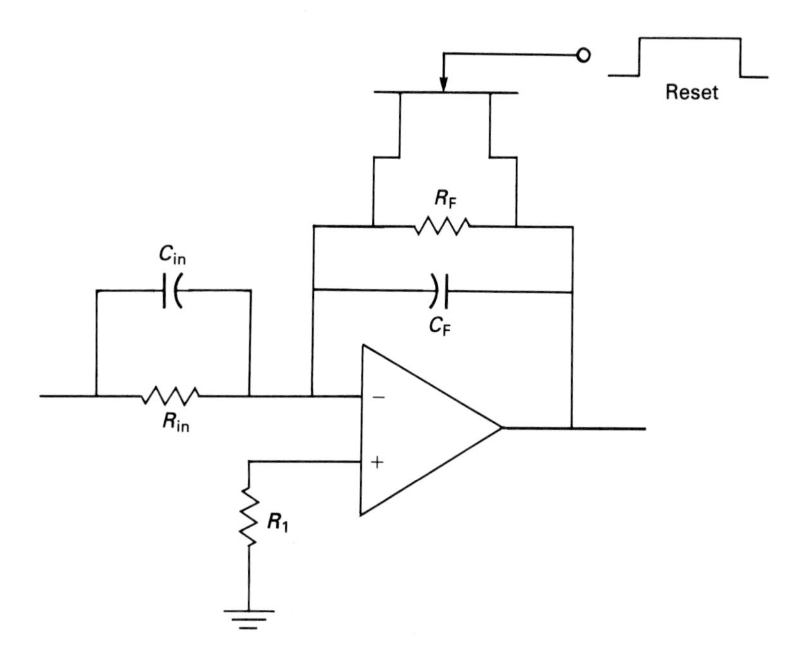


**Figure 15.9** A practical integrator circuit

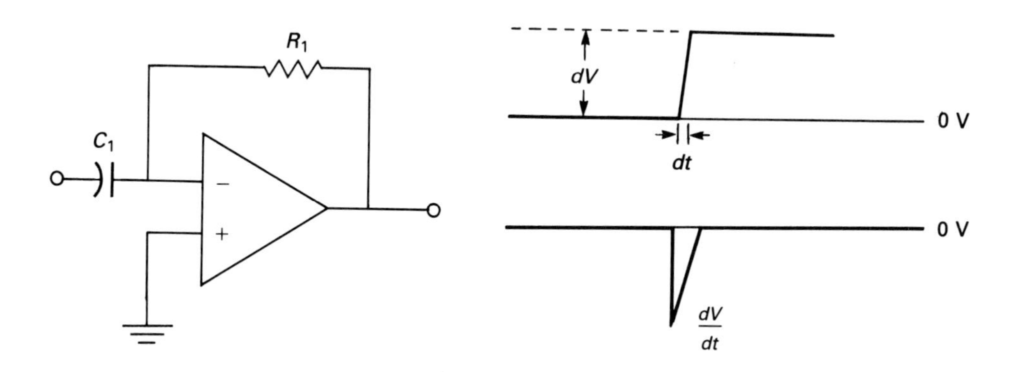


**Figure 15.10** An op amp differentiator

From the presence of $C_{in}$, you might guess that the differentiator will not respond to the steady dc voltage of steady-state error, which the integrator controls; rather it responds *only to changing voltages.*

We can calculate the output voltage:

$$V_{out} = \frac{dV}{dt}$$

$$\text{where } dV = \text{the change in voltage}$$

$$\text{and } dt = \text{the change in time}$$

This formula supposes that the time constant is much smaller than the time interval.

$$(RC << dt)$$

**Example 15.6**

A sudden change in a process variable causes a transducer voltage to change from 0.5 V to 0.6 V in 0.01 second. Calculate the resultant output voltage pulse. Also calculate the output voltage pulse for the same voltage change occurring in 0.1 second.

**Solution:**

Finding $dV$:

$$V_1 = 0.5 \text{ V}$$

$$V_2 = 0.6 \text{ V}$$

$$dV = V_2 - V_1$$

$$dV = 0.6 \text{ V} - 0.5 \text{ V}$$

$$dV = 0.1 \text{ V}$$

Finding $V_{out}$ (pulse):

$$V_{out} = \frac{0.1}{0.01}$$

$$V_{out} = 10 \text{ V (pulse)}$$

**Part 2:**

$$V_{out} = \frac{0.1}{0.1}$$

$$V_{out} = 1 \text{ V}$$

From this, we can see that a small change in input voltage can cause a large output pulse if it happens over a short time span, while more slowly changing voltages cause small output pulses. These rapidly changing voltages are said to have a *large differential*. Unfortunately, *noise* often has large differentials, so differentiator circuits must be used wisely.

Figure 15.11 shows the additional components you may find in practical differentiator circuits. $C_F$ limits the high-frequency gain of the circuit to improve stability. The addition of $R_{in}$ establishes a low-frequency cutoff point, and $R_1$ compensates for offset voltages.

Adding these extra components creates a stable differentiator circuit. It will differentiate input signals, ranging in frequency between two values, which are calculated as follows:

**Solution:**

Solving for $A_v$:

$$A_v = -\left(\frac{R_F}{R_1}\right)$$

$$A_v = -\left(\frac{5\text{ K}}{15\text{ K}}\right)$$

$$A_v = \frac{-1}{3}$$

Solving for $V_{out_1}$:

$$V_{out_1} = 3\text{ V} \times \frac{-1}{3}$$

$$V_{out_1} = -1\text{ V}$$

Solving for $V_{out_2}$:

$$V_{out_2} = 1.2\text{ V} \times \frac{-1}{3}$$

$$V_{out_2} = -0.4\text{ V}$$

Solving for $V_{out_3}$:

$$V_{out_3} = 2.7\text{ V} \times \frac{-1}{3}$$

$$V_{out_3} = -0.9\text{ V}$$

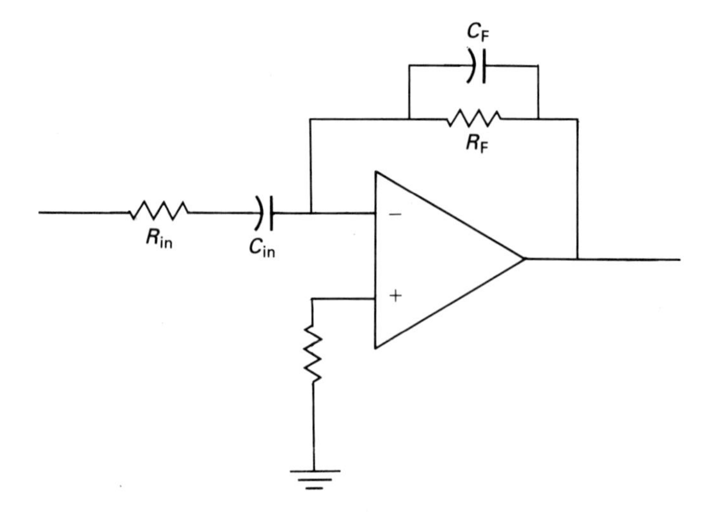


**Figure 15.11** A practical differentiator circuit

Solving for $V_{out}$:

$$V_{out} = V_{out_1} + V_{out_2} + V_{out_3}$$
$$V_{out} = (-1\text{ V}) + (-0.4\text{ V}) + (-0.9\text{ V})$$
$$V_{out} = -2.3\text{ V}$$

Averaging the three input voltages:

$$F_{lower} = \frac{1}{2}(R_F C_{in})$$

$$F_{upper} = \frac{1}{2}(R_{in} C_{in})$$

Thus, the slowest differential it will integrate is

$$dt_{lower} = \frac{1}{F_{lower}}$$

and the fastest differential it will integrate is

$$dt_{\text{upper}} = \frac{1}{F_{\text{upper}}}$$

At higher frequencies (shorter time differentials), the circuit is unstable and will actually behave like an integrator. Once again, never substitute parts values when repairing a differentiator.

## SUMMARY

In this chapter, we have seen that op amp circuits can be used to perform many different mathematical functions or *operations.* We have seen that the operations may be as simple as addition or as complicated as integration or differentiation and still work in simple op amp circuits. We have seen that:

1. The inverting amplifier is the heart of op amp math circuits.
2. The inverting input of an inverting amplifier is at ground potential, which allows us to combine signals at this *summing junction.*
3. The summing amplifier allows us to perform addition.
4. Adding an inverting voltage follower before a summing amplifier allows us to perform subtraction.
5. By making $R_F$ larger than the input resistor(s), we can perform multiplication.
6. By making $R_F$ smaller than the input resistors, we can perform division.
7. By making $R_F$ equal to $\frac{R_{in}}{n}$, we can average inputs.
8. By using different values for different input resistors, we can combine functions in one op amp circuit.
9. By substituting a capacitor for $R_F$, we can perform the calculus function of integration.
10. By substituting a capacitor for $R_{in}$, we can perform the calculus function of differentiation.

## SELF-TEST

1. The inverting input of an op amp inverting amplifier is at ______________ potential.
2. The inverting input of an op amp inverting amplifier is called the _______ ________________________________ junction.

3. Multiple inputs can be combined at this input because it is at __________.
4. When two inputs are combined at the summing junction and all of the input resistors are equal to the feedback resistor ($R_1 = R_2 = R_n = R_F$), the circuit forms a(n) ______________________________ amplifier.
5. When the feedback resistor in an inverting amplifier is larger than the input resistor, the circuit performs ______________________________.
6. When the input resistor is larger than the feedback resistor in an inverting amplifier, the circuit performs ______________________________.
7. By using different size input resistors in a summing amplifier circuit, the circuit can perform ______________________________.
8. When the feedback path includes a capacitor, the circuit performs ________ ______________________________.
9. When a capacitance is used in series with the input, the circuit performs ______________________________.
10. The ______________________________ output is a function of how long an input voltage has been present.
11. The ______________________________ output is a function of how rapidly an input signal changes.

## QUESTIONS/PROBLEMS

1. The resistances in a three-input adder are

$$R_1 = R_2 = R_3 = R_F = 15\ \text{k}\Omega$$

The input voltages are

$$V_1 = 1.5\ \text{V}$$
$$V_2 = 2.5\ \text{V}$$
$$V_3 = 4.5\ \text{V}$$

Solve for $I_1$, $I_2$, $I_3$, $I_F$, and $V_{out}$.

2. In the preceding circuit, an inverter is added to input $V_3$. Write the algebraic formula for the circuit.
3. Design a single op amp circuit to solve the formula

$$Y = -\left(A + 2B + \frac{C}{2}\right)$$

Prove that the circuit works by solving for

$$V_{out} = I_F \times R_F$$

4. Design a single op amp circuit that will average four different input voltages. Prove that the circuit works by solving for $V_{out} = I_F \times R_F$.
5. The steady-state error in a control system is 0.15 V. Design an op amp circuit that will output 3 V after the error signal has been present for 2 seconds.
6. The largest possible transient error in a control system causes a voltage change of 0.5 V in 100 ms. The *RC* time constant of the differentiator must be at least ten times smaller than the time interval. If $R_F$ must be 100 kΩ, what is the largest acceptable value of $C_{in}$? What will be the maximum output pulse from the controller?

# 16

# Transmitting Control Information

The industrial plant is often an electrically and mechanically hostile environment. Large electrical devices switch on and off, dc motor brushes arc, welders and even fluorescent light fixtures create *EMI* (*electromagnetic interference*). Oil fumes, steam, paint overspray, and heat threaten delicate electronic controllers.

For these reasons, electronic controls are often located in remote locations, away from the factory floor. Data from sensors must then be transmitted from the location of the process variable to the location of the controller. Control signals must be transmitted from the controller back to the actuators that adjust the process variables.

In this chapter, we will examine some of the techniques used to transmit sensor data and process control information.

## OBJECTIVES

You will have successfully completed this chapter when you can:

1. Explain the use of twisted-pair and shielded conductors to reduce the effects of EMI.
2. Describe a voltage-to-current converter circuit.
3. Describe a current-to-voltage converter circuit.
4. Explain how fiber-optic systems reduce interference in the industrial environment.

## 16.1 SOURCES OF EMI

EMI is created by virtually every electrical device. Perhaps the most familiar is the radio frequency interference (RFI) you can detect when you turn on your computer. The rapidly changing signals on the data and address lines create harmonics that can usually be heard on an AM radio. Switching regulated power supplies, which are growing in popularity, are yet another source of RFI from your computer. The magnetic field surrounding an electric motor, or even a fluorescent light ballast, can generate serious interference problems for a control system. When the device is switched on or off, sudden changes in electrical current flow generate large pulses of EMI, which are many times larger than the field surrounding the device while it is running. SCR and TRIAC speed controls add their own form of rapid on-off switching. Electric arc welding contributes to EMI, as do dc and universal motors which contain commutators and brushes with their attendant sparking. Devices that spark or arc act like old-fashioned "spark gap" radio transmitters.

When the wiring must run beside or across other wiring, we can add another problem: *two conductors separated by an insulator form a capacitor.* Rapidly changing signals such as the square waves of digital control or the sharp rise and fall times of analog motor speed controllers contain high-frequency components which can be capacitively coupled even through the tiny capacitances that exist between adjacent control wires.

Through this ever changing and very active field, we must run our control wiring, which carries tiny electrical signals between the sensor and the controller. From your early studies, you have seen that a moving magnetic field induces electrical currents in a conductor. That means that every change in the magnetic field within the plant can induce false signals into our control wiring. It is evident that we must protect our wiring from such induction and from capacitive coupling with adjacent wiring.

One form of protection is simply in the *routing of the wire.* Sensitive signal lines should be run *at least* one foot from other wiring, when possible. The active magnetic field surrounding fluorescent lighting should be avoided at all costs. We can use an AM radio tuned near the 600 kHz end of the band to identify the most active noise areas in our plant. Once identified, we can try to avoid them with our control wiring.

The electrical code requires that most wiring be run in metal conduit, which will provide some shielding from EMI. Even when it is not required, conduit is a good idea because it also protects wiring from physical damage and from chemical fumes, which may cause corrosion.

## 16.2 TWISTED PAIRS AND SHIELDING

The first approach to reducing or eliminating the effects of induced interference is to use *twisted-pair* wiring between the sensor and the controller.

Figure 16.1 shows how twisted-pair wire helps reduce the induced interference. In Figure 16.1a, we see that a signal induced in a twisted pair of wires is induced *equally in both wires.* The input to the controller, however, is usually a differential input.

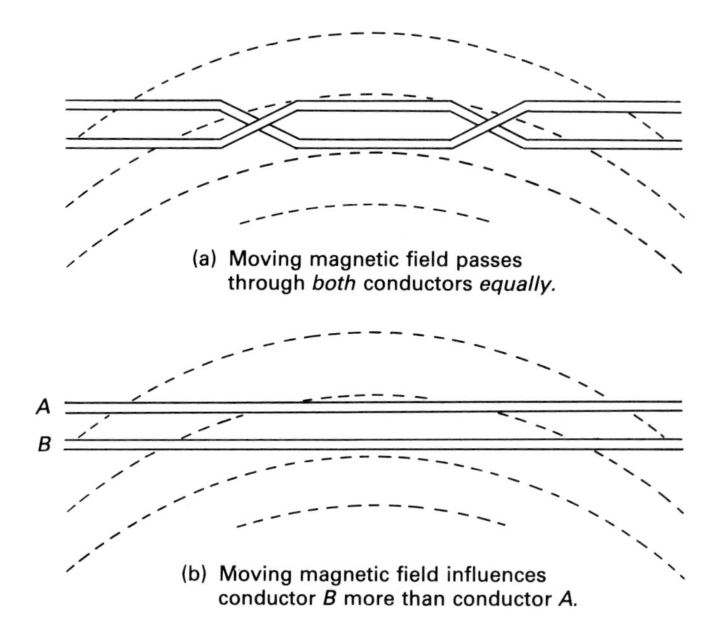


**Figure 16.1** Twisted-pair wire

Since a differential input amplifies the *difference* between signals at the two inputs, the induced signal is *canceled*.

Figure 16.1b shows that when the wires are not twisted, the induced signals may not be equal. If they are unequal, there will be some difference voltage. Thus, some portion of the interference will be amplified.

While twisted pairs can substantially reduce induced noise, they are not the perfect answer. The theory is fine, but in practice, the induced signals are *almost* equal; therefore, some portion of the interfering noise can be amplified. This is especially troublesome in systems with differential control.

In Figure 16.2, we see the idea behind shielded wire. The inside conductors shown here are a twisted pair, as described earlier. These conductors, however, are surrounded by a separate *shield*, a conductor that *completely surrounds the inner conductors*. The shield is grounded; therefore, any induced or capacitively coupled signal is shunted to ground by the shield and cannot reach the inner conductors.

Shielding may be a braided tube surrounding the inner pair, or it may be an aluminum-foil strip wrapped around the pair. In the case of the foil shield, there will be a third *drain wire* in the cable so that the shield may be grounded. It is important

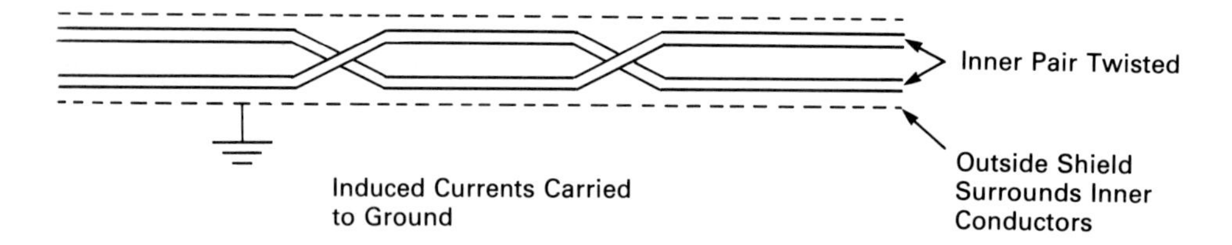


**Figure 16.2** Shielded wire

when splicing shielded wire to connect the drain wire. It is also important to arrange the shield to surround the splice.

Shielded wiring may be composed of an individual shielded conductor, a shielded pair, a shielded multipair, or a collection of many shielded pairs inside a single cable. When many shielded pairs are included in a single cable, there is often an extra shield surrounding the entire bundle. Once again, shields must be connected together when splicing this cable. The shield should be grounded *at one end only* (usually at the controller end), and *the shield should never carry signal.*

## 16.3 CURRENT LOOPS

One of the most effective techniques for reducing noise in signal lines is to convert the signals to *currents* and use low-impedance *current* amplifiers, rather than high-impedance voltage amplifiers, to process signals. Recall from our earlier work that the input impedances of voltage amplifiers are very high to reduce the losses that might appear from long lines and high source impedances. The problem with high-impedance inputs, though, is that very small induced currents can cause large error voltages in such systems.

In Figure 16.3, we can see why this happens. The noise source in this illustration causes only 5 microamps of induced noise current in the line. However, when those 5 microamps flow through the 500 kilohm input impedance of the amplifier, it causes a 2.5 V error signal!

Figure 16.4 shows the method used to reduce the effects of the induced currents in the transmission of information. Here, the output voltage of the transducer is converted into a *current* before being sent to the controller. At the controller, the current is converted back into a voltage for the signal processing amplifiers. The noise source induces the same 5 microamps of noise current into the line, but the effect is greatly reduced. The current-to-voltage converter in this circuit has an input impedance of only 1 kilohm. Then, 5 microamps of induced noise current are converted into a noise voltage of only 5 mV!

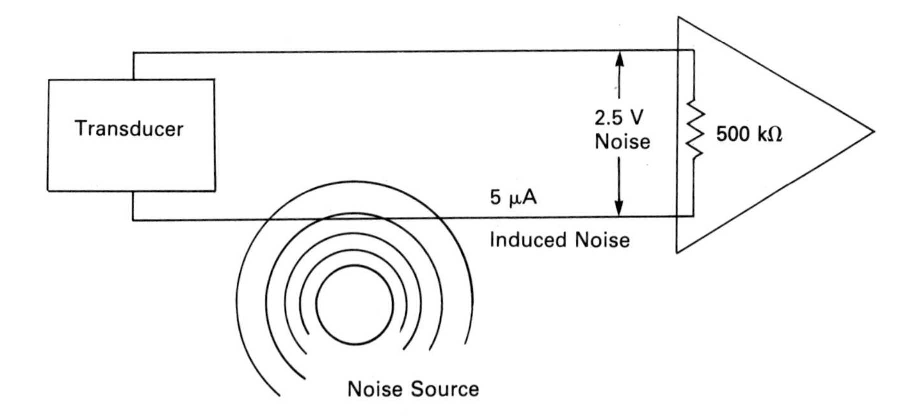


**Figure 16.3** Why voltage transmission accepts noise.

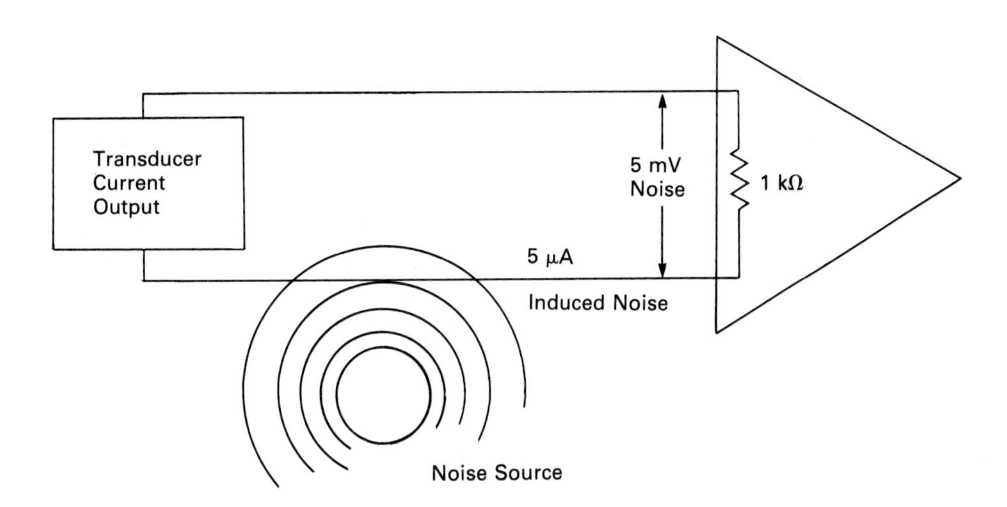


**Figure 16.4** The current loop data line

In Figure 16.5, we see the circuit of a simple voltage-to-current converter. In essence, the circuit is a noninverting amplifier. The difference is that the *output of the circuit is the feedback loop for the amplifier.* The output is a current, $I_F$. The line impedance for the circuit is now the low-impedance $R_F$, rather than the high-impedance $Z_{in}$.

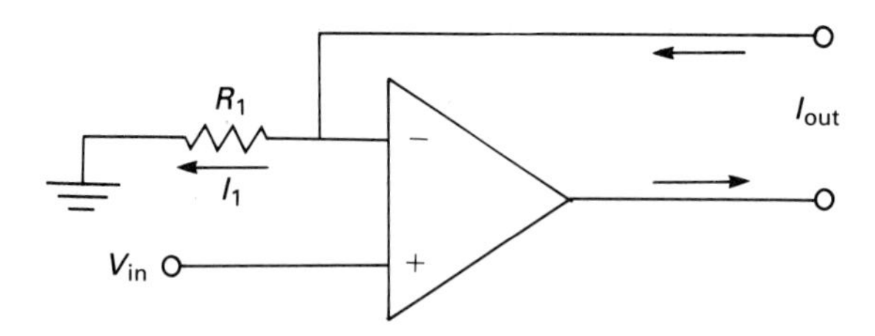


**Figure 16.5** A voltage-to-current converter

The current output is easily calculated:

$$I_{out} = \frac{V_{in}}{R_1}$$

The Instrument Society of America (ISA) has established standards for the transmission of data with currents, rather than voltages. The major standard is in the current range between 4 mA and 20 mA. There is also a secondary standard in the range between 1 mA and 5 mA, but it is much less often used.

In the 4–20 mA standard, the level of 4 mA represents the *minimum signal*, and the 20-mA level is the *maximum signal* for the system. At first, it may seem odd to have any level of current other than zero represent zero, but a moment of thought tells us why: the current through an open line is zero. That means a system would not be able to tell the difference between no signal and an open line. The ISA system makes sense.

In Figure 16.6, we see the design of a simple current-to-voltage converter. The output voltage of the circuit is easily calculated:

$$V_{out} = I_{in} \times R_F$$

Transducers and controllers are readily available with built-in ISA current-source transmitters and receivers. Most manufacturers offer a choice of either voltage output or current-source output in their transducer lines, and voltage input or current-source input in their controller lines. It is far more economical and reliable to use these ready-made devices than to attempt to fabricate them on the job.

Servicing current-source lines is generally a simple matter of inserting a milliammeter into the line. We must also beware of the effects of grounded wiring, which can cause erroneous currents.

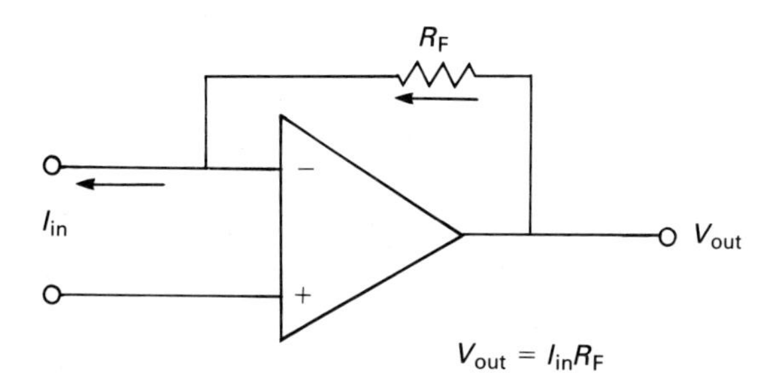


**Figure 16.6** A current-to-voltage converter

## 16.4 FIBER-OPTIC DATA LINKS

A new technology is emerging for data transmission. It is immune to the effects of RFI and EMI, allows more data paths per cable diameter, and is generally cleaner in overall operation than traditional wire coupling. This technology is the optical fiber.

An *optical fiber* is a light conductor made of glass or thin plastic fibers. The length of the fiber is highly polished to reflect light *back into the fiber*, rather than let it escape along the length of the path. This means that *most* of the light that is injected into one end of the fiber reaches the other end.

In Figure 16.7, we see the principle of the optical fiber. Notice that in Figure 16.7b, there is an abrasion in the outer wall of the fiber that allows some of the light

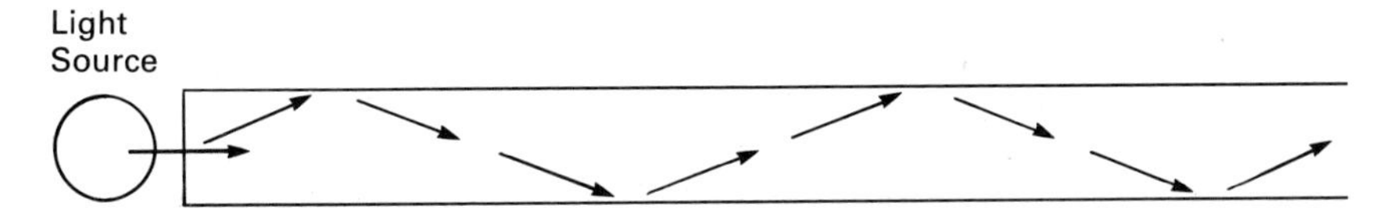


(a) Light reflects from polished sides of the fiber—it stays in the fiber.

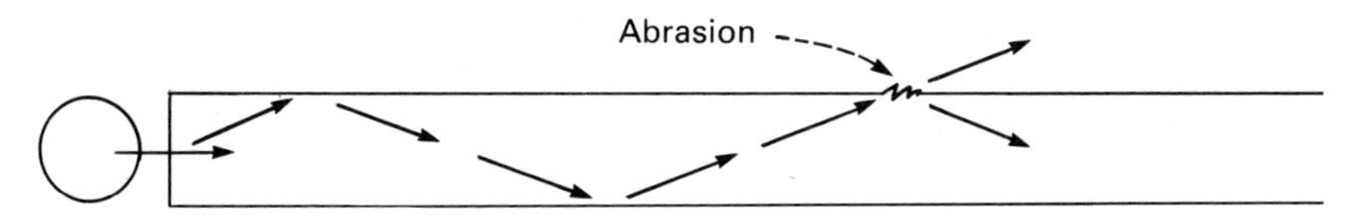


(b) Light escapes through the abraded portion of the fiber —it reduces light output.

**Figure 16.7** The principle of the optical fiber

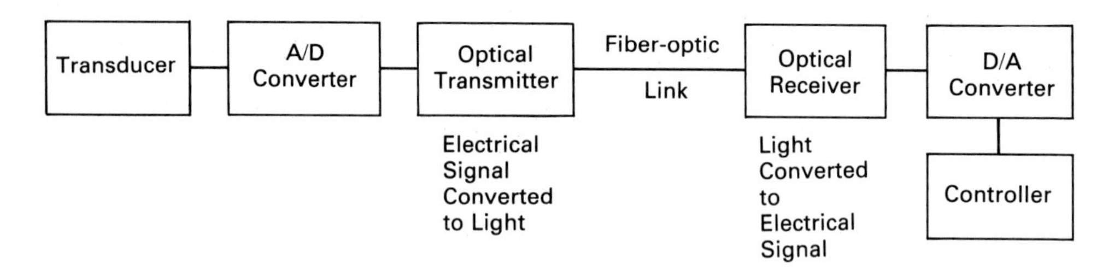


**Figure 16.8** A typical fiber-optic link

to escape. Obviously, we do not want that to happen because it will considerably reduce the effectiveness of optical fiber transmission. Care must be exercised in handling fiber-optic cables, but they usually have a strong plastic outer jacket, which tends to protect them.

Splicing fiber-optic cable requires special *butt-splice* connectors. The fiber-optic conductor must be cut off squarely to ensure maximum transfer of light from one fiber to the next. Special transparent cements are used in the joining technique.

Figure 16.8 shows the block diagram of a typical fiber-optic communications link. The transmitter end of the fiber consists of an analog-to-digital converter, which converts the analog signal into a digital signal. Digital signals are more suitable for transmission through optical fiber. An LED or laser diode then converts the electrical signals into light, which can be transmitted through the fiber.

On the receiving end, the digital light pulses are detected by a photodiode or phototransistor and converted back into electrical signals. The digital signal may then be converted back into an analog signal for analog controllers. Digital controllers may not need this additional conversion.

The transmitter and receiver units for fiber-optic communications are usually sealed units. The technician should make no effort to open such units, but should return them to the manufacturer for repair. A well-equipped factory shop will have spare units on hand to reduce downtime to a bare minimum.

## SELF-TEST

1. Electric motors turning on and off generate a source of ________________ in an industrial environment.
2. This "magnetic noise" can ______________________________ currents in control wiring systems.

3. In a system that employs voltage amplifiers, small induced ____________ can result in large false signals.
4. The differential input of the voltage amplifiers will ________________ equal signals that appear on both inputs.
5. Twisted-pair wires help ensure that induced signals will be ___________.
6. The level of induced signals can be substantially reduced by using ______ ___________________________ wires.
7. The shield in a shielded wire must be at _________________ potential.
8. Signal voltages can also be converted to __________________________ for transmission along wires.
9. Current amplifiers are inherently _____________________________ impedance devices.
10. The ISA has two standards for current transmission of data. The main one uses ___________ mA to represent the smallest signal and ___________ mA to represent the largest signal.
11. In the secondary ISA standard, the smallest signal is ___________ mA and the largest signal is ___________ mA.
12. Signal voltages can also be converted to light, which is transmitted along a(n) ___________________________ data link.
13. Optical signals are transmitted in ___________________________ form.
14. Optical data transmission is immune to both _____________________ and ___________________________.

## QUESTIONS/PROBLEMS

1. The transmission line for a sensor must run past several large dc motors. The sensor is available in voltage output or current-source output models. Explain which model you would select. Describe the other steps you would take to reduce the possibility of interference.
2. Which ISA current-source standard would you suppose to be less susceptible to induced noise? Why?

# 17

# A Summary of Signal Processing

In Section III, we have seen some of the signal-processing circuits used to amplify and clean up the output of measurement transducers. We have also seen some of the techniques used to transmit transducer signals safely through the electrically hostile atmosphere of the modern factory. In this chapter, we will summarize the main points of signal processing and transmission.

## OBJECTIVES

You will have successfully completed this chapter when you can:

1. Analyze the signal-processing section of various control systems.

### *17.1 AN INDUSTRIAL COUNTER SYSTEM*

The popularity of canned drinks has caused a boom in the can-manufacturing industry in recent years. Drink cans are manufactured in two pieces, the *can* and the *end*. The ends, or lids, complete with pop-top, are stamped from a sheet of metal and embossed in a single machine. The finished ends come out of older machines on edge in a V-trough.

Before the pop-top and the plastic liner used in some cans today, the ends were "counted" by measurement. A trough length of two feet was determined to be a specific number of ends. The ends were packaged in narrow bags and delivered to the drink-canning companies.

When the pop-top and plastic liner are added, the ends become "springy." The result is that the number of ends per foot varies with the strength and fatigue of the machine operator. The harder the machine operator presses the ends into the end of the trough, the more ends will be contained in a package. Some means was needed to count the required number of ends and insert a separator, so that the operator could bag exactly the right number of ends in a package.

It was determined that an optical sensor could detect the ends as they passed beneath a pinpoint light source and reflected it back to a phototransistor. Since this was an older, nonautomatic machine, the ends were moved beneath the sensor manually. That meant that the speed would be variable, depending on the operator.

In Figure 17.1, we see an illustration of the end-counter system as it was built. Since compressed air was readily available, an air solenoid was used to force a "blunt

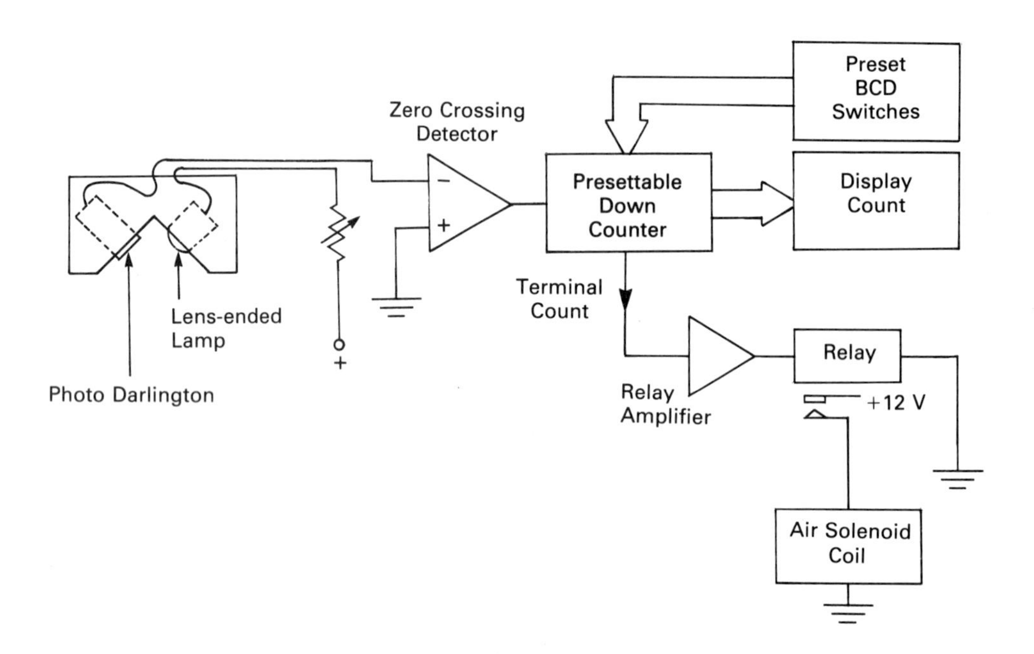


**Figure 17.1** The end-counter system

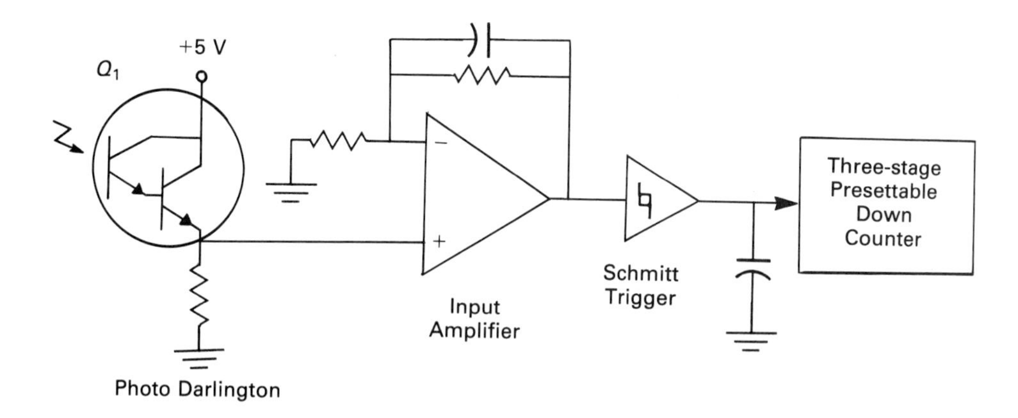


**Figure 17.2** The signal-processing section of the end-counter system

knife" between the ends when the proper count was reached. This solenoid was operated by a switch closure in the counter circuit.

Nearby equipment generated high levels of EMI, which had to be dealt with in the controller system. The varying speed of the ends passing beneath the light source caused different signal levels to be output by the phototransistor, compounding the problem.

Figure 17.2 is the circuit for the signal-processing section of the end-counter system. The input stage is a high-gain amplifier section, which brings the output level of the signal to a standard value. Almost any level of input signal from the phototransistor is amplified to the clipping level by this input amplifier. The capacitor in the feedback path adds low-pass filtration to the circuit, which eliminates most of the EMI from nearby machinery.

The second stage is a Schmitt trigger circuit, which "squares up" the edges of the signal for the CMOS counter chips. CMOS counters were selected because of their higher noise immunity. Note the capacitor on the input, which adds a second stage of low-pass filtering.

Although it is not shown, additional filtration was needed on the power supply. Ceramic disk capacitors were added across the power line and in parallel with the power-supply filter capacitors to help reduce power-line-carried noise that infiltrated itself into the system.

## 17.2 A ROBOTIC POSITION CONTROL

The industrial robot is perhaps the ultimate automatic machine. Programmable to perform a variety of tasks, the robot is a *generalist*, rather than a specialist. Still, when we analyze the operation of the robot, we find that it is really a collection of individually controlled variables, coupled with a computer which can be programmed for a variety of activities.

A *full servo* robot has transducers that report the position of each of the robot's axes of motion to the controller. Any variation in position between the desired position and the actual position can be controlled.

The circuit in Figure 17.3 is the feedback and control network from a popular light industrial robotic arm. Let's see how it works.

The control computer outputs an eight-bit code to the controller. This is clocked into the eight-bit latch, $U_1$, storing the desired position. The output of $U_1$ is connected to $U_2$, a digital-to-analog converter, which outputs an analog voltage. This analog voltage serves as one input to comparator $U_3$.

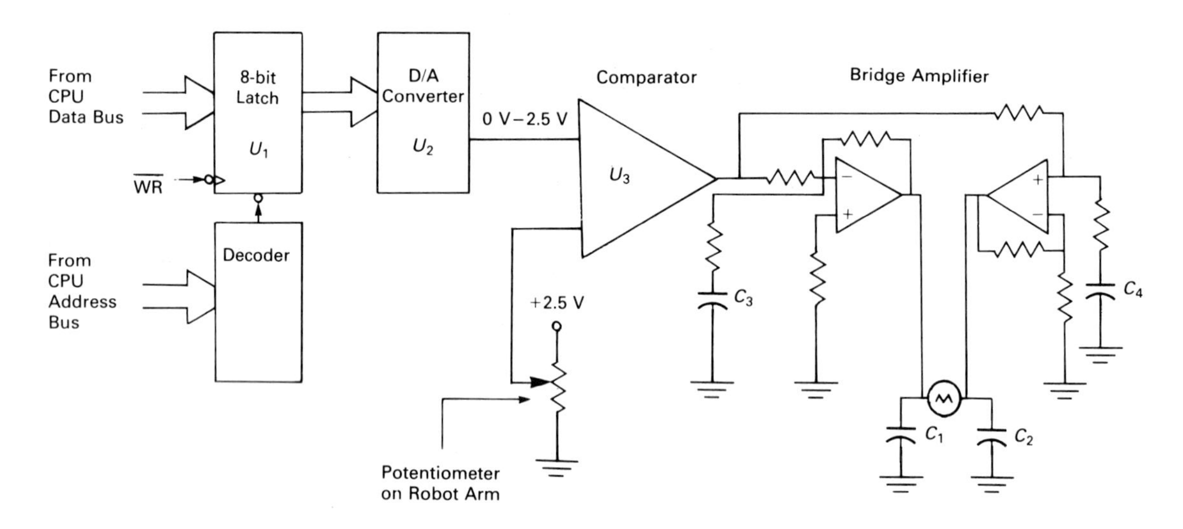


**Figure 17.3** A robotic-arm feedback controller

Located on the robotic arm is a potentiometer that is mechanically connected to the axis under control. As the axis rotates, it rotates the potentiometer, which reports back to the controller with a voltage representing the position of the axis. This voltage is the other input to the comparator.

The comparator output is a voltage whose *magnitude* is proportional to the distance the axis must move to arrive at the desired location. The *polarity* of the voltage depends on the direction in which the arm must move. The output of the comparator is fed to the input of a bridge amplifier, which drives the dc motor.

Dc motors contain brushes and commutators, which tend to arc. In other words, the motor controlled by this system generates its *own* EMI. To reduce this, capacitors $C_1$ and $C_2$ are mounted right at the motor, shunting much of the high-frequency energy to ground at that point. Additional capacitors $C_3$ and $C_4$ are used in an *RC* network acting as a low-pass filter on the bridge amplifier. Further filtration is provided by the addition of capacitors $C_5$ and $C_6$ on the power supply for the bridge amplifier and comparators.

## SELF-TEST

1. The sensitivity of the end counter system in Figures 17.1 and 17.2 can only be adjusted by varying ______________________________.
2. The input amplifier in Figure 17.2 is configured as a(n) ______________.
3. The Schmitt trigger in Figure 17.2 is used for ______________________.
4. The robotic arm position controller in Figure 17.3 indicates that this is a(n) ______________________________ servo robot.
5. What is the purpose of $C_1$, $C_2$, $C_3$, and $C_4$ in Figure 17.3?

## QUESTIONS/PROBLEMS

1. What effect will reducing the lamp voltage in Figure 17.1 have on the sensitivity of the end-counter?
2. When light strikes the base of the photo Darlington, what will happen at the noninverting input of the input amplifier?
3. In Figure 17.2, what will be the effect of increasing the value of the capacitor in the feedback loop? Explain your answer.

# Troubleshooting Signal-Processing Circuits

From an inspection of the systems in this section, it is clear that there is often no sharp line of demarcation between the signal-processing section and the controller section of analog control systems. The circuit board that filters the noise from a transmission line may also output a control signal to the process under control. With the exception of wire size, the input and output transmission lines are very similar. For these reasons, the remarks in this section also apply to analog control systems.

As stated in Chapter 10, one of the most common trouble spots in a system is in the wiring between the sensor and the controller and between the controller and the actuator. Before disassembling a controller for service, it is a wise move to perform a series of ohmmeter tests on the interconnecting wires. Refer to Chapter 10 for step-by-step instructions.

The *shielding* of the control wires is critically important. EMI and RFI must be prevented from reaching the controller as a matter of safety: unintended signals may cause unexpected movement or change in a system, which can be extremely hazardous to service personnel and system operators. When reconnecting the system wiring, be sure that the shield is properly connected.

If the system wiring is not at fault, begin servicing any circuit board with a careful visual inspection. An illuminated magnifying glass will be helpful in this step. Look for discolored, burned, or charred components, as well as small cracks in the circuit traces on the board. Also look for a dark ring around the lead of a component where it is soldered to the circuit board. These rings indicate a possible poor solder connection.

When a damaged component is replaced, the circuit must undergo further testing before it can be declared fixed. Every effort must be made to determine *what* caused the component to fail *before* restoring power to the circuit. A common cause of burned resistors, for example, is a failed electrolytic coupling capacitor, which allows dc current to pass. Test semiconductor junctions around the failed component with an ohmmeter; power surges often exceed the breakdown voltage of these junctions, causing them to short.

Monitor the circuit around the replaced component with voltmeters, ammeters, and oscilloscopes when power *is* applied to the circuit. Watch the meters for unusually large voltage change and excessive current flow.

Watch the oscilloscope for spikes, turn-on transients, and spurious oscillations. Something made the component go bad. The component itself may have been defective, or another defective component may have put a strain on it. It is important to determine which is the case; the alternative is more down time. An analog meter is best for this type of monitoring because the slow speed of the DMM will not be able to keep up with the changes.

Circuit trace cracks must be repaired with care. Do not attempt to repair a crack in a circuit trace by "flowing" solder over the break. The preferable method is to run a strand of insulated wire from a component lead on one side of the break to a component lead on the other side of the break. This wire must be soldered to the component leads, not to the PC trace itself. Measure the wire carefully and form it to follow the defective trace. If the board itself is cracked, it should be replaced.

If the board passes the visual inspection, the first section tested should be the power supply. Verify that all of the required voltages are present. Read the power-supply voltages with the power supply under load. Often a supply will test perfectly when it is *unloaded* but fail when load currents are drawn. Use an oscilloscope to check for excessive ripple. A modern, regulated power supply should have virtually no discernible ripple.

Low voltages and excessive ripple may be caused by failure of a filter capacitor, by the opening of a single diode in a full-wave rectifier circuit, or by excessive load currents being drawn by the circuit. The filter capacitor can be checked by *temporarily* bridging it with a good capacitor. If the capacitor is bad, it must be completely replaced, not bridged. The diodes in the rectifier can be tested with a simple ohmmeter check. Be sure that any replacement diode can handle the current and the voltage in the circuit. As a rule of thumb, the maximum current-handling capacity of the diode should be double the load current, and the PIV rating of the diode should be four times the rms applied voltage.

Virtually all modern controllers use regulated power supplies, often with a three-terminal voltage regulator. Generally, these regulators require a $V_{in}$ that is three volts larger than the rated $V_{out}$. If this is the case, but $V_{out}$ is not correct, the regulator must be replaced. Be sure that the heat sink is properly attached. One common error is caused by the fact that the "tab" on a negative regulator is $V_{out}$, rather than ground. If the regulator has a mica insulator between it and the heat sink, be sure to replace it — complete with heat-sink compound. Check with your ohmmeter to verify that there is no electrical connection between the case of the regulator and the heat sink *before applying power.*

These regulators may occasionally become oscillators, outputting a high-frequency sine wave, which can cause problems within a control system that will be nearly impossible to locate with ordinary troubleshooting techniques. Test for this with an oscilloscope. The only cure is to replace the regulator.

If the power supply has passed all of its tests, it will be necessary to troubleshoot the circuit of the controller. Probably the best technique is signal tracing.

Apply an input signal to the controller that is typical of what the controller might receive on the plant floor. Now use your DVM and oscilloscope to follow that signal through the circuit. If the circuit is small, simply trace the signal from input to output through the system. If the circuit is large, use a *binary search* technique: look for the processed signal *halfway through the circuit*. If it is there, divide the remaining circuit (between your test point and the output) in half and repeat. If the signal is not there, divide the circuit in half between your test point and the input. You will quickly isolate the defective portion of the circuit.

When the circuit under test contains op amps, check for signal on the *output* of the op amp. Remember that in an inverting amplifier configuration, the inverting input is at ground potential. You will not be able to measure any signal at that point.

Once the defect is isolated, you will find only a few components need to be checked. When these components include active components (transistors, op amps, or other ICs), check for power *right at the pins* of the component. You will occasionally find a bad solder joint.

When replacing components, it is important to use exact replacements. This is especially true when the circuit includes filters or PID control. The component values and tolerances were carefully selected in the engineering of the system. Use the same care in your repair.

# Section IV

# Modifying the Process Variable: Actuators

Thus far in the text we have looked at the devices used to *measure* the process variable and at the devices used to *process* the signal output of these transducers. Recalling the process-control loop, we also need both a *controller* and an *actuator*, one to make decisions about changes in the process and one to actually *make* those changes. In order to understand what kinds of decisions the controller must make, we will examine actuators in this section.

Actuators are actually the other end of the word *transducer*. Transducers are defined as devices that convert physical quantities into electrical signals, or electrical signals into physical quantities. For example, both a microphone and a loudspeaker are transducers.

In this section, we will look at some of the actuators used to control temperature, position, motion, speed, and fluid pressure.

# 18

# Controlling Process Temperature

Temperature is one of the most controlled of all parameters. From classroom to industrial process, close control of temperature is often vital to success. In this chapter, we will examine some of the techniques used to modify the process temperature.

## OBJECTIVES

You will have successfully completed this chapter when you can:

1. Describe the cartridge heater and its use.
2. Describe the band-type heater and its use.
3. Discuss the cast-in heater.
4. Describe the strip heater and its use.
5. Explain the use of fluid heat exchangers for both heating and cooling operations.

### 18.1 RESISTANCE HEATING ELEMENTS

The resistance heating element is one of the staples of industrial and home heating systems. Basically, their operation depends on the fact that when current passes through resistance, energy is given up as heat. This principle makes our morning coffee, bakes the breakfast doughnuts, and pasteurizes the cream for the coffee.

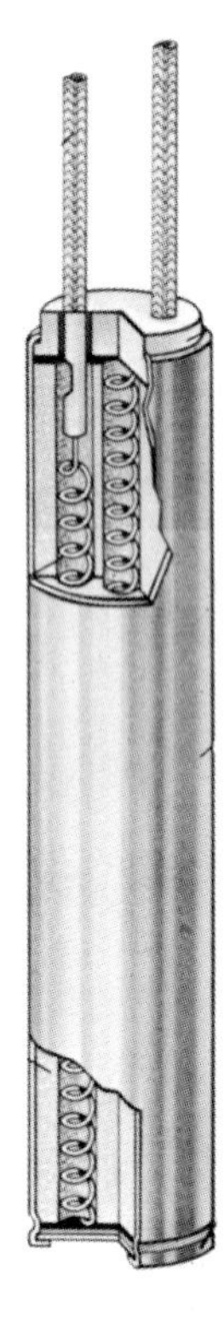

**Figure 18.1** Cutaway view of a cartridge heater *(Courtesy of Tempco Electric Heater Corporation)*

Resistance heating elements come in many different styles, shapes, and sizes. One of the most common is the *cartridge* element.

Figure 18.1 shows the internal construction of a typical cartridge heating element. Such elements are available in both round and square cross sections for various uses. Cartridge heaters are generally inserted in a hole in a metal plate. Heat is transferred from the metal plate to the object or material to be heated.

Cartridge heating elements are used in plastic injection molding, extrusion equipment, packaging equipment, food-service equipment, and a wide variety of other applications. They are available in *high-density* and *low-density* styles. *Watt density* is a measure of wattage per square inch of heater surface.

High-density units are used when high temperature is needed in small surface areas. An 80-watt unit only 1/4 inch in diameter and 1 inch long can generate a temperature of over 1000 degrees F in a metal part.

Low-density cartridge heaters are used when lower temperatures are needed or larger surface areas are available. They are more desirable because the high-density units burn out more often.

Immersion-type cartridge heaters are available with standard pipe fittings for easy insertion into tanks of liquids. It is always important when immersing heaters, thermocouples, or any other probe in a liquid that the proper sheath material is used to prevent corrosion. When replacing an immersion heater, always use the same type that was selected by the design engineers. If you are in doubt, manufacturers' engineering departments are more than happy to help in your selection.

Over a period of time, cartridge heaters tend to get bound up in their mounting holes. Replacement often begins by drilling out the old cartridge heater. Graphite or silicon lubricants may be used in the holes to facilitate insertion and (perhaps) extraction. Lubricants are conductive, and should be kept away from the terminal connections of the heaters. When mounting cartridge heaters in a solid block of metal, if possible, drill the mounting hole all the way through the block to make removal somewhat easier.

## 18.2 BAND HEATERS

*Band heaters* are used to apply heat to the outside of a container. They are used in applications varying from plastics manufacturing to food processing and vending machines. Most band heaters are designed for cylindrical containers, but they are available for rectangular and hexagonal containers as well. Most manufacturers will also make band heaters to your specifications to fit irregular shapes.

Band heaters are classified according to the type of insulation or the type of heating element used in their construction.

Figure 18.2 shows a mica-insulated band heater. The mica insulation and ribbon wire heating element allow this type of heater to be manufactured in small diameters.

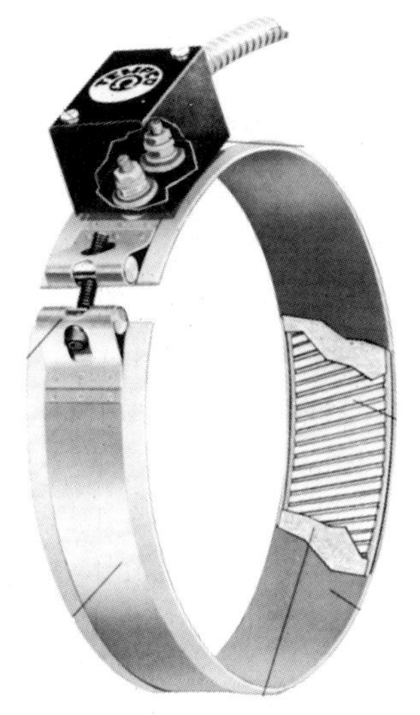

**Figure 18.2** Cutaway view of a mica band heater *(Courtesy of Tempco Electric Heater Corporation)*

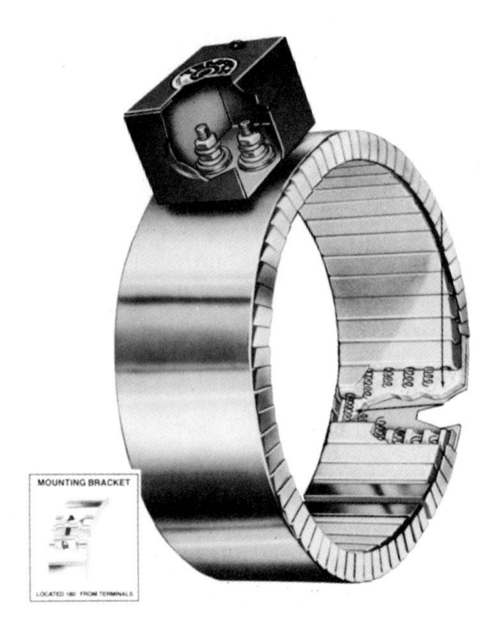


**Figure 18.3** Cutaway view of a ceramic band heater *(Courtesy of Tempco Electric Heater Corporation)*

Mica band heaters are available in diameters as small as 7/8 inch or as large as 16 inches, and in wattages from 300 W to 1000 W. Their width ranges between one and two inches. Their construction allows excellent heat transfer because the mica insulation can make good contact with the vessel.

Figure 18.3 shows a *ceramic-insulated* band heater. The ceramic-insulators limit these band heaters to diameters of three inches or greater. Ceramic-insulated heaters are made in widths from 1 1/2 inches to 16 inches, and in wattages from 500 W to 10,000 W.

The heating elements in the *tubular* band heater shown in Figure 18.4 are similar to those on your electric range at home. The heating wire is protected in a metal-clad ceramic tube. The tubular band heater is available in diameters from 3 1/2 inches to 60 inches, and in wattages from 300 W to 6000 W.

All of these band heaters are designed to be clamped onto the outside surface of a cylindrical vessel or tube of fluid. Before they are installed, the surface of the cylinder must be thoroughly cleaned to allow for efficient heat transfer. The surface must also be smooth, free of pits or protrusions. Air gaps beneath a band heater will cause *hot spots*, which will shorten heater life. When installing the units, the clamping screw must be firmly tightened until the band heater is in tight contact with the cylinder. After eight hours of operation, the clamping screw should be retightened to compensate for thermal expansion.

**Figure 18.4** Cutaway view of a tubular band heater *(Courtesy of Tempco Electric Heater Corporation)*

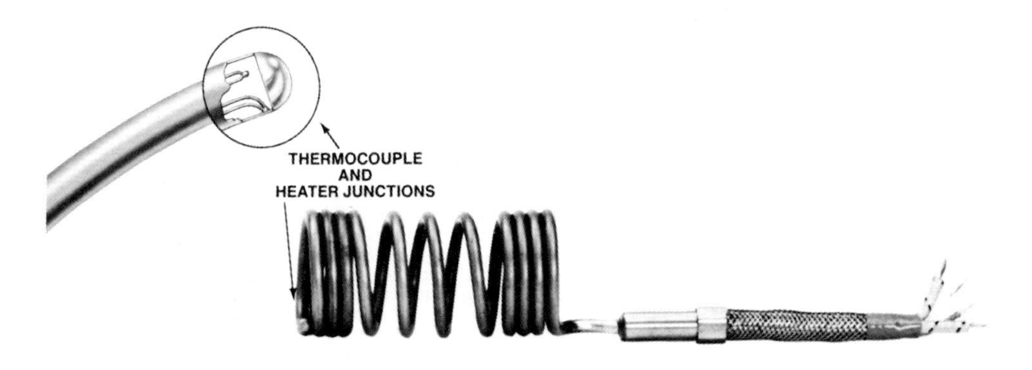


**Figure 18.5** Tubular band heater *(Courtesy of Tempco Electric Heater Corporation)*

The band heater in Figure 18.5 is specially designed for heating the nozzles used in plastic injection molding and extrusion. It is made without the outside clamp and it is slightly undersized for a tight screw-on fit over standard tubing, ranging from 1/2 inch to 2 1/2 inches in diameter. This type of band heater is available in wattages ranging from 150 W to 2500 W and in lengths from 1 inch to 6 inches.

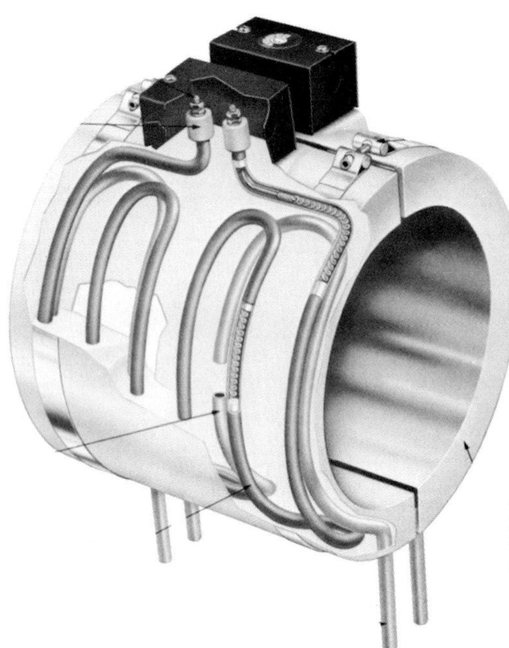

**Figure 18.6** Cutaway view of a cast-in heater *(Courtesy of Tempco Electric Heater Corporation)*

## 18.3 CAST-IN HEATERS

The cast-in heater, Figure 18.6, is made of a tubular heater element that is *cast* in a solid metal housing. A second tube is often installed for a cooling fluid, so that precise temperature control can be maintained. Most cast-in heaters are custom made for original equipment manufacturers, but a *cast-in band heater* is available for cylinders ranging from 3 inches to 16 inches in diameter.

## 18.4 STRIP HEATERS

The strip heater allows heat to be applied to flat surfaces for molds, platens, drying ovens, and so on. The strip heater is also used in bending and forming plastics because of its ability to concentrate heat in a narrow band.

Figure 18.7 illustrates the strip heater. The heating coil is supported by ceramic insulators inside a stainless steel outer sheath. Strip heaters are available in lengths from 6 inches to 6 feet and in wattages from 150 W to 6000 W.

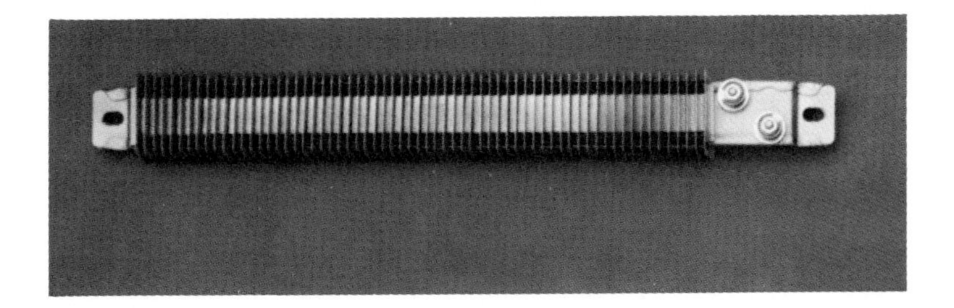

**Figure 18.7** A strip heater *(Courtesy of OMEGA ENGINEERING, an OMEGA Technologies Company)*

## 18.5 CALCULATING WATT DENSITY

Calculating the watt density for electrical heaters is a fairly simple task. Watt density is a shorthand way of saying *watts per square inch of heating surface* ($W/in^2$). To calculate watt density, then, we need to know two things: the wattage of the heater and the area of the *heated portion* of the heater. All electrical heaters have a heated area and a nonheated area. For example, a high-density cartridge heater usually has an unheated space at both ends — about 1/4″ at the butt end and 1/2″ at the wire end. We can calculate the watt density as shown in Example 18.1.

### Example 18.1

A 1/4″ diameter cartridge heater is 6″ long. The total unheated length is 3/4″. If the heater is rated at 500 W, find the watt density of the element.

### Solution:

Find the heated length:

$$HL = 6'' - 0.75''$$

$$HL = 5.25''$$

Find the circumference:

$$C = \pi \times D$$
$$C = 3.14 \times 1/4''$$
$$C = 0.785''$$

Find the heated area:

$$A = HL \times C$$
$$A = 5.25'' \times 0.785''$$
$$A = 4.12\ \text{in}^2$$

Find the watt density:

$$\frac{W}{\text{in}^2} = \frac{500}{4.12\ \text{in}^2}$$
$$\frac{W}{\text{in}^2} = 121.32$$

In general, it is far better to use low watt density heaters. While the initial cost may be greater, the lower watt density heaters burn out less often. The cost in replacement heaters will be lower, and downtime will be reduced. In the long run, it is downtime that is the determining factor. A process that is off line for repairs is not making any profits.

Downtime can be reduced by using very precise temperature controls near all heating elements. This is so important that some heating elements are made with built-in thermocouple wells. No heater should be operated without temperature control to guard against overheating and subsequent burnout.

Determining the watt density of a heater is simple. Determining the watt density needed for a particular process is much more difficult. This requires an intimate knowledge of the thermal characteristics of the process materials and some high-order mathematics. It is beyond the scope of this text and will not be covered here. As a technician, you should rely on the manufacturing engineers who design process control systems. The manufacturers of industrial heating equipment will be more than happy to help if your plant does not have an in-house engineer.

## 18.6 HEAT EXCHANGERS

A heat exchanger allows heat to be transferred from one medium to another. A heat exchanger is usually a closed, liquid-filled system. The liquid may be water or it may be some special composition of chemical elements. The heat exchanger you

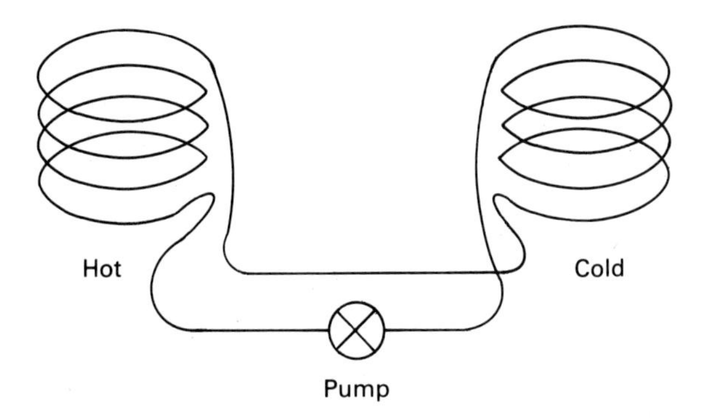


**Figure 18.8** A heat exchanger system

are most familiar with is your car radiator. The water or antifreeze in your radiator passes through channels in the engine block, absorbing heat from the block. It is pumped through the block and into the radiator where it gives up the engine heat to the air, cooling the engine block in the process.

Many solar heating systems also use heat exchangers. Here, an antifreeze liquid is heated by the sun's rays, then pumped through a heat exchanger in the hot-water storage tank. The heat in the antifreeze solution is thus transferred to the water in the holding tank.

The general idea behind a heat exchanger is shown in Figure 18.8. The liquid in the heat exchanger system is heated in one location and gives up its heat in the second location. Thus, it cools the first location and heats the second.

In our discussion of the cast-in heater, we mentioned that *cooling tubes* can be cast into the casting as well as the tubular heater coils. To warm the process, the heaters are activated. To cool the process, chilled water is pumped through these cooling tubes. Thus, cast-in systems can be used to maintain very precise temperature control, heating with electrical resistance heaters and cooling as a heat exchanger with chilled water.

## SUMMARY

In this chapter, we have seen some of the many types of electrical resistance heaters. We have seen:

1. The cartridge heater, which is designed to be inserted directly into a metal plate to be heated, or immersed directly in the fluid to be heated.

2. The band heater, which is designed to be clamped tightly around the outer diameter of a container.
3. The cast-in heater, which is usually custom made. This heater also offers the option of operating as a heat exchanger with chilled water.
4. The strip heater, which is designed to be bolted to flat surfaces.
5. How to calculate watt density.
6. The heat exchanger, which can be used to heat or cool a process.

## SELF-TEST

1. A hole is drilled in a steel plate and a heating element is installed directly into the hole. The heating element is most likely a(n) ____________________ heater.
2. Cartridge heaters tend to get bound up in their mounting holes. One way to facilitate removal is to ______________________.
3. When a cartridge heater must generate very high temperatures in a small space, ______________________ density heaters are used.
4. A(n) ____________________ heater is used to apply heat to the outside of a container.
5. For very small diameter containers, a(n) ____________________ band heater is selected.
6. For larger containers requiring higher heat, a(n) ____________________ band heater is selected.
7. When installing band heaters, the outside surface to which the heater will be clamped must be clean and smooth to avoid ______________________.
8. Heating elements that are built in to solid metal machine parts are called ______________________ heaters.
9. When it is necessary to apply heat to a flat surface, a(n) ______________ heater is selected.
10. These heaters also have the ability to cool because they have built-in ____ ______________________.

## QUESTIONS/PROBLEMS

1. A 1000 W strip heater has a width of 2″ and a length of 16″. The outer 1/4″ of each edge is unheated, and the last 2″ on each end is unheated. Find the watt density.
2. A 500 W cartridge heater is 5″ long and 1/2″ in diameter. It has an unheated space of 1/2″ on each end. Find the watt density.

# 19

# Controlling the Dc Motor

Motors are the muscle of industry. The electric motor can generate large amounts of force in an economical space. It is hard to imagine a world without electric motors, so widespread have they become.

Since the power companies transmit power from generator to factory as alternating current, it may seem surprising to find that industry makes wide use of direct current motors. In fact, within certain limits, the dc motor is very popular in industrial use. In this unit we will see some of the reasons for this popularity.

We will also see that controlling a large industrial motor involves more than a simple on-off switch. We will look at some simple motor starters and some simple motor speed controls to wind up the chapter.

## OBJECTIVES

You will have successfully completed this chapter when you can:

1. Describe the operating principles of the dc motor.
2. Describe the permanent magnet motor.
3. Describe the shunt wound motor.
4. Describe the series wound motor.

5. Describe the compound motor.
6. Explain the basics of starting dc motors.
7. Explain a simple speed controller for dc motors.

## 19.1 FORCE, WORK, AND POWER

Machinery is designed to perform work for us, to reduce the effort we must exert to do a particular task. *Webster* defines a *machine* as "any device consisting of two or more parts which may serve to transmit force and motion so as to do some desired kind of work."

These are all familiar words, but in the world of machinery they have special meanings.

*Force* is a directed effort that changes the motion of a body. Force can exist only if there is a body to *act upon*. Force can exist without producing a change in motion if there is opposition from an equal counterforce.

*Work* is the result of force overcoming resistance. Work is measured in *foot pounds*. If you lift a one-pound weight a distance of one foot, you will have done one foot pound of work.

*Power* is the *rate of doing work*. The power expended in lifting 550 pounds one foot in one second is *one horsepower*. When measured in metric units, power is measured in *watts*. One watt is the power expended to lift one kilogram one meter in one second. One horsepower is equal to 746 watts.

*Torque* is twisting, or rotary, force. A one-pound weight suspended from the end of a one-foot bar will exert one foot pound of torque on the other end of the bar.

## 19.2 PRINCIPLES OF THE DC MOTOR

An electric motor converts electrical energy into torque. There are electric motors that operate from direct-current power sources and there are electric motors that operate from alternating-current power sources. In this chapter, we will be looking at the dc motor.

In Figure 19.1, we can see the basic operating principle of a dc motor. Any dc motor consists of two magnets. The first magnet is a nonmoving magnet called the *field*. The second magnet is connected to a shaft and is allowed to rotate. It is called the *armature*. In some texts, the field is also called the *stator* and the armature is called the *rotor*.

The field in Figure 19.1 is a permanent magnet. The armature consists of two turns of very heavy wire wrapped in slots in an iron core. When direct current flows through this wire, it forms an electromagnet.

Since the armature must be free to rotate, power is connected to the armature windings through a split ring attached to the shaft. The halves of the ring are insulated from each other and also from the metal shaft of the motor. Power is connected to the split ring or *commutator* through *brushes* which slide on the commutator.

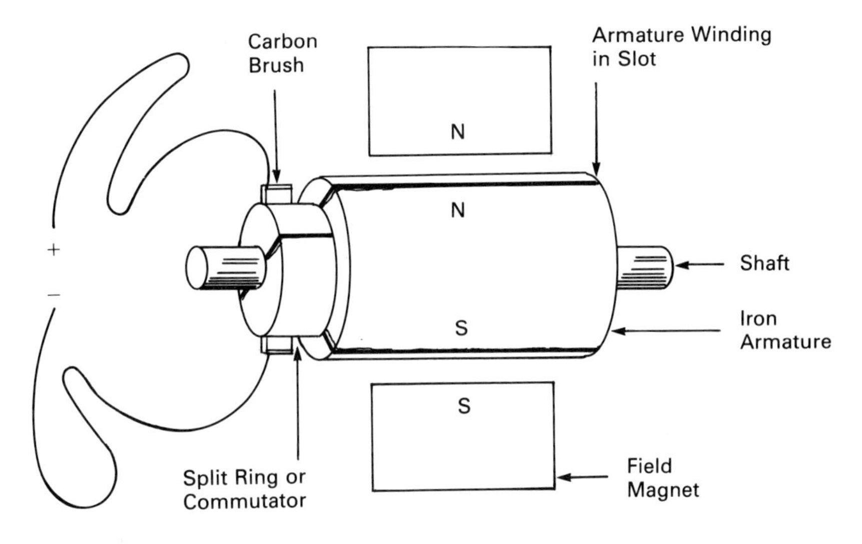


**Figure 19.1** The simple PM dc motor

When power is applied to the armature coil, current will flow from the negative battery terminal, through the coil, and back to the battery. The coil becomes an electromagnet. In Figure 19.1, the polarity of the electromagnet is such that the North pole of the armature magnet is near the North pole of the field magnet. Since two like magnetic forces repel each other, the armature magnet will be forced away from the field magnet, turning the motor shaft.

As the armature moves, its North pole begins to approach the South pole of the field magnet, and the attraction between the two unlike magnetic poles keeps the motion going until the North pole of the armature magnet is opposite the South pole of the field magnet.

Motion would stop at this point, except that the commutator has rotated with the armature. The result of this is that the polarity of the power connections to the armature coils is *reversed*. This, naturally, reverses the magnetic polarity of the armature magnet. What was a North pole on the armature electromagnet now becomes a South pole, which is repelled by the South pole of the field magnet, keeping the motion going. This is the principle behind all dc motors.

As the armature rotates, it rotates the armature coil through the field of the field magnet. When we move a wire through a magnetic field, a voltage is induced in that wire. The voltage induced in the coils of the armature is *of the opposite polarity to the applied voltage*. This induced voltage is called *counter EMF* (CEMF). In an unloaded motor, the CEMF will be almost equal to the applied armature voltage, reducing armature current.

The *base speed* of a motor is the speed the motor will attain with the rated armature voltage and the rated field flux. The *speed regulation* of a motor is a measure of how much the motor slows from the base speed when a load is applied to the motor.

In a practical motor, the armature is made of several separate windings, and the commutator has two separate segments for each winding. Commutators are made of copper segments, which are insulated from each other with mica spacers. The brushes are made of compressed carbon because it has good conductivity and is softer than the copper commutator. The softness means that the brushes will wear out, but not the commutator.

Brushes are held against the commutator by spring pressure. In some larger motors, the spring pressure is adjustable. Too much pressure causes rapid brush wear. Too little pressure causes arcing and carbonization of the armature. When adjusting brush pressure, always follow the manufacturer's recommendations to the letter.

## 19.3 THE PERMANENT MAGNET MOTOR

Dc motors are classified by the type of field magnet used in the motor. The motor described in Section 19.2 is a *permanent magnet* (PM) motor. Permanent magnet motors are primarily low-power motors, generally available in sizes below 1 hp. They are selected for tasks that need *high starting torque*. They are typically used in tasks where the load is connected to the motor before the motor is turned on. Pm motors also deliver high torque at low speeds, but have poor speed regulation.

## 19.4 THE SHUNT WOUND MOTOR

The second classification of dc motors is the *wound field* motor. In these motors, the field magnet is also an electromagnet. Wound field motors are classified according to how the field magnet is connected into the circuit.

In a *shunt* motor, the field magnet coil is connected in *parallel* to the armature coils. This is shown in Figure 19.2a. A variation of the shunt motor is shown in Figure 19.2b. This motor has a *separately excited field*, which allows for greater control.

The shunt motor is the most common dc motor. It has good speed regulation, and even though motor speed does decrease slightly with an increase in load, it is considered a constant-speed motor.

When the load is added to a shunt motor, the motor tends to slow down. But as the motor slows, the CEMF is reduced. This results in an increase in armature current, which provides more torque to compensate for the increased load. Thus, the shunt motor has a natural feedback system.

## 19.5 THE SERIES WOUND DC MOTOR

In the series wound motor, the armature and the field are wired in series with each other. This is shown in Figure 19.3.

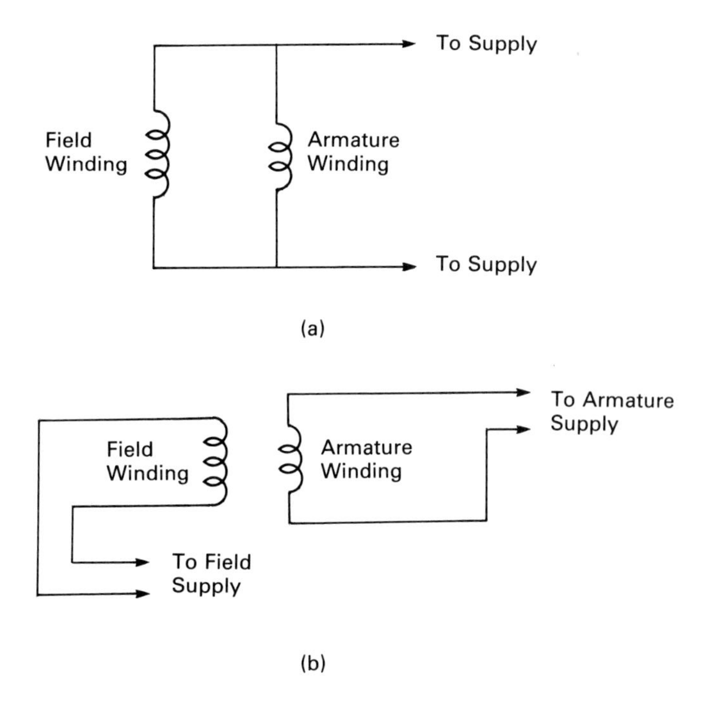


**Figure 19.2** The shunt wound motor

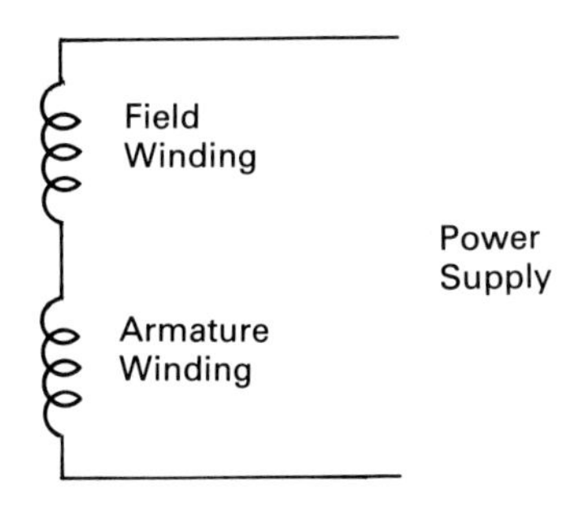


**Figure 19.3** The series wound motor

Any change in load causes a change in armature current. This will also cause a change in the field current. Thus, as the load changes, the speed changes.

As load is removed from a series motor, the motor speed increases; thus, the CEMF of the armature increases. This reduces current in *armature and field alike*, since they are in series. As the field current decreases, the speed increases. This causes an even greater CEMF, further reducing field current and increasing the speed even more. Soon the motor is running at dangerous speeds, which may cause damage to equipment and injury to personnel. For this reason, *a series motor should never be operated with no load.*

Series motors have high starting torque and are often used in starting heavy loads. You may find series motors in cranes and winches and in electric rail equipment, which must move heavy loads at low speeds.

## 19.6 COMPOUND MOTORS

A *compound* motor has two field windings, one in series with the armature and one in parallel with the armature.

As shown in Figure 19.4, there are two types of compound motors, the *long shunt* and the *short shunt.* Compound motors are designed to get the best features of both series and shunt motors in a single motor. Depending on design preferences, they usually offer good starting torque and good speed regulation.

The curves in Figure 19.5 show a comparison of the torque and speed characteristics of shunt, series, and compound motors.

## 19.7 MOTOR STARTING

Recall that the armature coils of a dc motor are made up of only a few turns of heavy wire. When power is first applied to the armature, the ohmic resistance of the coil is the only thing limiting current in the armature. In a typical motor, the ohmic

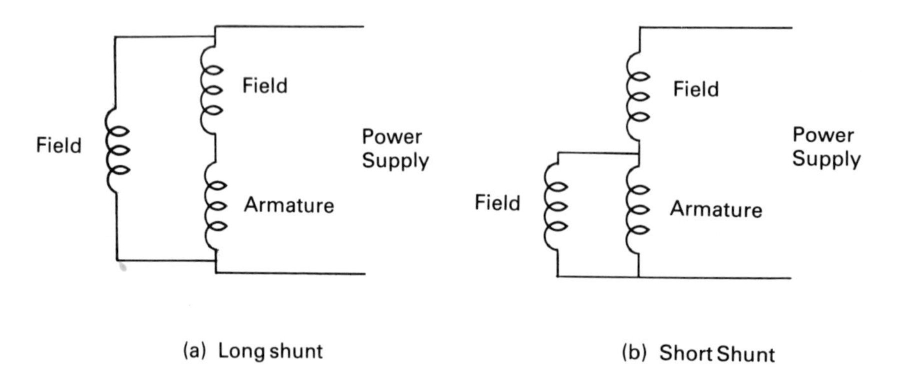


**Figure 19.4** Compound motors

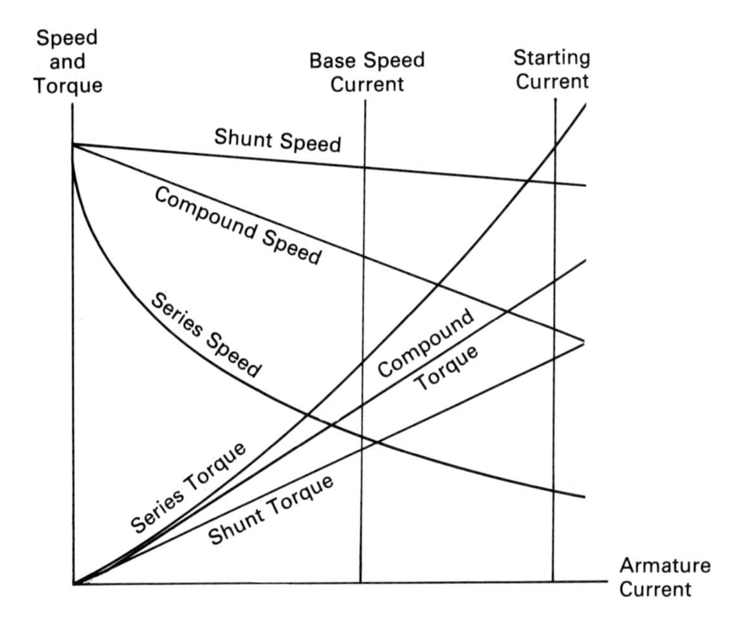


**Figure 19.5** Speed/torque comparison. Note that series motors have highest starting torque, lowest starting speed, and greatest speed variation. Shunt motors have lowest starting torque, highest starting speed, and smallest speed variation. Compound motors fall between series and shunt.

resistance is less than one ohm. Since large hp dc motors run from voltage sources ranging between 240 and 600 volts, you can see that the initial *starting current* can be very large. The motor and the supply lines to the motor must be protected from the flow of excessive current during the starting period. Once the armature is turning at or near its base speed, CEMF will keep the armature current flow within reason.

The simplest way to protect the system is by limiting the armature current until the motor has had a chance to start.

The circuit shown in Figure 19.6 is a simple *face plate starter*. Bare copper contacts are connected to the series string of resistors — $R_1$, $R_2$, $R_3$, and $R_4$. To start the motor, the contact arm is moved to the right so that it connects power to contact number one. Full field current will flow, but the armature current is limited by the four series resistors. The motor begins to turn. As CEMF builds, the armature current falls. When the motor speed is no longer rising, we move the contact arm to the next contact. This switches one of the resistors out of the armature circuit and allows the armature current to increase. Note that resistor $R_1$ is now switched in series with the field. This reduces the field strength. The combination of reduced field strength and increased armature current causes the motor speed to increase rapidly. The sequence is continued until the contact arm is pulled fully to the right, in contact with the *holding coil*. At this time, the motor is running at its full speed.

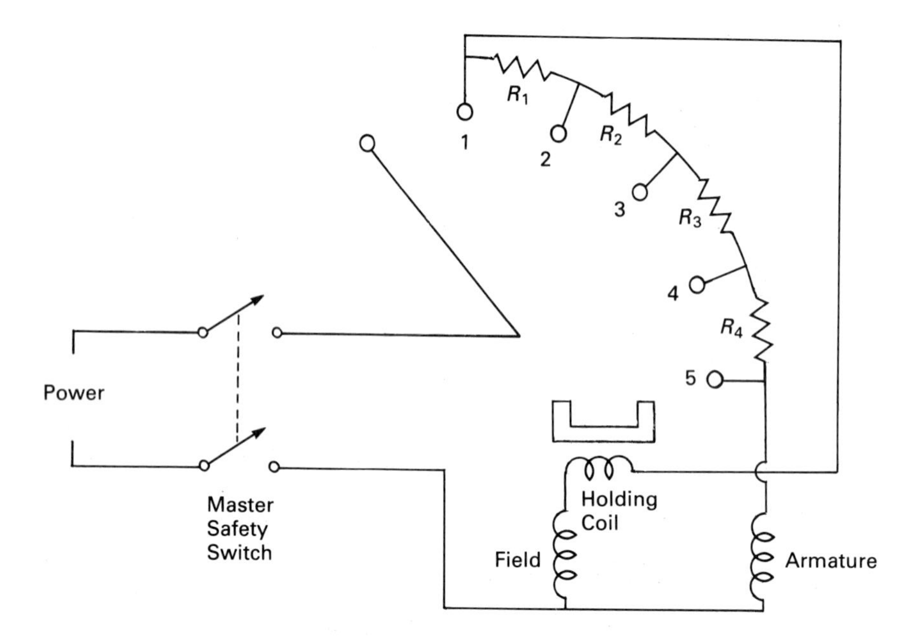


**Figure 19.6** The face plate starter

Note that the field current is now running through the four resistors and the holding coil. The holding coil is a small electromagnet which holds the contact arm in place. In the event of a low-voltage condition or some circuit failure that would cause field current to become too low, the holding coil will not have sufficient strength to hold the arm. The return spring will pull the arm back to the off position, protecting the motor from running too fast, which would destroy the motor.

This *low-field protection* is essential on shunt and separately excited motors. If the field flux is too low, the motor will over-rev and will most likely destroy itself.

The resistors in a face plate starter are obviously not the resistors you are used to seeing in a stereo chassis. They are large coils of resistance wire that look like heater coils. They *act* a lot like heater coils, too, which is why they are housed in a perforated metal housing.

With the high voltage and the high temperatures, there is serious potential for injury. Use extreme care when working on motor starters. Large machinery always has a safety switch to disconnect the power from the machine. This switch *must be turned off and tagged* before you begin working on the machine. The tag is a warning to prevent someone from turning the safety switch back on while you are working on the machine. Don't neglect this simple precaution.

The face plate starter is being replaced by automatic starters. We will examine them later in Section IV of the text.

## 19.8 BRAKING

Turning off a motor may seem to be the simplest thing in the world — simply disconnect the power. While that approach is effective in small machinery, a large machine may coast for over an hour after the power is removed. That is usually not acceptable; therefore, some means of *braking* must be included.

The simplest braking method is the one you use in your car: your car's brakes convert the forward momentum of your car into heat by using friction. It is rather crude, but it works. You may also see this method in use on some older machines.

In *dynamic braking*, Figure 19.7, power is removed only from the armature; the field is left excited. When power is removed from the armature, a resistance is switched into the armature circuit. The spinning armature generates a CEMF in the field, forcing current through the resistor. This acts like a load on the armature and brings the motor to a rapid, smooth stop. When the motor has stopped, the field supply is also disconnected. In practice, the resistance is chosen so that the initial braking current is about twice the rated motor current.

We can stop the motor even faster by *plugging*, Figure 19.8. Basically, this consists of suddenly reversing the supply polarity to the armature. Because the CEMF and the reversed supply voltage are now in the same polarity, plugging circuits also requires some series resistance to limit the current. Plugging a motor without the added resistance will destroy the commutator and brushes almost instantly. Once again, the resistance in the plugging circuit is selected to give an initial braking current that is about double the motor's rated running current.

With this technique, a reverse torque is present even after the motor has stopped. To prevent running the motor in reverse — a calamity in most machinery — all power must be removed as soon as the motor has stopped.

## 19.9 SPEED CONTROL

There are two different ways to control the speed of a wound field dc motor. As we have seen, reducing the armature voltage will slow the motor, and reducing the field current will speed up the motor.

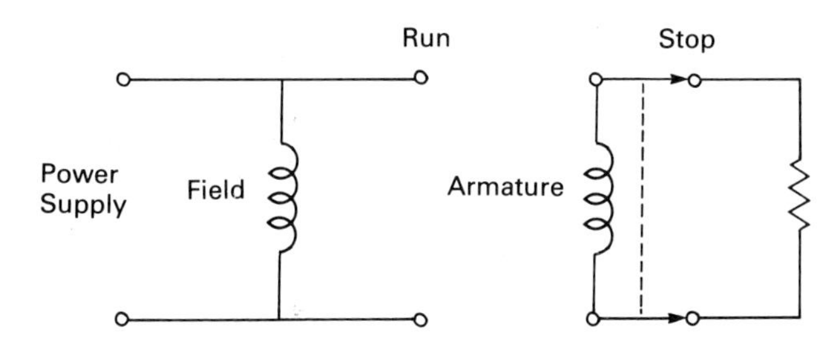


**Figure 19.7** Dynamic braking

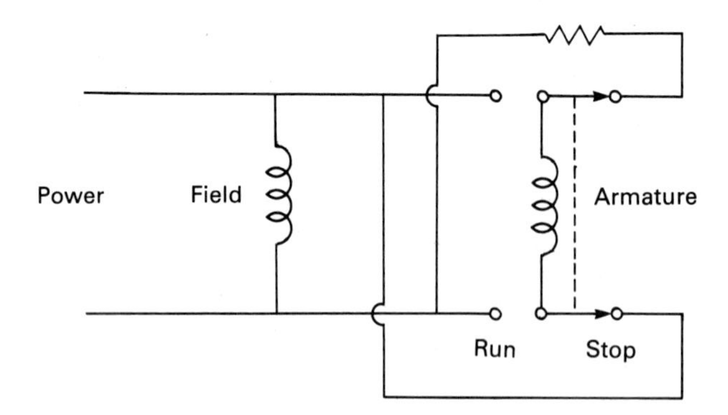


**Figure 19.8** Plugging the dc motor

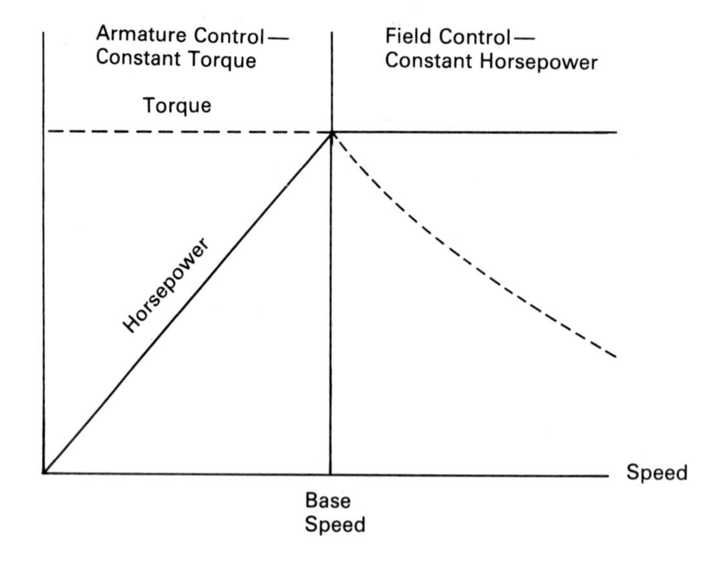


**Figure 19.9** Motor speed control curves

The portion of the curve to the left of the base speed in Figure 19.9 shows the relationship between armature voltage control, motor speed, and horsepower output of the motor. This graph shows us that operating the motor at reduced armature voltage provides constant torque but reduced hp at speeds below the motor base speed. This is the area in which the motor operates during starting. It may also be operated in this area when low-speed operation is desired.

The curve to the right of the base speed in Figure 19.9 shows the relationship between field current control, motor speed, and the motor's power output in hp or kW. This part of the graph shows that the motor may be operated above the base speed by reducing the field current. Here the hp output of the motor remains constant, while motor torque is reduced. This is not as much of a disadvantage as it might first seem because the greatest torque is needed when the motor is starting, to overcome the inertia of a stopped machine.

Figure 19.10 shows a simple circuit for motor speed control below and above the base speed. The armature potentiometer will adjust for operation below base speed, and the field potentiometer will adjust for operation above the base speed. Care must be taken to ensure that field current is not reduced below a certain point to prevent overspeeding the motor.

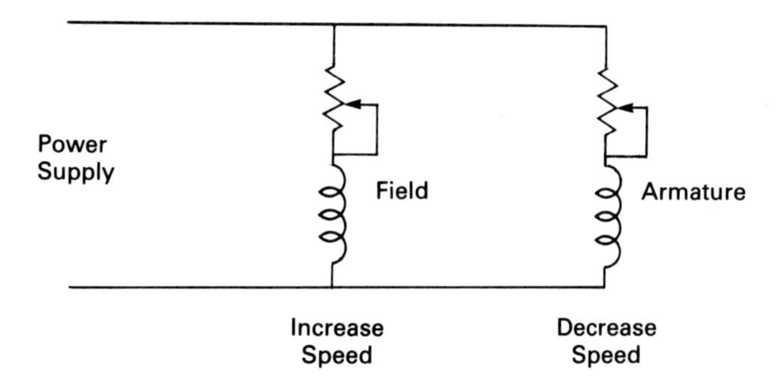


**Figure 19.10** Armature and field speed control

Most motors can be operated without problems at speeds below the base speed. The range of speed available above the base speed will depend on whether the motor was designed for variable-speed operation. For a motor not designed for high-speed operation, we should limit speed to about 150% of the base speed. If the motor is designed for high-speed running, we may increase the speed up to about 600%. Running a motor above its maximum design speed may cause excessive arcing and carbonization of the commutator as the brushes attempt to supply rated armature current while the commutator races by.

## SUMMARY

In this chapter, we have seen that there are different types of dc motors.

1. The permanent magnet motor is used for low-power, high-torque applications. It has poor speed regulation.
2. The shunt motor is the most used of the dc motors. It is considered a constant-speed motor because of its excellent speed regulation.
3. The series motor is used for its high starting torque and high torque at low speeds. It has poor speed regulation.
4. The compound motor is a compromise between series and shunt motors.
5. Dc motors are selected for the characteristics required by a particular task. The selection of motors should be left to the manufacturing engineer. Manufacturers will provide help in selection.
6. Starting large dc motors requires control of armature voltage to prevent excessive current flow.
7. Large motors must usually have a braking system to prevent coasting.
8. Dynamic braking inserts a load resistor into the armature circuit. Current from the armature's CEMF puts a load on the armature and brings the motor to a safe stop.
9. Plugging a motor can result in faster stopping. When a motor is plugged, the polarity of the armature voltage is reduced.
10. The speed of a dc motor can be controlled by controlling either armature voltage or field current.
11. Reducing armature voltage causes the motor to run below the base speed.
12. Reducing field current causes the motor to run above the base speed.

## SELF-TEST

1. An effort that changes the motion of a body is called ________________.
2. An effort that raises a one-pound weight a distance of one foot is ________ ____________________________ of work.
3. The rate of doing work is called ____________________________.
4. Power is measured in ____________________________.

5. In metric measurement, power is measured in ______________.
6. A twisting force is called ______________.
7. The rotating part of a motor is called the ______________.
8. The stationary part of a motor is called the ______________.
9. In a permanent magnet dc motor, the ______________ is a permanent magnet.
10. Power is connected to the armature electromagnet through brushes and the ______________.
11. Motor brushes are made of ______________.
12. The speed of a motor with rated armature voltage and field flux density is called the ______________.
13. The main advantage of a permanent magnet motor is that it has very high ______________.
14. When the field magnet of a motor is an electromagnet, the motor is called a(n) ______________ motor.
15. When the field magnet is wired in parallel with the armature winding, the motor is called a(n) ______________ motor.
16. When the field magnet is wired in series with the armature winding, the motor is called a(n) ______________ motor.
17. When the field magnet is wired partly in series and partly in parallel with the field, the motor is called a(n) ______________ motor.
18. The shunt motor is known as a(n) ______________ speed motor.
19. The serial motor is used for its ______________ starting torque.
20. The dc motor with the best speed regulation is the ______________ motor.
21. When a dc motor is started, there must be protection in the circuit against high ______________ current in the armature.
22. The ______________ starter adds resistance in series with the armature while the motor is starting.
23. Controlling the armature voltage allows the motor to run ______________ the base speed.
24. Controlling the field current allows the motor to run ______________ the base speed.
25. Dynamic braking switches a(n) ______________ into the armature circuit in place of the armature voltage.
26. Plugging connects the armature to ______________ for motor braking.

## QUESTIONS/PROBLEMS

1. The 50-hp dc motor in a gantry crane has burned out. The name plate, which gives the specifications for the motor, is missing. What type of dc motor would you select for the job? Explain your answer.
2. A separately excited motor must have dynamic braking added to its circuit. The motor is rated at 90 VDC and 25 amps. What size resistance must be switched into the armature circuit for maximum safe braking current?
3. You must operate a motor at a point near its full rated torque, but at a reduced speed. Would you use armature voltage control or field current control for the job? Explain your answer.

# 20

# Stepper Motors

In addition to the continuous rotary motion provided by regular electric motors, we often need a *partial rotary motion*, a turn of a shaft of a few degrees or less. The *stepper motor* can offer rotary position control with a high degree of accuracy.

## OBJECTIVES

You will have successfully completed this chapter when you can:

1. Describe the operating principles of the permanent magnet stepper motor.
2. Describe the operating principles of the variable reluctance stepper motor.
3. Explain the control logic for stepper motor operation.

### 20.1 THE STEPPER MOTOR

Stepper motors are used in a wide variety of positioning tasks. The read/write head in a disk-drive unit is probably positioned by a stepper. Steppers are used to position the joints of some lightweight robotic arms. Stepper motors are used in *X-Y* plotters and in *X-Y* positioning tables for heavy machinery.

The stepper motor is a dc excited device in which the dc supply is switched to each of the motor's coils in turn, causing the rotor to turn in a series of equal steps. The number of steps is directly proportional to the number of pulses. Steppers that take steps as small as 1.8 degrees are available. Other steppers take steps as large as 45 degrees. By using *half-step* techniques, a 1.8-degree stepper can be made to take 0.9-degree steps.

The stepper motor output can be modified by coupling it to gears or lead screw assemblies. This modification can make the stepper capable of extremely fine control of rotary or linear motion.

## 20.2 THE PERMANENT MAGNET STEPPER

The rotor of a permanent magnet stepper motor is a permanent magnet. The stator consists of sets of windings, which are grouped into pairs.

Figure 20.1 is a simplified PM stepper motor. Note that the rotor is a permanent magnet with two poles. Normally, the rotor would contain many magnetic poles, which would look like teeth on the rotor.

The stator of this simplified motor is made of two pole pieces. Each pole piece has two windings, which are connected at a center point. These two windings are wound in opposite directions from each other. This is a *two-pole* stepper. Steppers are available with two-, three-, and four-pole designs.

In Figure 20.1a, switches 1 and 2 activate windings *A* and *C*, respectively. The direction of current flow and windings causes poles *W* and *X* to become North magnetic poles of the electromagnet poles. This causes the South pole of the rotor magnet to align itself directly between the two North poles. As long as current flows through the stator magnets, the rotor will remain in this position.

In Figure 20.1b, switch 1 has been moved to activate coil *B*. Switch 2 is unchanged. The change in switch 1 causes pole *Y* to become a South pole and pole *W* to become a North pole. The South pole of the rotor is repelled by pole *W*, and moves counterclockwise to position 2, between poles *X* and *Y*. The rotor has turned 90 degrees. As long as the current is maintained through coils *B* and *C*, the rotor will remain in this position.

In Figure 20.1c, switch 2 has been moved to activate coil *D*. Switch 1 is unchanged. The change in switch 2 causes pole *Z* to become a South pole and pole *X* to become a North pole. The South pole of the rotor is repelled by pole *X* and moves in a counterclockwise direction to position 3, between poles *Y* and *Z*. As long as current is maintained through coils *B* and *D*, the rotor will remain in this position.

In Figure 20.1d, switch 1 has been moved to activate coil *A*. Switch 2 is unchanged. The change in switch 1 causes pole *Y* to become a South pole and pole *W* to become a North pole. The South pole of the rotor is repelled by pole *Y* and moves to position 4, between poles *W* and *Z*. As long as current is maintained through coils *A* and *D*, the rotor will remain in this position.

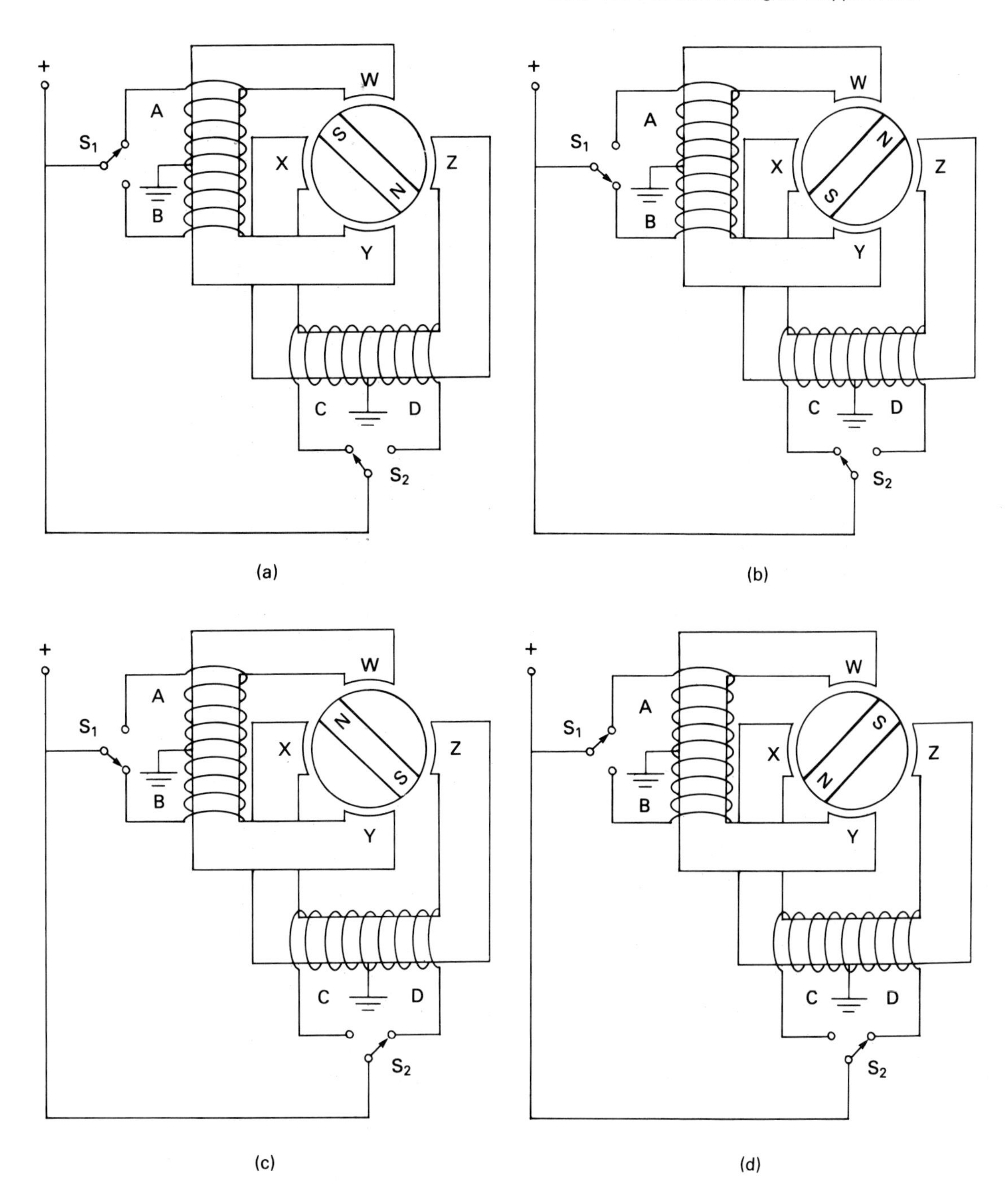


**Figure 20.1** A simple PM stepper motor

The next step would return the switches to the positions shown in Figure 20.1a, and the rotor back to position 1. The motor is a 90-degree stepper motor. Each step is 90 degrees. It can be stopped in any of the four positions, and it will return to this position if some external force tries to make it rotate further. The ability to remain in a position is called *detent torque*. Detent torque is an important parameter of stepper motors.

Because the rotor is a permanent magnet, there will also be a tendency for the motor to remain in position when there is *no current in the coils*. This is called *holding torque*.

The direction of rotation can be reversed at any time by reversing the switching sequence. Start at position 1 in Figure 20.1a and move to the switch settings in Figure 20.1d. Follow the sequence through Figure 20.1c, then Figure 20.1b. The rotation is now clockwise.

The mechanical switches of Figure 20.1 are used to simplify the diagram. The stepper switch is normally driven by a digital circuit, and solid-state switching is afforded by transistors or Darlington transistor pairs. If we take the coils in *DCBA* order (most significant to least significant) and assume a *1* to be an activated coil and a *0* to be a nonactivated coil, we can specify the binary sequence for operation.

Using these specifications, the counterclockwise sequence will be 1010, 0110, 0101, 1001. Clockwise operation will be 1010, 1001, 0101, 0110.

We can take *half steps* with a stepper motor if we are willing to suffer a decrease in starting torque and detent torque. Rather than activate two coils for each step, we can activate the coils in a different sequence. Using the poles in Figure 20.1 as an example, start with coils *A* and *C* energized. Now *turn off coil A*. Neither pole *X* nor pole *Y* will be energized, and the rotor South pole will align itself directly with pole *W*, a step of 45 degrees. Now energize coil *B*. The rotor will rotate another 45 degrees to position itself between poles *W* and *Y*, our original position 2. The half-step sequence for counterclockwise operation is 0101, 0100, 0110, 0010, 1010, 1000, 1001, 0001. For clockwise operation, reverse the sequence.

Sequences other than those given will result in unpredictable operation and should be avoided. The result will often be that the rotor does not move, or moves back and forth between two positions.

## 20.3 THE VARIABLE RELUCTANCE STEPPER

The *variable reluctance (VR) stepper* does not have a permanent magnet rotor. The rotor is made of soft iron, and has teeth in its periphery. The stator coils are arranged in groups of four. Each group of four is called a *phase*. The pole pieces of the stator are also toothed.

The VR stepper shown in Figure 20.2 is a *three-phase* stepper. There are 12 stator coils in three groups of four. In Figure 20.2a, we can see that phase *A* is wired so that its four poles are 90 degrees apart around the stator. When coil *A* is energized, all four of its poles will become North poles. The remaining poles become induced South poles.

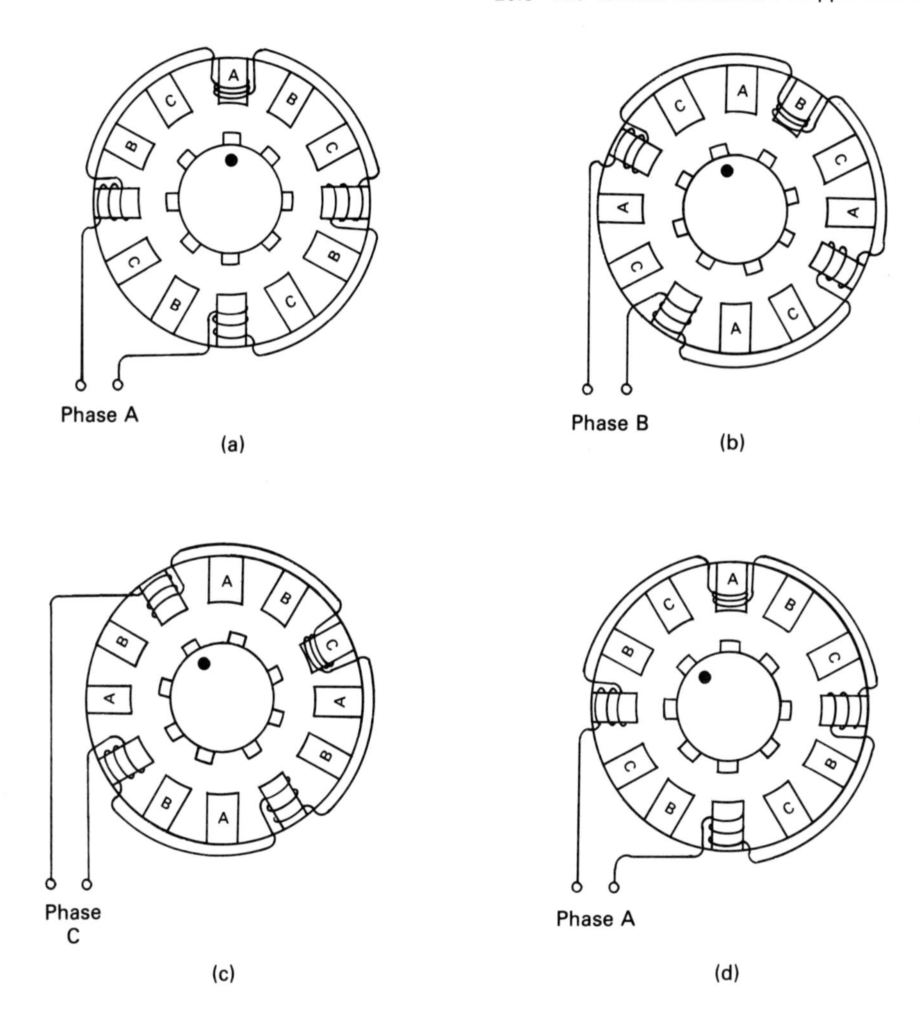


**Figure 20.2** Four steps of a VR stepper

The rotor in Figure 20.2 has eight teeth. When coil *A* is energized, the teeth must align themselves as shown in Figure 20.1a, with the path of least magnetic reluctance.

When coil *A* is de-energized and coil *B* is energized, the rotor will step counterclockwise, aligning the second set of teeth with the four poles of coil *B*.

When coil *B* is de-energized and coil *C* is energized, the rotor will make another counterclockwise step, now aligning the first set of rotor teeth with the four poles of coil *C*.

The VR stepper can be made with smaller stepping angles than the PM stepper, but it does not have the same holding torque with all coils de-energized.

This motor has three step sequences, one for each phase. The relationship between the number of step sequences and the number of rotor teeth tells us how many steps per revolution this motor will take:

$$\text{Steps per revolution (SPR)} = (SS)\ T_r$$

**Example 20.1**

Calculate the steps per revolution of the motor in Figure 20.2.

$$SPR = 3 \times 8$$

$$SPR = 24$$

Calculate the degrees/step.

$$DPS = \frac{360}{SPR}$$

$$DPS = \frac{360}{24}$$

$$DPS = 15 \text{ degrees}$$

## 20.4 CONTROLLING THE STEPPER

When a stepper motor has a steady dc current applied to one winding, the rotor will be lined up with that stator field. If an external force attempts to turn the shaft, a torque, which attempts to restore the shaft to the original alignment, develops. That torque is at its maximum when the shaft is one step away from the original position. It is called the *holding torque*, and it is specified by the manufacturer.

The *pull-out torque* is similar to the starting torque of a dc motor. Pull-out torque is the maximum torque the motor can develop from a stopped position. It determines the maximum load that the stepper can move. Pull-out torque varies with the step rate, decreasing as the step rate increases.

There is a maximum step rate for an *unloaded* stepper motor from a stopped condition to the stepper's maximum response time. As load is added to the motor, the maximum step rate decreases. One of the things that controls the pull-out torque and maximum step rate is the *current rise time* in the stator winding. This is because the strength of the magnetic field is a function of the current, not of the applied voltage.

Recall the formula for current rise time:

$$\text{Current rise time} = 5T$$

where:

$$T = \frac{L}{R} \text{ seconds}$$

$$R = \text{winding resistance in ohms}$$

$$L = \text{winding inductance in henrys}$$

The time constant can be reduced (and the response time improved) by adding resistance in series with the windings. When this is done, however, the applied voltage must be increased so that the windings still receive their rated current. This is a very common method for improving the maximum step rate of a stepper motor.

If the stepper is pulsed at a slow enough rate, it comes to a rest between steps. This provides reliable operation where the shaft position can be stated with certainty simply by counting steps.

As the motor step rate is increased, its operation changes from the start-stop mode to a continuous operation called *slewing*. In the slew mode of operation, the motor does not come to a stop between steps.

Figure 20.3 shows the relationship between torque and step rate for a PM stepper motor in both start-stop mode and slew-mode operation. Any load below the start-stop line in Figure 20.3 is safe for start-stop operation. The motor can be operated at that rate without skipping any pulses.

Note that the stepper will actually generate higher torque in the slew mode at a given step rate. This picture is a little deceptive, because the torque levels shown are for *continuous operation*. That means that the load is *already moving at that rate*. Changes in step rate must be made carefully so that the stepper's peak torque is not exceeded. To change from a full stop to a high step rate, the pulse rate must be increased gradually. To change from the high step rate to a full stop, the step rate must be decreased gradually. This is called *ramping*.

Figure 20.4 illustrates this idea. To go from full stop to maximum velocity (step rate), a certain amount of *ramp up* time is needed. To get from maximum velocity to a full stop, a certain amount of *ramp down* time is needed. If these times are not allowed for, the motor will skip steps. Many stepper motor controllers contain dedicated microprocessors that automatically deal with the ramp up and ramp down requirements. These are highly recommended for new installations and replacement work.

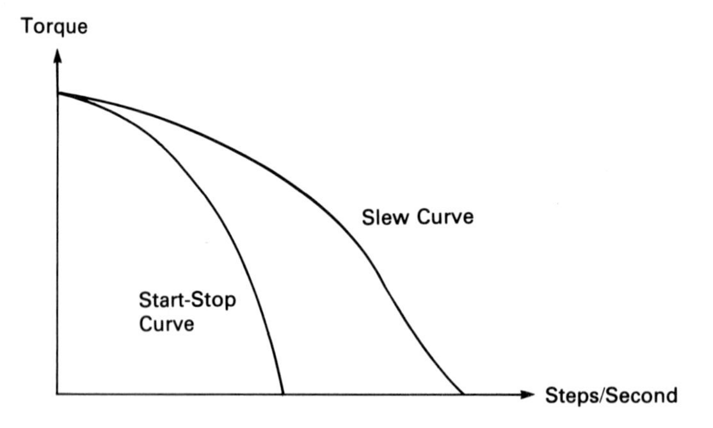


**Figure 20.3** Speed/torque curves for a PM stepper

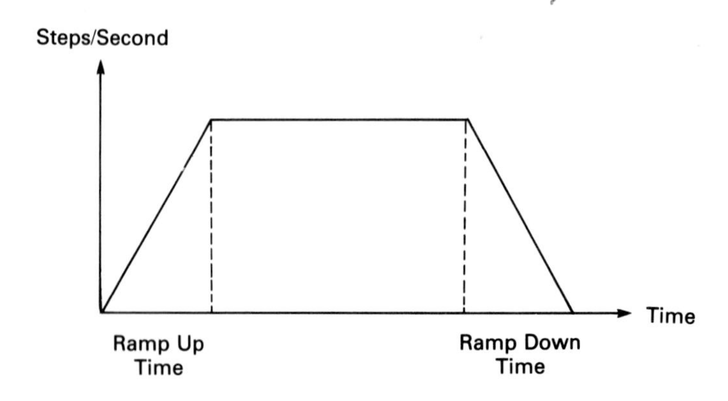


**Figure 20.4** Velocity control of a stepper

With careful attention to ramp up and ramp down times, the stepper motor can be operated *open loop* — without positional feedback to the controller. The controller determines the position of the shaft based on the number of steps the motor has been given.

When the load is variable or when the mechanical components of the system make load position less certain, shaft encoders or other positional feedback devices are required. When in doubt, err in the direction of closed-loop systems, rather than open-loop systems.

## SUMMARY

In this chapter, we have looked at the stepper motor. We have seen that:

1. The stepper motor is used to move a shaft in small steps, rather than continuously.
2. The permanent magnet stepper motor has permanent magnets in the rotor.
3. The variable reluctance motor has a toothed rotor made of soft iron.
4. Stepper motors require careful attention to the step rate.

## SELF-TEST

1. The stepper motor is operated from a(n) ______________________________ power source.
2. By using half-step techniques, a 15-degree stepper motor can take steps as small as ______________________________ degrees.
3. In a PM stepper motor, the rotor is a permanent magnet and the stator is two or more ______________________________.
4. In a VR stepper, the rotor is made of ______________________________.
5. A PM stepper will remain in position even when power to the coils is turned off. This is called ______________________________ torque.
6. A stepper motor will remain in position as long as power is maintained on the stator winding(s). This is called ______________________________ torque.
7. The starting torque of a stepper motor is also called ____________________ torque.
8. Another name for the stator windings is ______________________________.
9. As additional load is added to a stepper motor, the step rate of the motor must be ______________________________.
10. As the step rate is increased, the pull-out torque of the motor is __________ ______________________________.
11. One way to increase the maximum step rate is to reduce the current rise time of the stator coils by adding ______________________________ in series with the windings.
12. When a stepper motor is operated at a low step rate, the rotor comes to a stop between pulses. When the step rate is increased so that the rotor does not stop, the motor is operating in the ______________________________ mode.
13. Gradually increasing the step rate of the motor is called ____________________.

## QUESTIONS/PROBLEMS

1. A stepper motor takes 7.5-degree steps. The motor must take __________ steps to complete one rotation.
2. You need a stepper motor that takes 1.8-degree steps. The motor will step ______________________________ per revolution.
3. A 7.5-degree stepper motor is being operated in the half-step mode. Each step is now ______________________________ degrees.
4. A four-phase VR stepper has twelve rotor teeth. This motor will take ______________________________ degree steps.
5. The step rate of a stepper motor is fixed at 150 steps per second. What is the clock frequency for the steps?

# 21

# Ac Motors

Since ac power is the power that is delivered to homes and factories worldwide, it is reasonable that we use ac motors when the special characteristics of dc motors are not required. Ac motors do not require rectifier circuits to correct the line voltage before it is sent to the motor. As we shall see, most ac motors also do not require carbon brushes and commutators, reducing the cost of preventive maintenance. For these reasons, ac motors are the muscle of industry, far more prevalent than dc motors in home and factory alike.

In this chapter, we will take an overview of the many different types of ac motors and their characteristics. We will look at the similarities and differences among them, and at the different types of motor starters and controllers.

## OBJECTIVES

You will have successfully completed this chapter when you can:

1. Describe the basic operating principles of the ac induction motor.
2. Describe the basic operating principles of the ac synchronous motor.
3. Explain the importance of power-factor correction for industrial users.

## 21.1 AC POWER

The power supplied to your home is single-phase ac power, which is characterized by the familiar sine wave. This is the waveform you would see if you connected an oscilloscope directly across the ac line in your home. Single-phase ac power, however, is of a pulsating nature. In a purely resistive circuit, the power delivered to the load will actually be zero twice during each cycle.

We can see this relationship in Figure 21.1. In a resistive circuit, current and voltage are in phase. Note that the power is always positive because current and voltage are negative at the same time. Note, also, that the power delivered to the load is zero when current and voltage cross zero in step.

In Figure 21.2, we see the relationship between voltage, current, and power in an inductive circuit. Recall that in a pure inductance, the voltage leads the current by

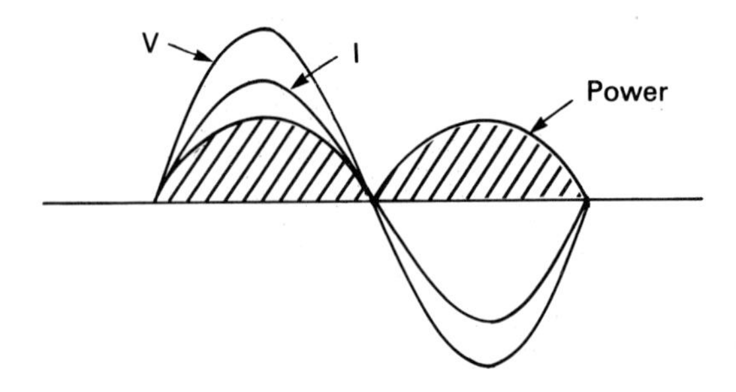


**Figure 21.1** Single-phase ac power in a resistive circuit

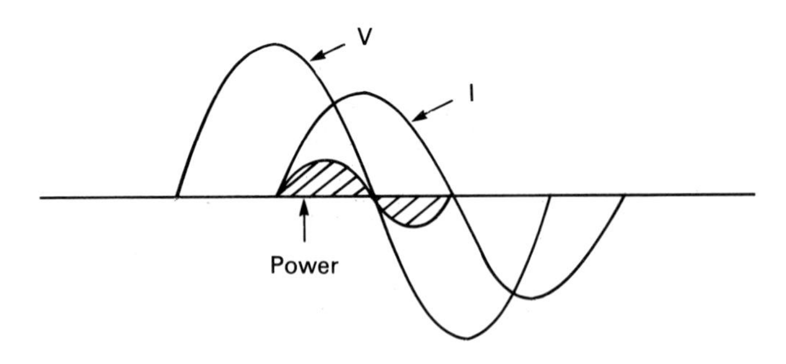


**Figure 21.2** Single-phase ac power in an inductive load

90 degrees. The relationship between voltage, current, and power in a pure inductive circuit is shown in Figure 21.2. Note that the power curve is sinusoidal and at a frequency *twice the line frequency*. Notice also that the power is *negative* twice in each cycle of line voltage, and zero four times in each cycle. Because the power is negative for half of the time and positive for half of the time, the resultant power in a pure inductance is always zero.

By convention, the negative power is called the *reactive power*. It represents the power returned to the line when the magnetic field around the inductance collapses. In effect, an inductance absorbs power from the line during one half-cycle and returns it to the line during the next half-cycle.

There is, of course, no such thing as a pure inductance, and that could be no more true than in an electric motor. Rather, the circuit of a motor is composed of inductance and resistance — an *RL* circuit. Figure 21.3 shows the relationship between voltage, current, and power in an RL circuit. Because of the addition of the resistance into the circuit, the phase angle between the voltage and current is less than 90 degrees. As shown in Figure 21.3, the resultant power curve is negative for less time than it is positive.

It is possible to draw sine-wave diagrams for any combination of resistive and reactive elements in a circuit, but it is a tedious process. Vector diagrams make a much better tool for analysis of such circuits.

Figure 21.4 shows the use of vector diagrams to show the relationship between voltage and current in circuits containing reactance. In Figure 21.4a, we can see the relationship between voltage and current in a purely inductive circuit. The voltage is shown along the horizontal axis at three o'clock, and the current is shown along the vertical axis at six o'clock — later in time.

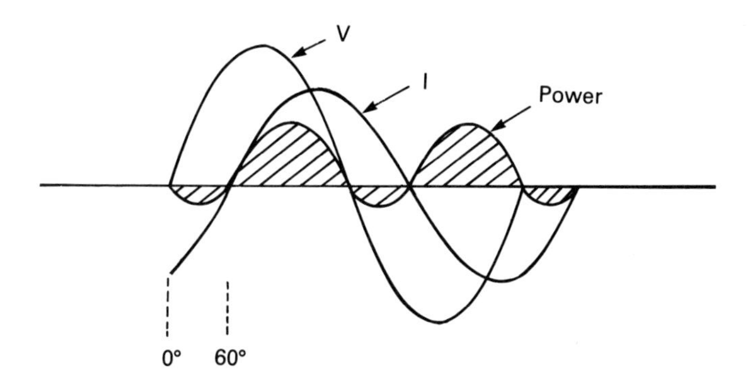


**Figure 21.3** Single-phase ac power in an RL circuit

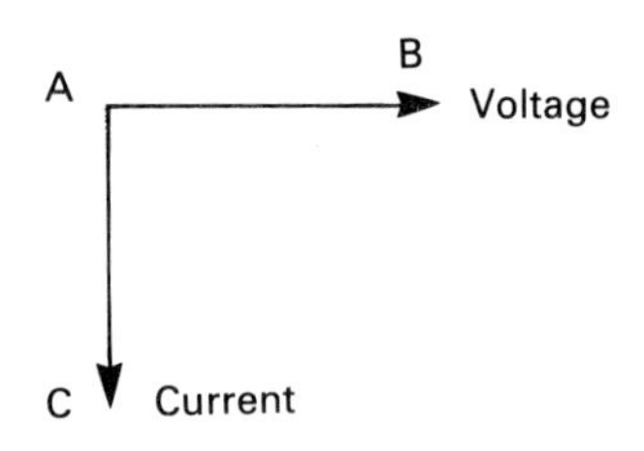


(a) Vector relationship between voltage and current in a pure inductance.

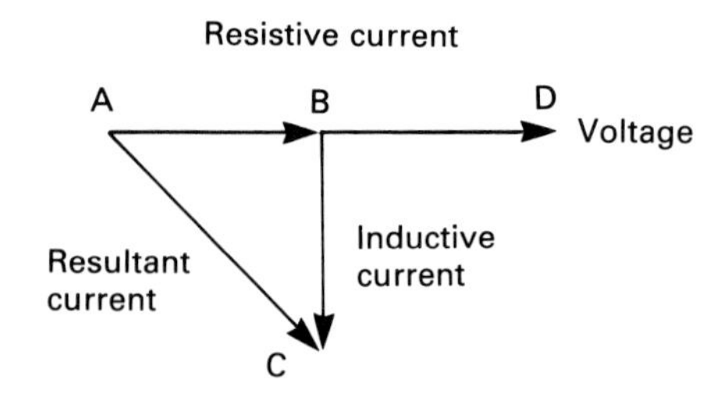


(b) Vector relationships between voltage and current in an RL circuit

**Figure 21.4** Vector diagrams

In Figure 21.4b, we see the phasor diagram for an RL circuit. Once again, the voltage is shown drawn toward six o'clock. Now, however, there are *two* current vectors, one for the reactive current and one for the resistive current. Since the resistive current is *in phase* with the applied voltage, it is drawn along the same axis as the applied voltage, shown here as line *AB*. The reactive current is drawn 90 degrees lagging and is represented by line *BC*. The *resultant current* is shown along line *AC*, the hypotenuse of the right triangle *ABC*.

The resultant current is called the *apparent current* — the current absorbed by the load and carried by the plant wiring. The reactive portion of the current is the current that is absorbed by the inductance in one half-cycle and returned to the line in the next. The only portion of current in the line that performs useful work is the *resistive portion*, which is called the *effective current*.

The phase angle between the voltage and the apparent current is an important measurement, for it allows us to calculate the effective current, and, as we shall see, add enough reactive current of the opposite polarity to reduce or eliminate the reactive portion of the current.

Canceling the reactive component in motor circuits can lead to a substantial savings in the cost of power for an industrial user. The power company offers discounts to industrial users who reduce the VAR in their plants. The reason is simple: although the VAR does no useful work, power lines and generator wiring must be large enough to carry this portion of the current, as well as the effective portion. By using power-factor correction, industrial users can reduce the wiring capacity for the utility, as well as for their own plants. Most utilities *require* that the overall power factor be 0.8 or better.

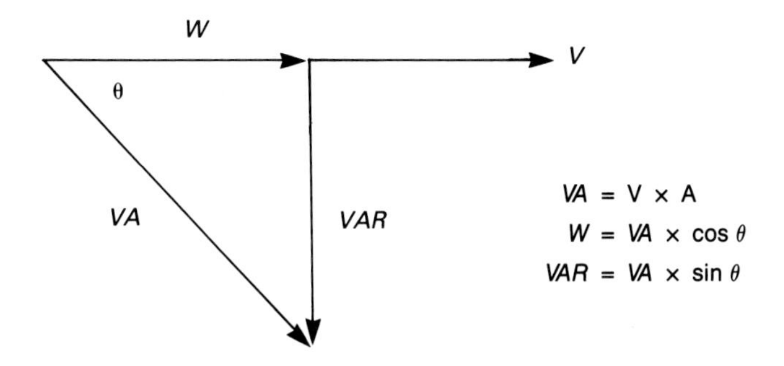


**Figure 21.5** The vector relationship of kV, VA, and kVAR

Figure 21.5 is a phasor diagram showing the relationship between the various types of power consumed in a reactive circuit. The apparent power, measured in *volt amps*, or *VA*, is the power that determines wire sizes and circuit breaker sizes in your plant. The effective power, which is measured in *watts* or *W*, is the power that is actually used to perform useful work in your plant. The reactive power, which is measured in *volt amps reactive*, or *VAR*, is the power in the reactive portion of the circuit. It is first consumed and then fed back to the power lines by the inductance in the circuit.

The ratio between effective power and apparent power in a circuit is the *power factor* of the circuit:

$$PF = \frac{W}{VA}$$

The power factor of a given circuit is a direct function of the phase angle relationship between current and voltage:

$$PF = \cos\theta$$

And knowing *VA* and *PF*, we can readily calculate effective power:

$$W = VA \times \cos\theta$$

This math is important because motors are often rated in terms of *VA* and *PF*. Given those quantities, we can determine the effective power in watts and the reactive power in volt amps reactive.

### Example 21.1

An ac induction motor is rated at 1 kVA and has a power factor of 0.8. What is the power consumed by the motor in watts? What is the reactive power in VARs?

**Solution:**

Solving for effective power:

$$W = VA \times PF$$
$$W = 1000 \times 0.8$$
$$W = 800 \text{ W}$$

Solving for the phase angle:

$$\theta = \arccos 0.8$$
$$\theta = 36.87°$$

Solving for VAR:

$$VAR = VA \times \sin \theta$$
$$VAR = 1000 \times 0.6$$
$$VAR = 600 \text{ VAR}$$

The owner of this plant must wire this circuit for 1 kW of electrical consumption, although he receives the benefit of only 800 W of work.

## 21.2 POWER FACTOR CORRECTION

In a large plant, the extra cost of wiring for apparent power consumption can be substantial. The saving in power cost from electric company discounts is an even greater incentive to use *power factor correction.*

Adding the correct amount of capacitance to the circuit of an induction motor can totally cancel the effect of inductive reactance.

The basic wiring of power factor correction is shown in Figure 21.6. The added capacitance is wired in parallel with the motor. If the capacitive reactance is exactly equal to the inductive reactance, the power factor will be one. In that case, the circuit appears to be a pure resistive circuit because the reactive power that is returned to the line by the inductance is absorbed by the capacitance. When the capacitance returns its reactive power to the line, it is absorbed by the inductance. In effect, the motor inductance and the added capacitance become a resonant circuit.

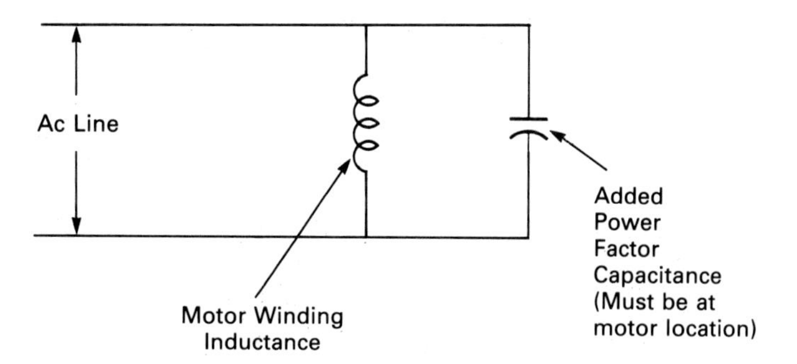


**Figure 21.6** Power factor correction

Fortunately, we do not have to calculate the inductance of the motor in order to find the correct capacitor for power factor correction. Manufacturers of capacitors produce special power factor correction capacitors for use with industrial induction motors. These capacitors are electrolytic oil-filled units with screw terminals for easy connection into the circuits. They are rated in kVAR!

Power factor correction capacitors are readily available in ratings from 1 kVAR to 40 kVAR. Properly selected power factor correction can raise the power factor of a motor to 0.95 or better.

## 21.3 THREE-PHASE AC POWER

As we saw in the beginning of this chapter, power in a single-phase ac circuit is pulsating in nature. In a pure resistive circuit, the power is *zero* twice in each cycle, while in a reactive circuit, the power is actually *negative* during part of each cycle. Such pulsating power, when converted to rotary motion by a motor, results in pulsating or *vibrating* motion.

Industry consumes power in such quantities that a more efficient and smoother delivery of power is essential. Polyphase (more than one phase) ac power offers an economical solution.

Figure 21.7 is a simplified diagram of a three-phase generator. It has three stator windings: $A\text{-}A'$, $B\text{-}B'$, and $C\text{-}C'$, which are located 120° apart. As the rotor magnet rotates within these windings, each winding produces its own sine wave. The output of winding $B\text{-}B'$ is 120° *behind* the output of winding $A\text{-}A'$; it *lags* $A\text{-}A'$. The output of winding $C\text{-}C'$ lags the output of winding $A\text{-}A'$ by 240°.

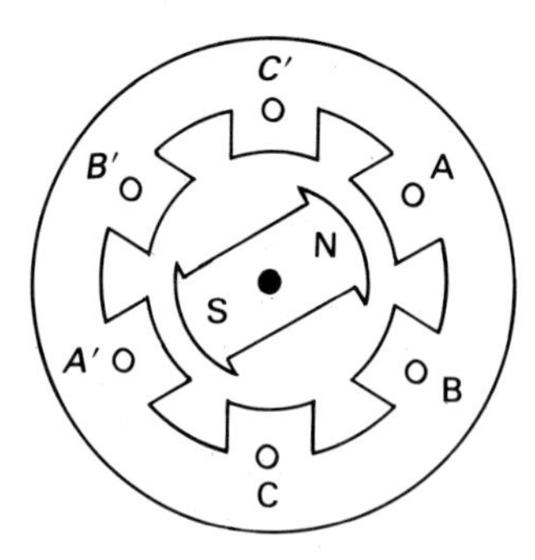


Figure 21.7 A three-phase generator

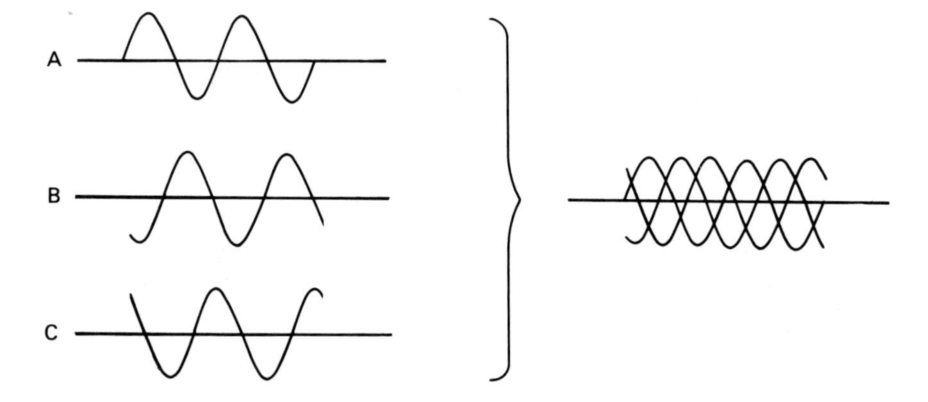


Figure 21.8 Three-phase power

Figure 21.8 illustrates these time relationships among the three output voltages. From observing this figure, we can see that the *total* application of power is smoother than the application of power in the single-phase circuit.

There are two possible connections in three-phase circuits, the delta connection and the wye, or star, connection.

Figure 21.9 illustrates both types of connection. The three-phase generator uses only the wye connection because it suppresses harmonics, which cause distortion in the sine wave under load. The center node of the wye connection is the neutral lead, which is often labeled *N*. When the three loads of a wye connection are equal, the neutral lead carries no current, and it is often eliminated.

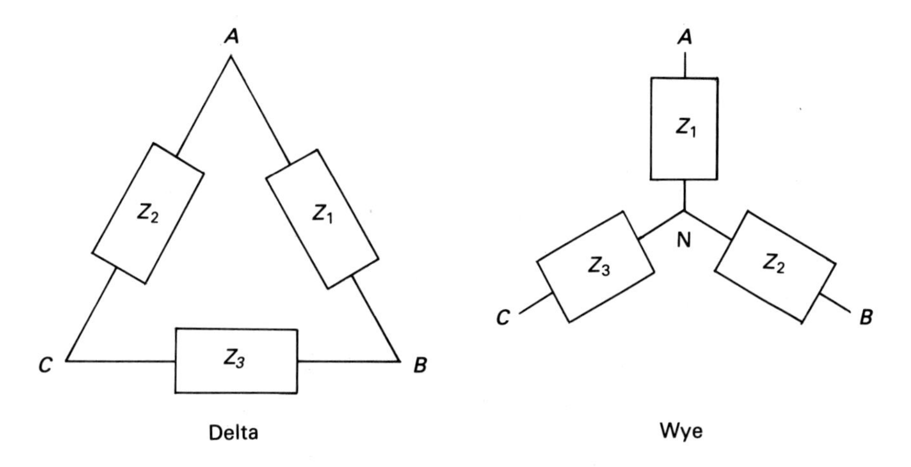


**Figure 21.9** Three-phase connections

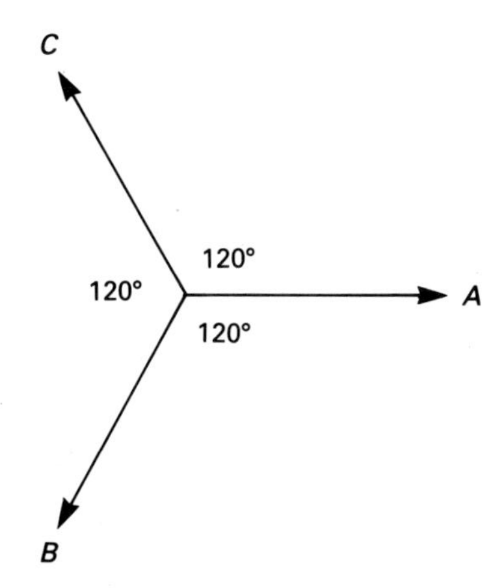


**Figure 21.10** Phasors for three-phase voltage

Figure 21.10 shows the phasor diagram of a three-phase voltage. Note that voltage *B* lags voltage *A* by 120°, and voltage *C* lags voltage *B* by 120°. This illustrates the phase relationship that we expect from three-phase power sources.

The load on a three-phase power connection may be wired in either a wye or a delta configuration. Because the effect of the applied voltages is different for each of them, we shall examine them separately.

We will first consider the wye connection. By examining the sine waves in Figure 21.8, we can see that when voltage *A* is at its positive peak, voltage *B* has some negative value. The voltage measured between $V_A$ and $V_B$, then, must be greater than $V_{peak}$. We can use vector analysis to determine the maximum voltage between these two lines.

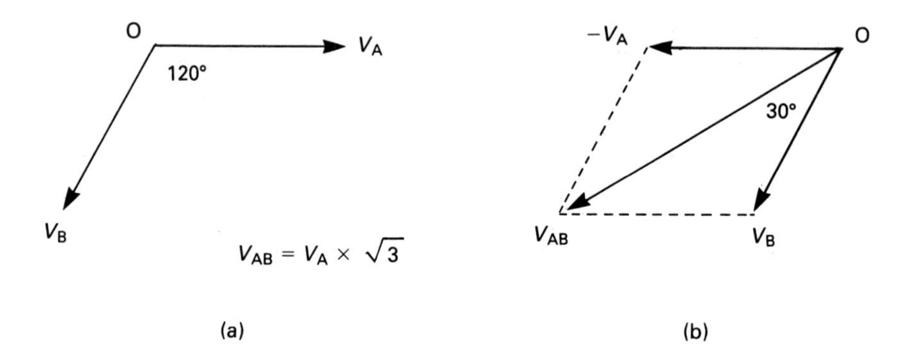


**Figure 21.11** *VA* and *VB* phasors for a wye-connected circuit

Figure 21.11a shows the phasors for $V_A$ and $V_B$ in a wye connection. The magnitude of the resultant voltage ($V_{AB}$) could be found by measuring the distance between the end points to $V_A$ and $V_B$. However, since we are also concerned with the phase relationship of $V_{AB}$, we must use the diagram of Figure 21.11b so that $V_{AB}$ has the origin in common with $V_A$ and $V_B$. From this, we can determine that

$$V_{AB} = V_A \times \sqrt{3}$$

The phasors for the voltages developed across the three separate loads of a wye-connected circuit are shown in Figure 21.12. Note that these three resultants are also separated by 120°.

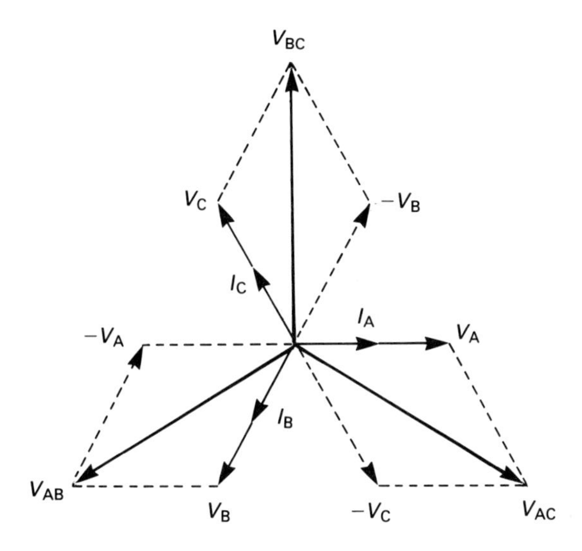


**Figure 21.12** Voltage phasors for the wye connection

An examination of the delta connection in Figure 21.9 indicates that the voltage between phases is equal to the line voltage. In this connection, because there is no neutral point, the current through $Z_1$ is affected by all three voltage sources. It is the *current* we must solve with phasors.

The current phasors for a delta connection are shown in Figure 21.13. Here we find that

$$I_{AB} = I_A \times \sqrt{3}$$

Table 21.1 summarizes the important relationships between voltage, current, and power in three-phase circuits.

## Example 21.2

A three-phase heater (resistive load) is connected in a wye configuration to a 208 V power source. The measured current in phase $A$ is 3 amps. Find:

a. The power dissipated by each heating element.

b. The total power consumed by the heater.

**Solution:**

Solving for power in one phase:

$$P = \frac{EI}{\sqrt{3}}$$

$$P = \frac{624}{\sqrt{3}}$$

$$P = 360.27\ \text{W}$$

Solving for total power:

$$VA = \frac{3EI}{\sqrt{3}}$$

$$VA = \frac{1920}{\sqrt{3}}$$

$$VA = 1108.51\ \text{W}$$

## 21.4 THREE-PHASE AC MOTORS

Electric motors perform the bulk of the work in modern industry. They move products on assembly lines, mix chemicals, provide compressed air for pneumatic tools, operate the pumps for hydraulic tools, drive cranes and drag-lines — the list is almost endless. Because the application of power is more continuous and efficient with three-phase motors, most of the large electric motors in industry are three-phase motors.

The motor in Figure 21.14 is a three-phase industrial motor. This particular motor style is available in horsepower ratings from 3/4 hp to 500 hp. Inside the main frame, we can see the field coil windings for the motor. The connections to the field coils are shown behind the open conduit cover. Note the eye bolt on top, for ease of handling the larger motors.

In Figure 21.15, we can see the way the windings are configured in a three-phase stator. The stator in the illustration is wye connected, but the winding pattern would be the same for a delta-connected motor. Each phase in Figure 21.15 is wound on two poles of the stator, providing the two poles of an electromagnet for each phase. This arrangement gives the motor the name *two-pole motor.* Note that the two poles from each phase are directly opposite each other in the stator.

The number of poles in a stator is not limited to two. A four-pole motor has *two* electromagnets per phase, a six-pole motor has *three* electromagnets per phase, and an eight-pole motor has *four* electromagnets per phase. The added poles are arranged so that the sequence of poles is always *ABC*, *ABC*, and so on. Three-phase motors typically have four, six, or eight poles. The added poles provide a smoother application of power, but it is done at the cost of rotational speed.

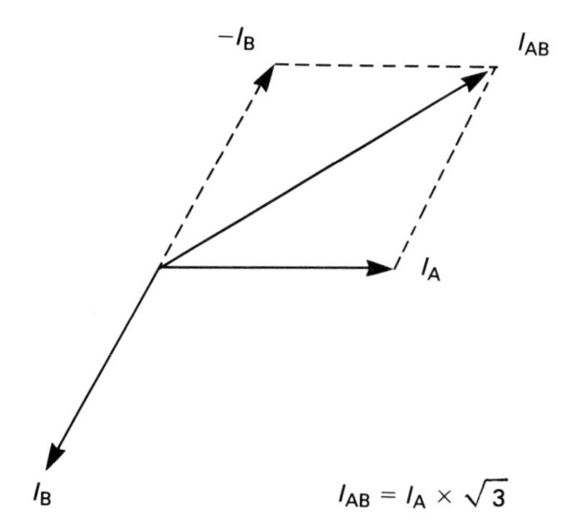


**Figure 21.13** Phasor diagram for the delta-connected circuit

**Figure 21.14** Exploded view of a three-phase industrial motor *(Courtesy of Reliance Electric)*

| | WYE CONNECTION | DELTA CONNECTION |
|---|---|---|
| **Current** | $I_{\text{line}}$ | $\frac{I_{\text{line}}}{\sqrt{3}}$ |
| **Voltage** | $\frac{E_{\text{line}}}{\sqrt{3}}$ | $E_{\text{line}}$ |
| **Phase** | The voltages across the resistors are 120° out of phase. | |
| | The currents through the resistors are 120° out of phase. | |
| **Power** | The power in any single resistance is: $P_R = \frac{EI}{\sqrt{3}}$ | |
| | The power in any single reactance is: $P_x = \frac{EI \cos \theta}{\sqrt{3}}$ $VA = P^2 - VAR^2$ | |
| | The total power in the circuit is: $VA = \frac{3EI}{\sqrt{3}}$ | |

**Table 21.1.** Voltage, current, and power in three-phase circuits.

Figure 21.15 shows the poles protruding from the frame, and some older motors actually have this type of construction. In modern electric motors, the stator is smooth and the windings are pressed into slots in the stator and held in place with wedges.

Figure 21.16 shows the effect of connecting three-phase power to the stator windings. Current flow in the windings builds up a magnetic field. As the current in each

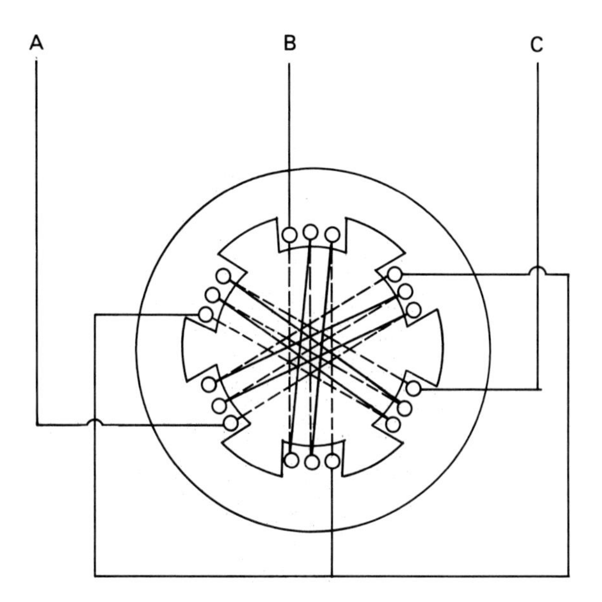


**Figure 21.15** The windings in a three-phase stator

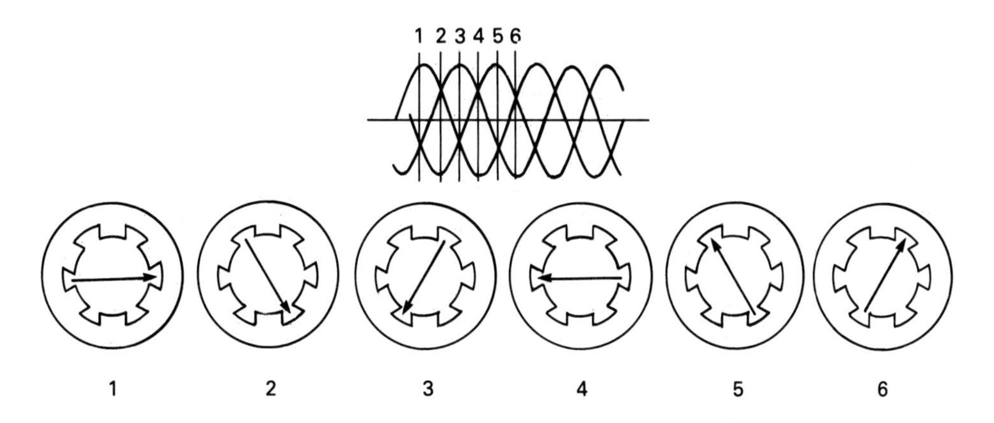


**Figure 21.16** The rotating magnetic field

phase winding varies, the magnetic field *rotates* within the stator. A magnet placed within this rotation will rotate with the magnetic field — the basis of motor action.

The speed at which the field rotates within the stator is a function of the line frequency and the number of poles per phase. It is easily calculated:

$$ns = \frac{120f}{p}$$

where:

$ns$ = synchronous speed in RPM
$f$ = line frequency
$p$ = poles per phase

**Example 21.3**

Calculate the synchronous speed of a two-pole motor if the line frequency is 60 Hz.

**Solution:**

$$ns = \frac{120 \times 60}{2}$$

$$ns = \frac{7200}{2}$$

$$ns = 3600$$

The motor in Figure 21.17 has a permanent magnet rotor. This rotor will follow the rotating magnetic field of the stator *exactly*, so that the rotor speed is the same as the synchronous speed calculated in Example 21.3. For that reason, it is called a *synchronous motor.* Synchronous motors are used when exact speed is desired.

For large motors, however, the cost of a permanent magnet rotor is prohibitive. In such cases, electromagnets are used for the rotor. The dc power is applied to the rotor through slip rings and brushes on the armature shaft.

This technique also has drawbacks. Both slip rings and brushes are subject to wear and need periodic maintenance. The newest innovation in synchronous motors is the *brushless synchronous motor.*

The diagram in Figure 21.18 shows how the brushless motor works. The parts of the circuit within the dotted lines are *inside* the rotor of the motor, and rotate with it. On the right side of the diagram, we see a winding labeled *exciter field.* This field is connected to one phase of the ac power line. Within the rotor portion of the diagram, we see another winding labeled the *exciter.* These two components make up a *transformer.*

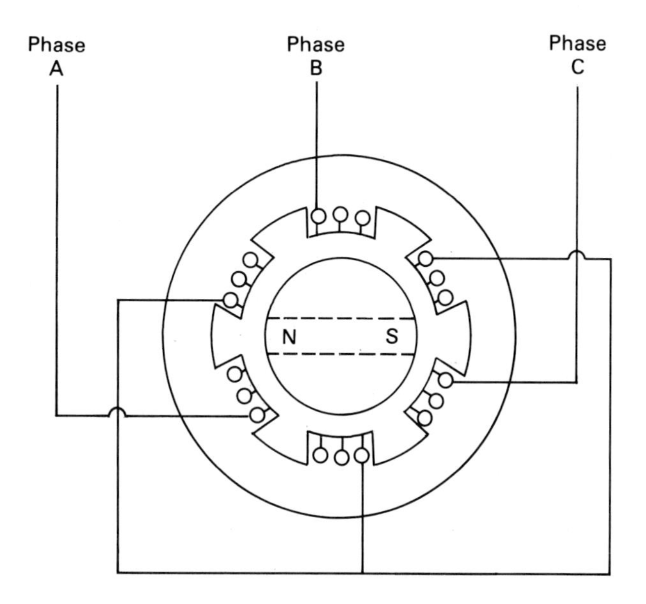


**Figure 21.17** A synchronous motor

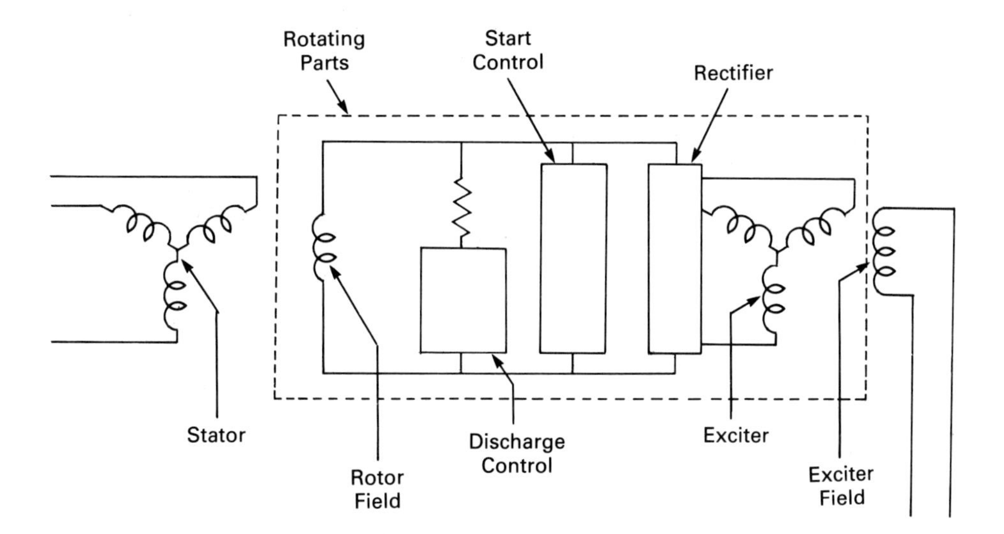


**Figure 21.18** A brushless synchronous motor

As the rotor turns, the exciter windings output three-phase ac voltage, which is converted to dc by the rectifier. This dc voltage is applied to the rotor field, which then becomes an electromagnet.

The start control circuit shunts the rotor windings with a resistance during start-up to assure that dangerous voltages are not built up during the start sequence as the rotating magnetic field of the stator cuts through the windings of the rotor field. This resistance, the *discharge resistance*, is switched out of the circuit when the rotor nears synchronous speed.

Synchronous motors offer a second advantage, beyond the exact speed control. Synchronous motors can be designed to have either leading or lagging power factors. Thus, it is possible to use a synchronous motor for power factor correction.

Most electric motors, however, do not require the exacting speed of the synchronous motor, and the extra cost is not justifiable. In these cases, an *induction motor* is used. Induction motors make up the majority of electric motors used in industry.

The rotor in Figure 21.19 is a *squirrel cage rotor.* The main body of the rotor is made of laminated iron to reduce the effects of eddy currents, which are induced when the rotor turns in a magnetic field. Each of the thin laminations is insulated from its neighbors so that these induced currents remain very small. Wide slots are left in the outer surface of the rotor, and aluminum or copper *rotor conductors* are pressed into these slots. The ends of these rotor conductors are joined by *end rings* to form complete electrical circuits.

As the rotating magnetic field cuts through these rotor conductors, a large current is induced in them. The rotor currents, in turn, create a magnetic field of their own. The magnetic field in the rotor attempts to follow the rotating field of the stator, generating considerable torque.

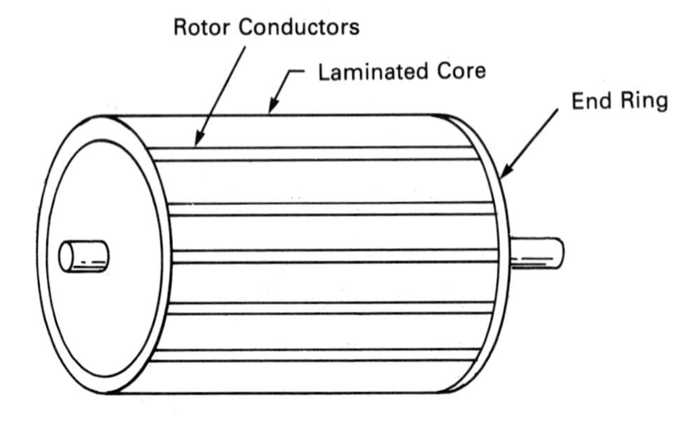


**Figure 21.19** A squirrel cage rotor

The rotor, however, can never quite rotate at the synchronous speed. If it were to do so, the rotating field would no longer cut through the rotor conductors, and the magnetic field of the rotor would be lost. The difference between the synchronous speed and the rotor speed is called *slip*. The slip in an induction motor is quite small, ranging from about 3% in small motors to less than 1% in larger motors. Since this slip is essential to the operation of the squirrel cage rotor, it is not an undesirable trait.

## 21.5 SINGLE-PHASE AC MOTORS

While the three-phase motor is the standard in industry, many smaller tasks are performed with single-phase motors. The single-phase motor follows many of the same principles as the three-phase motor, but some special considerations are required.

Figure 21.20 will serve to illustrate the problems in a single-phase system. The stator windings induce a magnetic field into the stator poles, but the field does not rotate. Rather, as the voltage changes polarity, so does the magnetic field. The rotor conductors are cut by this magnetic field as it increases, collapses, and reverses. A magnetic field is built up in the rotor by the current through these conductors. The rotor magnetic field, however, is aligned with the stator field, and causes no torque. This motor is *not self-starting*. If we give the rotor a twist, the magnetic field in the rotor will be shifted and the motor will continue to turn.

Figure 21.21 shows one of the most common ways to make a single-phase motor self-starting. A second set of stator windings, known as the *start winding*, is added to the motor. This winding is connected to the line voltage through a centrifugal switch and some type of impedance. This extra impedance causes a phase shift in the current through the start winding. Now as the main winding field begins to collapse, the start winding field is building, and attracts the induced field in the rotor, causing it to start

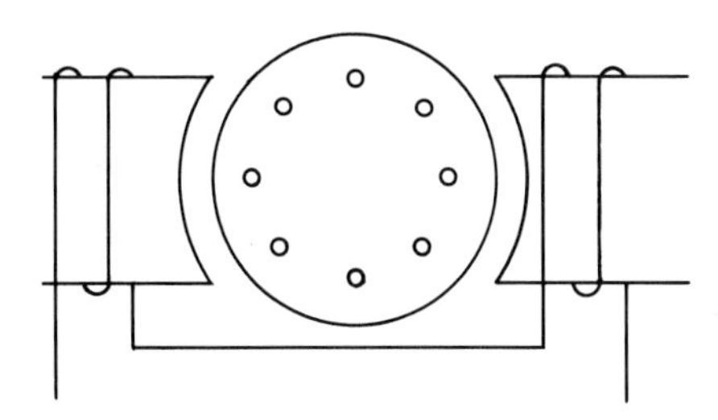

**Figure 21.20** A single-phase stator

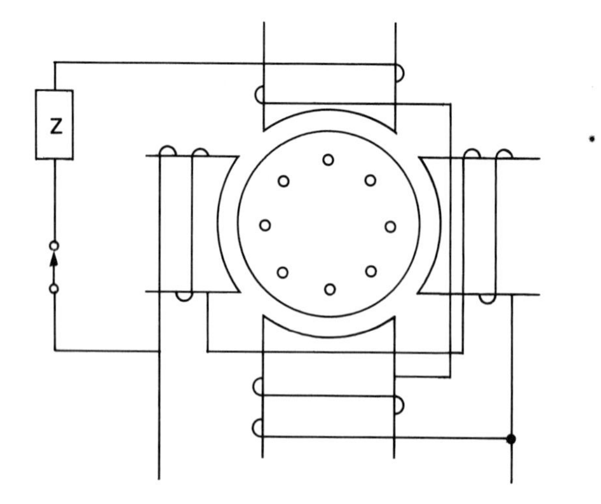


**Figure 21.21** A split-phase stator

turning. Once the rotor speed is high enough for the motor to continue running, the centrifugal switch disconnects the start winding. This type of motor is called a *split-phase motor.*

Figure 21.22 shows two different types of split-phase motors. Figure 21.22a shows a *resistive-start* motor, while Figure 21.21b shows a *capacitive-start* motor.

The resistive-start motor has two sets of windings, a run winding and a start winding. The start winding is made up of many turns of relatively small wire. The

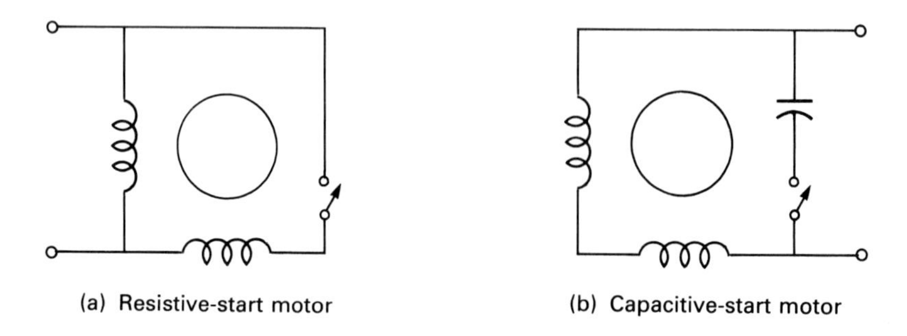


**Figure 21.22** Split-phase motors — resistive and capacitive starting

added resistance of this wire causes the phase shift required to make the motor self-starting.

More common is the capacitive-start motor of Figure 21.22b. Here a capacitor is used in series with the start winding to create the phase shift. The capacitor is selected so that the current through the start winding will lead the voltage. The starting capacitor must be large enough to allow a substantial starting current to flow in the start winding. These are most often electrolytic capacitors, which are not well-suited to provide this kind of current for any length of time. Thus, a centrifugal switch must disconnect the capacitor once the motor has reached its running speed.

A *permanent split capacitor motor* has a capacitor permanently connected into the circuit. This is most usually an oil-filled capacitor. Its value is selected as a compromise between the ideal size for starting and the ideal size for running current. Permanent split capacitor motors have a lower starting torque, but are more efficient and run more quietly.

A split-phase motor can be easily reversed, as shown in Figure 21.23. With the capacitor connected as shown in Figure 21.23a, the motor will run counterclockwise. In the connection shown in Figure 21.23b, the motor will run clockwise. Figure 21.23c shows a split-phase motor with a reversing switch.

Figure 21.24 shows a *shaded pole motor.* The shaded pole motor is characterized by a heavy *shading coil,* a single turn of heavy copper conductor on each pole of the stator. Large currents are induced in the shading coils, creating distortions in the magnetic flux of the stator. The effect is as if the magnetic field *shifted* across the face of the pole. This is sufficient to start the rotor turning. Shaded pole motors are limited to very low torque, low power applications, such as projector fans, drives in inexpensive record players, and the fans in small appliances like microwave ovens.

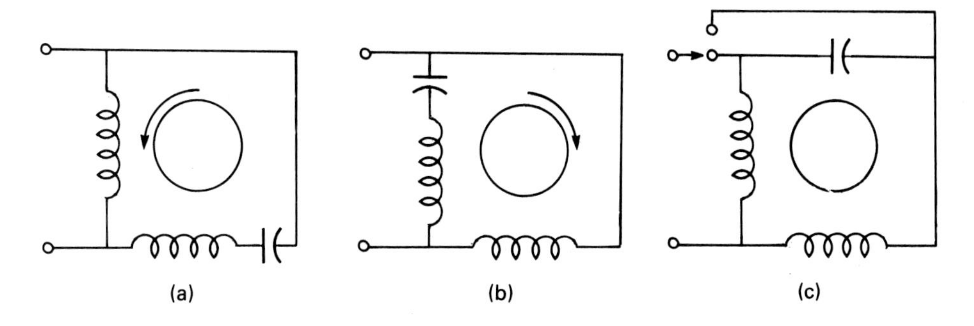


**Figure 21.23** Reversing a split-phase motor

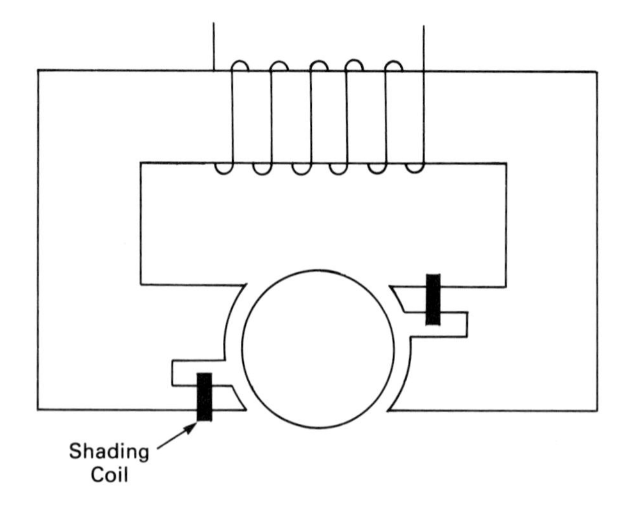


**Figure 21.24** A shaded pole motor

## 21.6 SPEED CONTROL

Synchronous motors run at exactly the synchronous speed of the rotating magnetic field. Induction motors run within a few percent of the speed of the rotating magnetic field. From this, it should be obvious that we cannot control the speed of an ac motor by controlling field current, nor by controlling the current in wound rotor synchronous motors.

There are two basic methods of speed control for ac motors. The first provides two- or three-speed motors by adding extra field windings and switching so that we can switch extra sets of poles in or out of the motor circuit. This approach is effective, but it does not give us a *continuous* control of speed.

In the second approach, power for the motor is supplied by a variable frequency source. There are two approaches to the variable frequency source: inversion and cycloconversion.

In the first approach, the ac power source is first rectified, and then *inverted* into ac again, this time at a different (and variable) frequency.

The single-phase *bridge inverter* in Figure 21.25 shows one approach to inversion. In this illustration, $SCR_1$ and $SCR_4$ are gated on for a time that amounts to 180 degrees of the desired line frequency. The current path through the load is then from

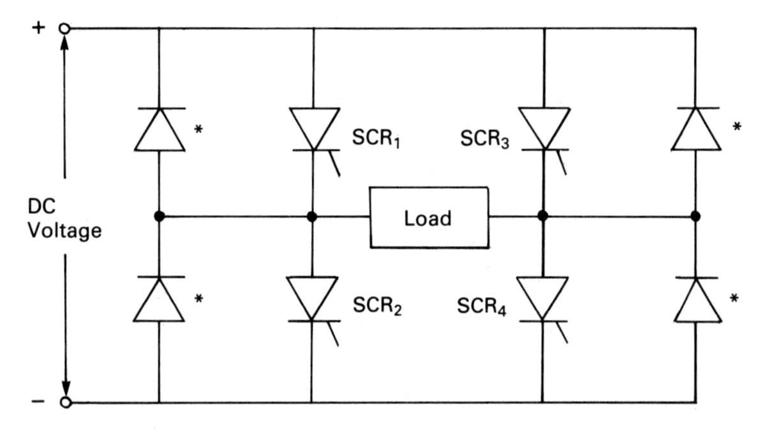


**Figure 21.25** Bridge inverter for ac motor speed control

the minus line, through $SCR_4$, through the load, and through $SCR_1$ to the positive line. At the end of the time period, $SCR_2$ and $SCR_3$ are gated on for the same time period. Now the path for current is from the minus line, through $SCR_2$, through the load, through $SCR_3$ to the positive line. Note that the direction of current through the load reverses — a definition of ac. The power applied to the motor is an ac square wave of the desired frequency.

An important variation of the bridge inverter reduces the large harmonic voltages and currents that can be induced in the lines by the simple on-off operation of the circuit. When these harmonics flow in the windings, they cause fluctuations in the torque delivered by the motor. At high speeds, these fluctuations are smoothed by the flywheel effect of the armature and its load. At lower speeds (and lower frequencies of operation), the torque variations become objectionable.

In *pulse width modulation (PWM) controllers*, each of the half-cycles is composed of a *series* of pulses, with the pulse widths narrower at the beginning and end of the half cycle. The *average* of the pulse voltages approximates a sine wave. PWM control is finding increased acceptance today because of the improvements made in solid-state control devices such as SCRs and TRIACs.

The *cycloconverter* can produce an output variable frequency ranging from zero Hz to one-third of the line frequency. Its main use is operating slow-speed motors. The

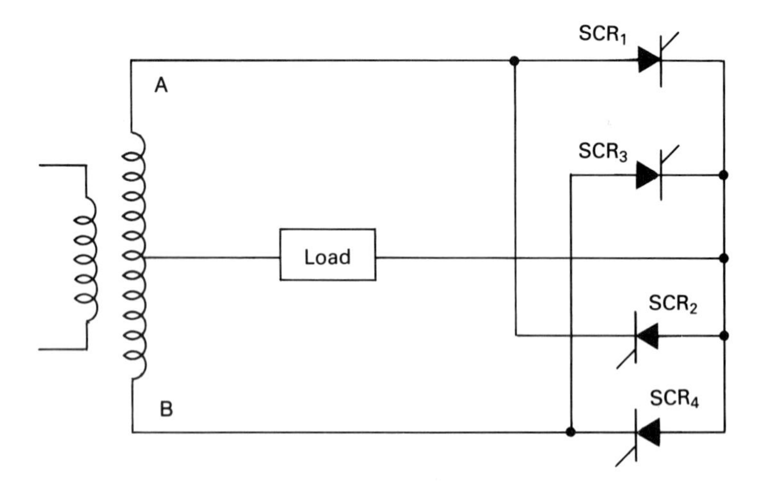


**Figure 21.26** A single-phase cycloconverter

cycloconverter converts ac at one frequency directly to ac at another frequency without the intermediate dc conversion.

The single-phase cycloconverter shown in Figure 21.26 illustrates how the principle works. It uses four SCRs in two groups: a positive group and a negative group. At time $t_0$, when point *A* is positive with respect to point *B*, $SCR_4$ is conducting. When the voltage changes polarity at time $t_1$, the return of energy from the inductive load keeps $SCR_4$ turned on. At time $t_2$, $SCR_2$ is turned on and $SCR_4$ turns off. At time $t_3$, $SCR_4$ is turned on and $SCR_2$ turns off. And at time $t_4$, $SCR_3$ turns on and $SCR_4$ turns off. The *average* voltage applied to the load is a sine wave. By varying the firing point of the SCRs around 90 degrees, the frequency can be controlled.

The A-C VS Drive shown in Figure 21.27 is an example of the variable frequency PWM type of controller. This unit is available for three-phase 460 V motors, ranging from 5 hp to 125 hp. The output frequency is continuously variable from 6 Hz to 120 Hz. The unit may be controlled from a separate operator controller, which is shown on the bottom left, or from virtually any type of process control system. It offers dynamic braking and automatic electronic control of reversing.

Variable frequency controllers provide levels of control that are not possible with multiple windings. Controlled acceleration and deceleration, usually called ramp up and ramp down controls, are an important part of motor control. Ramp up control allows the motor to be started at 0 V, 0 Hz, with the drive voltage and frequency being increased over a predetermined period. This allows motor starting without the

**Figure 21.27** A-C V-S Drive ac motor speed controller *(Courtesy of Reliance Electric)*

usual motor starter circuits. Ramp down allows the system to be brought to a gradual stop, protecting it against damage from sudden stops and eliminating the need for braking motors. Such controllers also often have presettable *jogging* speeds, which allow the load to be brought to a particular part of its operating cycle before full power is applied.

Scientists have shown that the internal torque developed in an ac motor is constant when the ratio of applied voltage to frequency (V/Hz) is constant. This means that when we vary the frequency, we must also vary the applied voltage in order to maintain a constant torque output from the motor. In other words, as we increase the frequency, we must also increase the applied voltage or the motor will lose torque. This constant V/Hz ratio is maintained in the variable frequency speed controllers discussed in this chapter.

At frequencies below 10 Hz, the stator appears more resistive than reactive to the ac line. This reduces the magnetic flux in the air gap between the rotor and the stator, causing a subsequent reduction in torque. In some cases, the loss in torque is not objectionable, and a constant V/Hz ratio can be maintained. Where this is not possible, provisions can be made to adjust the V/Hz ratio to improve the torque at lower speeds. A-C VS drives have an adjustable V/Hz ratio, ranging from 3.4 V/Hz to 15.1 V/Hz.

## SUMMARY

In this chapter, we have seen that:

1. Single-phase ac power is pulsating in nature.
2. Three-phase ac power offers a smoother transfer of power.
3. Some portion of the current in a load containing reactance is not effective in producing useful work.
4. The effective power consumed in a reactive circuit is measured in watts.
5. The apparent power consumed in a reactive circuit is measured in volt amps.
6. The reactive power in a reactive circuit is measured in volt amps reactive.
7. The ratio of effective power to apparent power in a reactive circuit is called the *power factor* of the circuit.
8. The power factor of a circuit is related to the phase angle in the circuit.
9. Power factor correction can reduce the wiring costs for both the industrial user and the power company.
10. The power company offers discounts to industrial users who reduce the VAR in their plants.
11. A synchronous motor rotates at the synchronous speed of the field coils because the rotor is a magnet.
12. An induction motor rotates at a speed slightly below the synchronous speed. The rotor magnet in this motor is the result of induced current in the rotor conductors.
13. A single-phase ac motor requires some additional means to make the magnetic field rotate. This is usually accomplished with a start winding in the stator field.
14. Continuous speed control for an ac motor can only be accomplished by varying the line frequency.

## SELF-TEST

1. A single-phase ac line is connected to a pure resistive circuit. The power consumed by the circuit will be ______________________________ twice in each cycle.
2. When an ac line powers a reactive circuit, the power will be ____________ during some part of each cycle.
3. The power factor in an ac circuit is the ratio of effective power to _______ ______________________________ power.

4. The power factor in an inductive circuit can be reduced by adding ______ to the circuit.
5. Correcting power factor can provide savings in both the wiring cost of a plant and ______________________ because the power company offers discounts for power factor correction.
6. Industrial users prefer three-phase ac power because the transmission of power is ______________________.
7. The magnetic field induced in the stator coils of a three-phase motor rotates at the ______________________ speed.
8. The difference between the synchronous speed of an induction motor and the rotational speed of the rotor is called ______________________.
9. A single-phase induction motor requires a start winding to get it rotating. When the motor speed is adequate, this winding is usually switched out of the circuit by a(n) ______________________ switch.
10. Motor speed control can be effected in an ac motor by switching extra stator poles into the circuit or by varying the ______________________ of the line voltage to the stator.

## QUESTIONS/PROBLEMS

1. An ac induction motor is rated at 500 VA. The motor has a power factor rating of 0.87. What is the reactive power in the circuit? What is the effective power in the circuit?
2. An ac induction motor is labeled *6 pole*. What is the synchronous speed of the motor on a 60 Hz power line?
3. A four-pole induction motor is found to rotate at 1725 RPM on a 60 Hz power line. What is the percentage of slip for this motor?

# 22

# Controlling Fluids

Much of the work that is done in modern industry is done by *fluid power* — either hydraulic or pneumatic power sources. Fluid power is used in many of today's industrial robots, and in mobile and fixed applications within the plant.

Fluid power has advantages over electric motors for certain applications. With a fluid power system, the source of power can be located remotely from the actuator. This eliminates the weight of the power source from the load on the actuator. The power can be transferred to the actuator through hoses and can thus be easily transmitted around corners and through small spaces. This eliminates drive belts, gear trains, or articulated drive shafts from the power transmission path.

With the advent of automation, electrical and electronic means of controlling the application of fluid power have been developed. In this chapter, we will examine the basics of fluid power and its electrical control.

## OBJECTIVES

You will have successfully completed this chapter when you can:

1. Explain the basic application of fluid power.
2. Define hydraulic and pneumatic, and explain the basic differences between them.
3. Describe the techniques used for controlling fluid power with electronic controls.

## 22.1 FUNDAMENTALS OF FLUID POWER

A *fluid* is a substance that is capable of flowing and that takes the shape of its container. There are two types of fluids: *liquids* and *gases.* The molecules in a liquid are more closely bound together than those in a gas. As a result, gases tend to expand to *fill* their container, and to be *compressible.* Liquids, on the other hand, are relatively *incompressible.* When we speak of fluid power, we must differentiate between *pneumatic* systems, which use gas (usually air), and *hydraulic* systems, which use liquid (usually oil). Our discussion in this chapter will concentrate on hydraulic systems. We will point out the occasional differences between hydraulic and pneumatic systems in basic operation when they occur.

When a force is applied to a confined liquid, the force is transmitted equally throughout the liquid in the form of hydraulic pressure. When a force is applied to a confined gas, the gas first *compresses,* then the force is transmitted equally throughout the gas. This compressibility is one of the fundamental differences between hydraulic and pneumatic systems. Pneumatic systems tend to have a "cushiony" effect, while hydraulic systems operate more abruptly.

This distribution of power throughout a fluid is called *Pascal's law* in honor of the French mathematician, Blaise Pascal, who discovered it. The mathematical expression that describes Pascal's law is

$$\text{PRESSURE (PSI)} = \frac{\text{FORCE (lbs.)}}{\text{AREA (sq. in.)}}$$

Figure 22.1 illustrates Pascal's law. Here a force of 100 pounds is exerted on a piston that has a surface area of 1 square inch. The pressure on the piston, then, is 100 pounds per square inch (PSI). This same pressure is exerted on the entire surface of the container, as indicated by the arrows in the drawing.

In Figure 22.2, we can get an idea of how to use Pascal's law to our advantage. We are using the same piston and the same 100-pound force in this illustration, but we have added a second piston, connected through a narrow pipe, to the system. This second piston, called a *load* piston, has a surface area of 20 square inches. The pressure on the load piston, then, is

$$100 \text{ PSI} \times 20 = 2000 \text{ pounds}$$

When the load on the load piston is 2000 pounds, it will be perfectly balanced by the force on the first piston. The system is in *equilibrium.* If the load is less than 2000 pounds, however, the second piston will be moved upward.

This is the magic of fluid power — we can step up the effects of our effort. This is analagous to a lever in mechanics, or to a transformer in electricity. Of course, we pay a price for our mechanical advantage: what we gain in effort we must sacrifice in distance. In our example, the first piston must be pressed into the cylinder ten inches

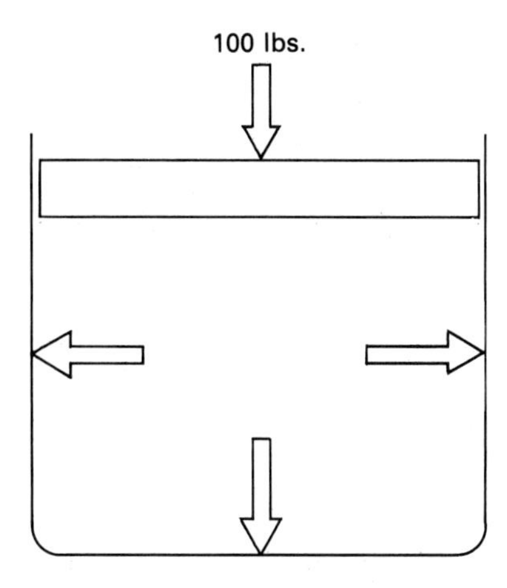


**Figure 22.1** Pascal's law

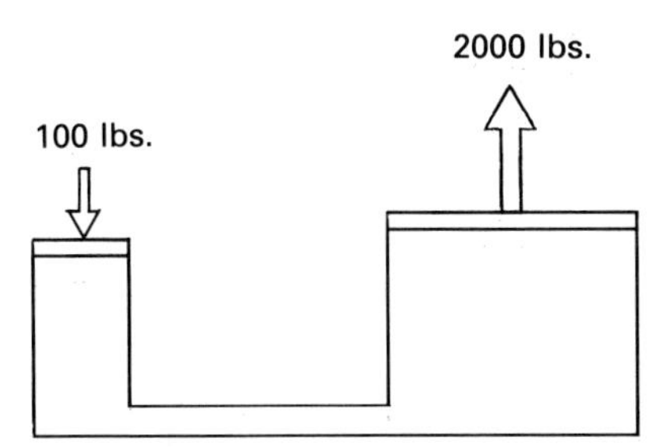


**Figure 22.2** Transmitting power with Pascal's law

to raise the load piston a distance of one inch! The total work done will be the same on both ends of the system (ignoring friction).

## Example 22.1

A cylinder with an area of 0.5 square inches has a force of 20 pounds applied to it. A load cylinder with an area of 2 square inches is connected into the system. Find the pressure in the system in PSI, and the load on the load cylinder for equilibrium.

**Solution:**

$$\text{PSI} = \frac{\text{force}}{\text{area}}$$

$$\text{PSI} = \frac{\text{20 lbs.}}{0.5}$$

$$\text{PSI} = \text{80 lbs./square inch}$$

$$\text{load} = \text{area} \times \text{PSI}$$

$$\text{load} = 2 \times 80$$

$$\text{load} = \text{160 lbs.}$$

## 22.2 CREATING THE FLUID ENERGY

While we may create a mechanical advantage by applying external force in a hydraulic jack, the energy used in hydraulic and pneumatic systems in industry is created using *compressors* for pneumatic systems and *pumps* for hydraulic systems.

A compressor converts mechanical energy from an electric motor into compressed gas, usually air. The compressed air is stored in a tank and it is usually distributed throughout the plant through pipes and hoses.

Hydraulic systems use a pump to create a flow of hydraulic fluid through pipes and hoses to the actuators on the machinery. Accumulators, devices that are similar to the tanks of a compressed air system, are used to absorb pulses in the pressure and smooth the application of power.

Hydraulic and pneumatic systems must be guarded against the buildup of too much pressure in the system. A pressure sensor is used to control the motor for the pump or compressor. When the pressure falls below the setpoint, the motor is turned on. When the pressure reaches a second setpoint, the motor is turned off.

Safety requires a second level of protection, as well. This is met with a pressure control valve.

In Figure 22.3, we see a cross-sectional view of a pressure control valve. The primary or inlet port is connected to the pressure side of the system. System pressure is felt at this port by the spool piston inside the cylinder. When the system pressure becomes greater than the pressure applied by the spring, the valve opens and the fluid is allowed to pass from the inlet to the outlet of the valve.

When used as a pressure relief valve in a hydraulic system, the hydraulic oil is returned through lines into the reservoir. When used in a pneumatic system, the excess compressed air is simply allowed to escape into the atmosphere. These relief valves are very important parts of a system. Their function should never be interfered with.

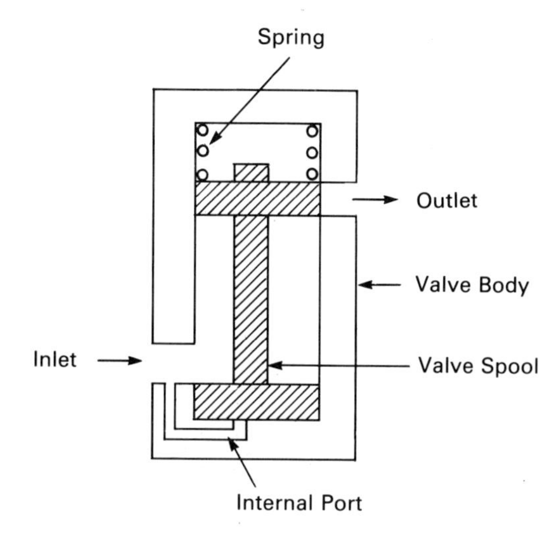


**Figure 22.3** A pressure control valve

## 22.3 ACTUATORS

The *actuators* in the system convert the fluid energy into mechanical energy which performs the actual work. There are two main types of actuators: those that create linear motion and those that create rotary motion.

The cylinders in Figure 22.4 convert system pressure into linear motion. When pressure is applied to the inlet port in Figure 22.4a, it forces the movable piston to the right. This, in turn, moves the piston rod. The vent behind the piston allows air to escape.

The cylinder in Figure 22.4b has a *return spring*, which forces the cylinder back into its inactivated position when system pressure is removed.

The cylinder in Figure 22.4c is a *double-acting cylinder*. This port has two inlet valves, port $A$ and port $B$. System pressure is applied at port $A$ to move the piston to the right, and at port $B$ to move the piston to the left.

The motor in Figure 22.5 will convert fluid pressure into rotary motion. Fluid pressure at the inlet port applies pressure on the upper vane, forcing it to move. The vane's motion turns the shaft clockwise, which brings the second vane up into the flow of fluid. In this manner, continuous rotation is maintained.

Thus, using fluid power systems, we can provide a machine with linear or rotary motion, as needed.

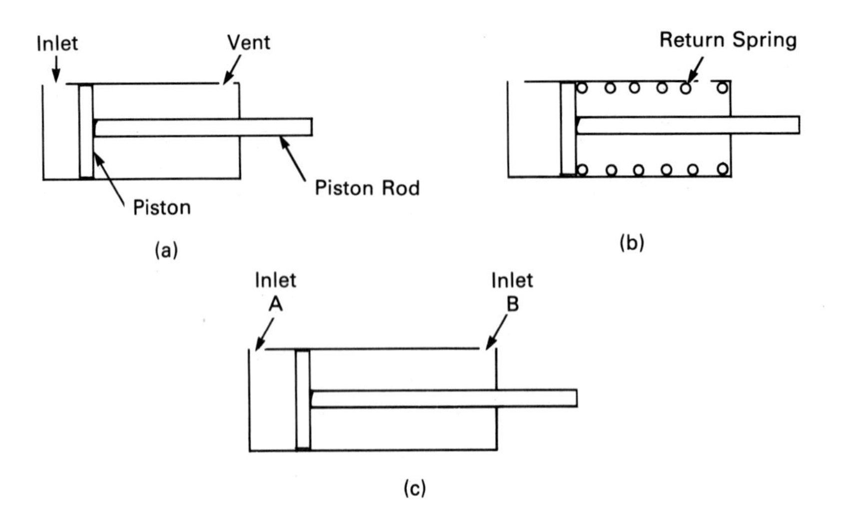


**Figure 22.4** Actuator cylinders

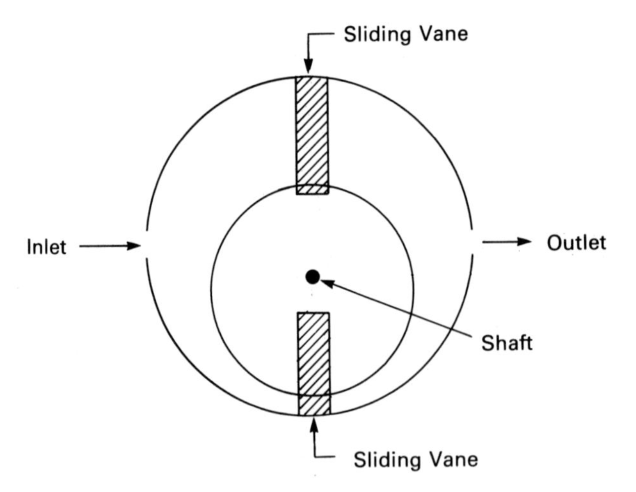


**Figure 22.5** A simplified fluid motor

## 22.4 CONTROLLING FLUID POWER

Fluid systems contain many valves for the control of the flow of the fluid drive power. While it is not the task of an electronic technician to repair these valves, it is useful to be familiar with them.

Figure 22.6 shows the cutaway view of a *check valve*. The check valve is the fluid analog of a diode: it allows the flow of fluid in only one direction. The spring in this check valve is rated at a low pressure, simply to hold the valve closed. When fluid pressure on the inlet side of the valve is greater than the pressure on the outlet side, the valve opens and fluid can flow. When the pressure on the outlet side is greater, the valve closes. At no time can fluid flow *backward* through a check valve.

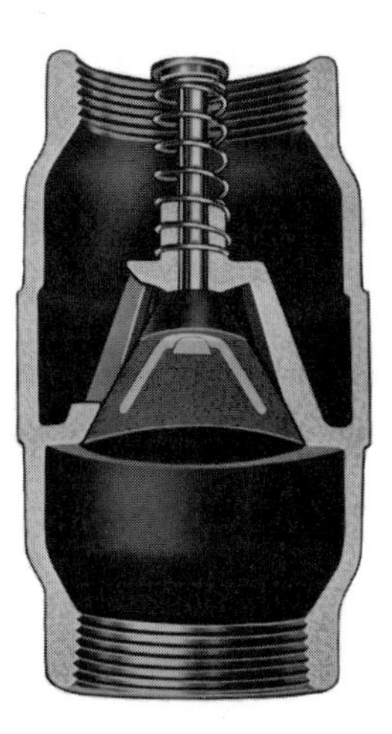

**Figure 22.6** Cutaway view of a check valve

*Direction control valves* control the direction of fluid flow in a system. The valve illustrated in Figure 22.7 is a pilot-operated four-way direction control valve. The actuator spool is moved to the right or left in this valve by *pilot pressure* from the fluid power supply.

When the spool is moved to the left, system pressure from the inlet port is directed to outlet port *A*. The fluid from port *B* is allowed to return to the reservoir. When the spool is moved to the right, system pressure is directed to port *B* and port *A* is allowed to drain to the reservoir.

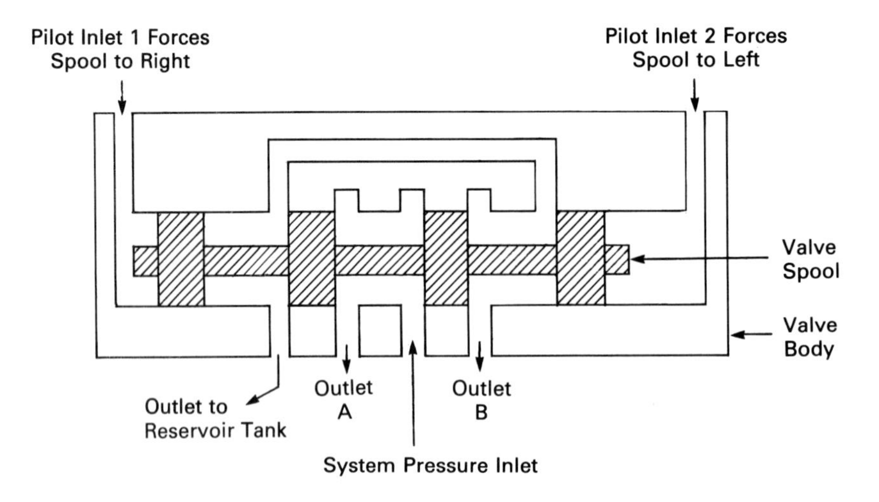


**Figure 22.7** A pilot-operated four-way valve

The valve shown in Figures 22.8a and b accomplishes the same task using solenoids to operate the valve. The system controller supplies the electrical signal that operates the solenoid.

The pilot-operated valve was shown first because solenoid valves are limited in their ability to control great amounts of pressure such as may be found in industrial systems and larger robots. In this case, the electronic controller controls a small solenoid valve, which, in turn, supplies pilot-operating pressure to the pilot-operated valve. This two-step operation is illustrated in Figure 22.9.

**Figure 22.8a** A solenoid-operated four-way valve *(Courtesy of Schrader Bellows Div.)*

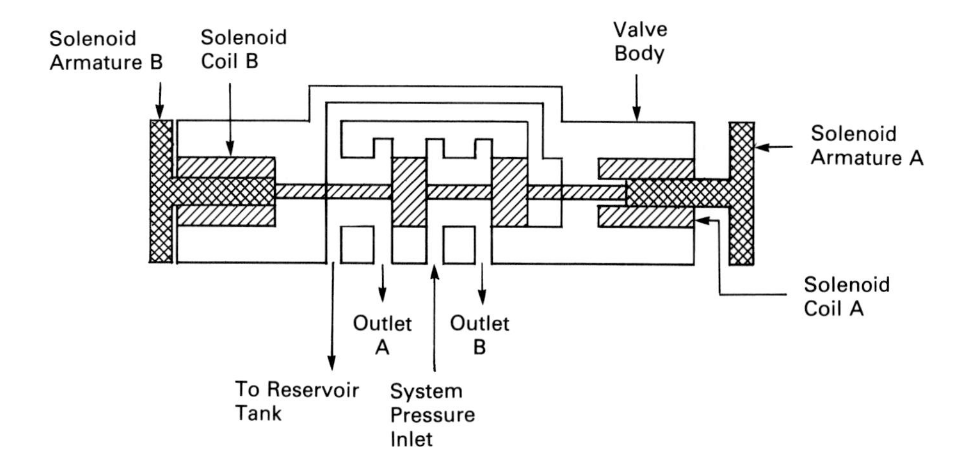


**Figure 22.8b** Diagram of the solenoid-operated four-way valve

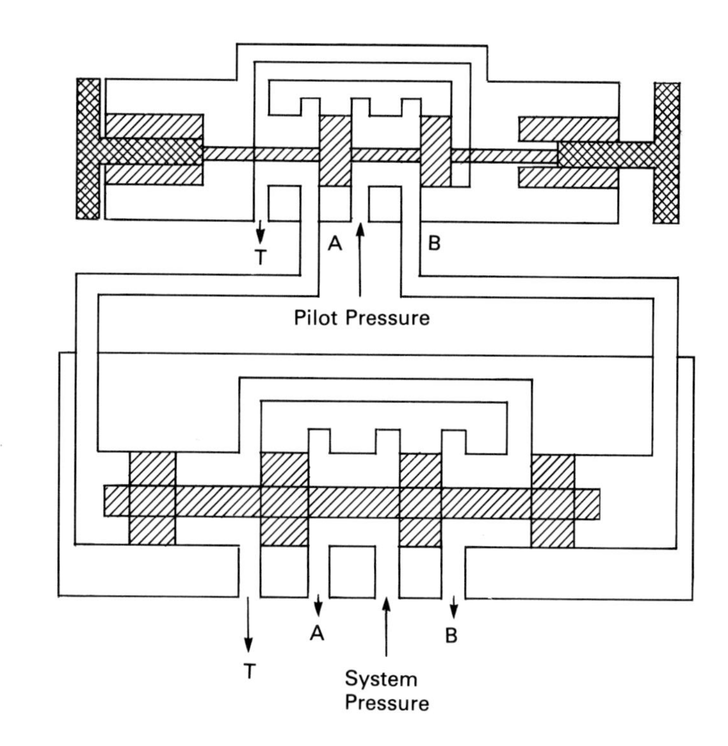


**Figure 22.9** Using a solenoid four-way valve to control a high-pressure pilot-operated four-way valve

## SUMMARY

In this chapter, we have looked at a few of the basics of fluid power. We have seen that:

1. Fluid power is convenient because the source of the power can be located remote from the application.
2. In pneumatic systems, the power is transferred through compressed air.
3. In hydraulic systems, the power is transmitted through hydraulic oil.
4. Hydraulic systems are used for heavier work because the hydraulic fluids are *incompressible.*
5. Electrical control of hydraulic power is accomplished by using solenoid-operated valves.

## SELF-TEST

1. The working system pressure in a pneumatic system is provided by a(n) ______________________.
2. The working system pressure in a hydraulic system is provided by a(n) ______________________.
3. A(n) ______________________ valve is used to prevent excessive system pressure.
4. A(n) ______________ converts system pressure into linear movement.
5. A(n) ______________________ valve allows fluid to flow in only one direction.
6. A(n) ______________________ valve allows fluid to be sent to one of two different ports.
7. A(n) ______________-operated valve is controlled by fluid pressure.
8. A(n) ______________-operated valve is controlled by electrical signals.

## QUESTIONS/PROBLEMS

1. A cylinder with an area of 1.5 square inches has a force of 100 pounds applied to it. The load cylinder in the system has an area of 10 square inches. Find the pressure in the system in PSI and the load on the cylinder for equilibrium.
2. A pneumatic system is operated at 150 PSI. What size load cylinder will be required to support a load of 3000 pounds?
3. A hydraulic system must apply 1200 PSI of pressure to operate a press. The load cylinder is eight inches in diameter. Determine the pressure setting for the pressure control valve.

# Troubleshooting Actuators

## DC MOTORS

Much of the troubleshooting and repair of the actuators described in Section III is beyond the scope of the technician. However, the equipment is controlled by electronic controllers, and the technician is often the one who discovers the problem. As an industrial electronics technician, you should be able to recognize problems and potential problems with these devices.

One key to the longevity of all rotating machinery is proper lubrication. The large motors used in industry, ac, dc, hydraulic, or pneumatic, require periodic lubrication. Many of them are fitted with *grease fittings* much like those on your car. They require greasing with a pressure gun. Find and remove the pressure relief plug on the bottom of the enclosure before greasing the bearings on these motors. Clean the grease fitting and the surrounding area before connecting the grease gun. The grease should be added one shot at a time with the motor running. Stop when grease begins to come out of the pressure relief valve. Allow the motor to run for five or ten minutes before replacing the valve plug. Avoid overgreasing! Too much grease in a bearing will increase friction.

Smaller motors will have *sleeve bearings* instead of the ball bearings of the large motors. These motors will have a ring-oiling system, a yarn-packing oiling system, or a porous-bearing oiling system. Both the ring-oiling system and the yarn-packing oiling system require periodic addition of oil. These motors will have oil filler caps and oil reservoirs. The reservoirs should be filled with the type and grade of oil recommended by the motor manufacturer. The porous-bearing system is used on small motors. There is no oil filler cap and no reservoir for adding oil. *Do not lubricate these bearings.* The motor end bells contain a wick that is saturated with a special oil for these bearings. The supply in the wick will last the life of the motor. Added lubrication will cause the bearings to clog and will destroy the lubrication.

Belt-driven machinery can generate static electricity, which can cause a current flow between the motor shaft and the bearing. This current flow will break down the oil film between the bearing and the shaft and can cause rapid wear to both. Proper grounding of the motor will prevent this problem.

The armature of a dc motor is subject to more abuse than any other part of the motor. Any overload on the motor will cause the armature to draw more than its rated current. Every time a motor is overloaded, the armature will be damaged to some extent. Frequent overloading will result in a burned-out armature. Less frequent overloading may result in a shorted coil. This condition can be discovered by applying a voltage across a pair of commutator bars which are several bars apart. Use a DMM to read the voltage drops across adjacent pairs of commutator bars. All of the voltages should be nearly equal. A wide variation in voltage drop indicates a shorted coil.

The commutator and brushes of a dc motor are another source of problems. Recall, the commutator is a group of copper wedges separated from each other by mica insulation. The mica should be below the surface of the bars by a distance about equal to the thickness of the mica. If the copper wears, the mica may protrude above the bars. In that case, the mica must be *undercut*. Undercutting is not difficult, but a poor job can ruin a motor. If you have not been trained for this, do not do it.

The interface between the commutator and the brushes should not be lubricated. The only suitable lubricant in this area is the copper oxide film which forms naturally through the use of the motor. The film may be any color from light tan to dark brown. A good film will have a soft, satin-like sheen. Oil or grease trap dust and other contaminants that will cause a high resistance glaze on the commutator. This glaze is hard and shiny, and is not hard to distinguish from the natural copper oxide film. The glaze must be removed. It causes brushes to chatter and break. This is a job for a motor specialist.

The brushes must be changed as they wear. This is more than a matter of pulling out the old ones and popping in new ones. The brushes must be *faced* so that their curve matches the curve of the commutator exactly. Install the brush in its holder and draw a piece of coarse sandpaper between the brush and the commutator. Finish with finer sandpaper. Set the brush pressure according to the manufacturer's recommendations.

Some sparking is inevitable between the brushes and the commutator. Excessive sparking, however, can cause serious problems. Worn brushes, glazed commutator bars, incorrect brush pressure, and high mica are among the causes of excessive sparking. If you notice excessive sparking in any motor, it is a sign of trouble. Report it to the appropriate supervisor for repair.

## AC MOTORS

A single-phase capacitor start ac motor that becomes sluggish in starting, or will not start unless the shaft is twisted, probably needs a new starting capacitor. The size of the original capacitor was selected to match the motor's characteristics. Replace the capacitor with one of the same type and capacitance. Do not substitute electrolytic capacitors for oil-filled capacitors, or oil-filled capacitors for electrolytic capacitors. Electrolytic capacitors are used for starting only, and must not be used as run capacitors on two value capacitor motors or permanent-split capacitor motors.

Both capacitive start and resistive start motors contain a centrifugal switch which may occasionally fail. The failure may be caused by an accumulation of dirt or by contact welding. A thorough cleaning and a burnishing of the contacts will probably fix the problem.

The most common three-phase ac motor is the induction motor. It has no brushes to wear, no commutator or slip rings, and needs no starting capacitor. Short of severe overload without protection, it is hard to damage a three-phase induction motor. The only regular maintenance needed is periodic lubrication. The heaters in thermal overload protectors may need replacement if the overload frequently turns off the motor.

Wound rotor ac motors have slip rings and brushes that require maintenance. The slip rings are smooth, without the mica spacers of a commutator. Brush replacement is the most common electrical maintenance needed for these motors.

## HYDRAULIC AND PNEUMATIC SYSTEMS

Hydraulic and pneumatic cylinders have seals which must be replaced when they wear. Wear is detected by leakage beyond the expected norm and a drop in system pressure when the cylinder is actuated. Hydraulic leakage is obvious by the oil that collects around the cylinder. Pneumatic leakage past cylinder seals is less obvious because the hiss of air is lost in the sound of the cylinder operating. Many shops replace cylinder seals according to a schedule of preventive maintenance.

The vanes in vane-type motors are subject to wear, and will eventually fail to make a good seal with the housing. This is shown by a loss of motor power. The vanes must be replaced.

Hydraulic and pneumatic systems must be checked frequently for leaks in hoses, pipes, and fittings. Remember that a pinhole leak jets a stream of fluid which cuts like a sharp knife. A Navy veteran related that aboard his ship the crew searched for leaks with a broom handle. A pinhole leak would cut the broom handle in half! Do not search for leaks with your fingers.

Cavitating pumps send a penetrating squeal throughout the system. The pump must be turned off at once to avoid damage. If the system was working before the cavitation occurred, check for leaks in the intake line or the pump gaskets. The purists will tell you this is *pseudo-cavitation*, but the effect is the same — the pump is destroyed. If the cavitation occurs during start-up of a new system, it may be true cavitation because the pump was not selected correctly. Check for the same leaks, then call the engineer who designed the system. The only repair for true cavitation is the correct pump.

Once again, most of the repair and troubleshooting ideas in this section are beyond the normal duties of an electronics technician. A large plant will have maintenance people who are trained to do this work, or will employ outside vendors for the jobs. A smaller plant may require that some of these jobs be performed by the electronics staff. Use manufacturer's maintenance manuals when available, and seek advice and guidance from more experienced coworkers.

# Section V

# Controllers

In the preceding sections of this text, we have seen how the results of processes are measured, how the signals from the measuring devices are processed and transmitted, and how an actuator can be used to vary the process variable. In this section, we will come full circle and close the process control loop, which we opened in Section I.

The process controller may vary from a simple arrangement of relays to a mainframe computer. In all cases, however, the idea is the same: allow a measurement of the process to control the operation of the process. The process controller makes the decisions. It decides when to alter the process, by how much, at what rate, and for how long. The process controller is the system brain, in other words.

We have already seen some examples of process controllers in Section III. The operational amplifiers in that section are used not only to process the error signal, but often also to output the correction signal, which corrects the error. In this section, we will examine the concept of the controller more deeply.

# 23

# Relay Controllers

The electromechanical relay is one of the most often used control devices in industry. While at first glance the relay may seem too simple to make decisions about process control, we will see that in its many forms, the relay can be a sophisticated controller.

In this chapter, we will examine many different types of relays and see how they are used in control circuits.

## OBJECTIVES

You will have successfully completed this chapter when you can:

1. Describe the basic operating principles of the relay.
2. Explain the different ways that relay manufacturers describe the contact configurations of their relays.
3. Read and decipher a ladder diagram.

### *23.1 THE RELAY*

The relay is best defined as an electromagnetic switch. In its simplest form, a relay is a solenoid-operated switch.

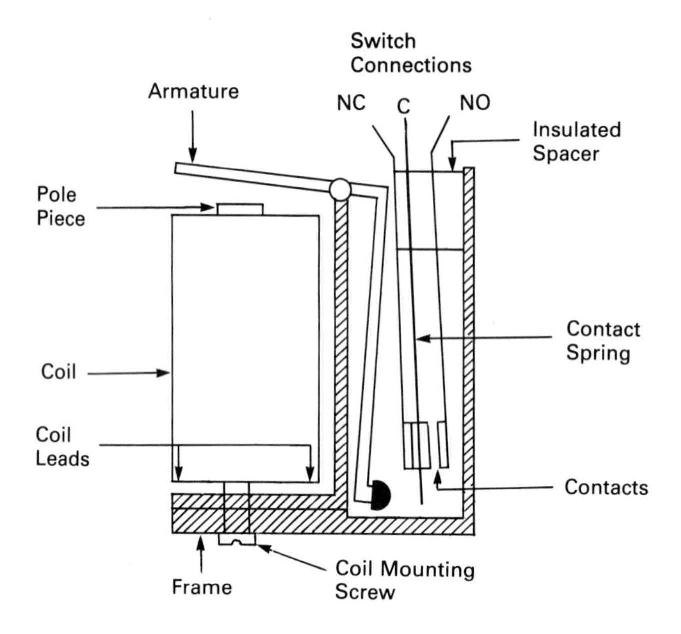


**Figure 23.1** An open-frame relay

Figure 23.1 illustrates the basic parts of a relay. The coil is mounted in the frame of the relay. The armature of the relay is attracted to the electromagnet when the coil is energized. As the armature moves toward the pole piece, it operates the relay contacts. When the coil is deenergized, the return spring pulls the armature back to the off position. It also deactivates the switch contacts.

There are many different ways to make a relay. The type of relay shown in Figure 23.1 is the common *open-frame relay*. There are also relays that are operated by a solenoid plunger. These relays are used for heavier loads. Regardless of construction, the operation is the same.

Figure 23.2 shows some of the common schematic symbols used for relays and their contacts. Figure 23.2a shows the symbols commonly used in electronics and communications diagrams. Figure 23.2b shows the symbol used in telephone circuit diagrams. Figure 23.2c shows the symbols used in industrial control diagrams.

Note that the contact symbols in Figure 23.2c look like one version of the electronics schematic symbol for a capacitor. While this could be a cause for concern, you will rarely encounter both symbols in the same diagram.

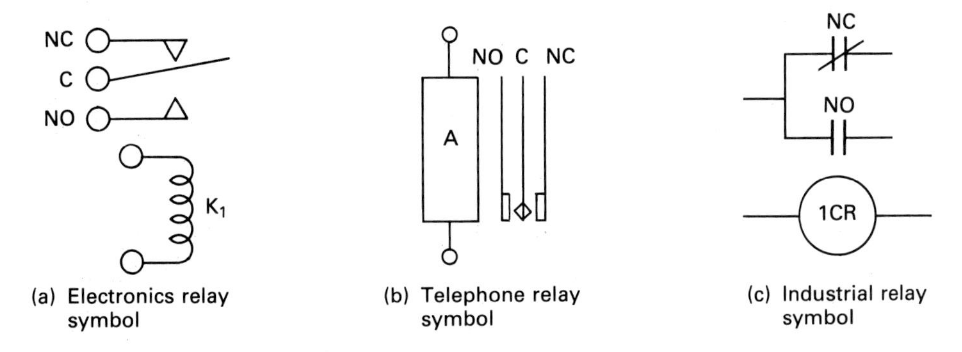


(a) Electronics relay symbol

(b) Telephone relay symbol

(c) Industrial relay symbol

**Figure 23.2** Relay schematic symbols

Relays in industrial control circuits most often plug in or have screw terminal connections, which makes changing them fairly easy. Most control relays are enclosed in plastic or metal cases to help reduce contamination of the contacts by airborne oil and dust. Dirty contacts interfere with reliable operation of the relay. Relay contacts may be cleaned with denatured alcohol and a relay burnishing tool, but in cases of doubt, replace the relay. Industry is more interested in reliability in control systems than in saving a few dollars.

When they carry large currents, the contacts have a tendency to arc. They will occasionally even weld themselves closed. When this happens, the only cure is replacement of the relay. No amount of filing and burnishing can make the relay reliable after its contacts have burned or welded. With some industrial relay types, it is possible to replace the contact set instead of the whole relay. This can be a big time saver.

When replacing relays, it is important to match the relay with the task at hand. Relays are rated according to contact voltage and the current-carrying capacity of their contacts. A 5-amp relay will not do the job of a 10-amp relay, even for a short period of time. Most systems use NEMA-rated contacts at either 300 V or 600 V, at 10 A. Larger current ratings are available under the name *contactor.* Another important rating for a relay is the *coil voltage.* A plant will usually standardize on one single voltage for control systems, but an odd voltage occasionally shows up in turnkey equipment. The plant maintenance department should keep a stock of replacement relays on hand to avoid long downtimes.

## 23.2 LADDER DIAGRAMS AND SYMBOLS

The *ladder diagram* has become the standard diagram used in relay control circuits. This type of diagram is found in control systems ranging from an automatic washing machine to a railroad locomotive. At first, it may look strange to the electronics technician, but the symbols used are simple and the reading of the diagram is equally simple.

The symbols found in Figure 23.3 are some of the common symbols used in ladder diagrams. Many of them, you will notice, are pictorial drawings of the device they represent. Others are identified by the letter symbols that appear with them. Most of these symbols are self-explanatory, but a few require some explanation.

Any motor used in industry requires protection from overload. The *thermal overload protector* protects the motor from long-term overload, which would cause the motor to overheat. It does not, however, provide protection against *short-circuit current overloads*, which might occur. In this type of overload, the motor would suffer extensive damage before the thermal protector began operating. Such overloads are readily detected by the *magnetic overload protector.* Both types of protectors have two elements: a *sense* element and a *switch* element. The sense element symbols are shown in Figure 23.3. The switch elements are NC switches that open when the fault condition is sensed.

The many different types of relays and line contactors are identified by the letter symbols that appear with them. A standard control relay is identified by the letters *CR*. For example, *CR*3 represents control relay number three in the circuit. Line contactors are simply heavy-duty, high-current relays. They are identified by the letters *LC* or *AC* for line contactor or auxiliary contactor.

The ladder diagram in Figure 23.4 will introduce us to the basics of reading ladder diagrams.

At the top of the diagram, we find a 480-volt three-phase power source on lines *L*1, *L*2, and *L*3. Connected to that power source is a *control transformer.* Note the terminal designations of the transformer. These are standard designations. The transformer has two identical primary windings. The terminals for the first winding are *H*1 and *H*2. The terminals for the second winding are *H*3 and *H*4. *H*1 is the winding start for the first primary. *H*3 is the winding start for the second primary. For the 480-volt connection, the two primaries are wired *in series.* In the next ladder diagram, we will see how to connect the same transformer to a 240-volt power source.

The control transformer serves to step the supply voltage down to 120 volts, which is standard *rail voltage* in industrial controllers. It is possible that you will also find other voltages used for the rail voltage. For example, either a 12-volt system or a 24-volt system is used in your central heating unit. Control relays are available in coil voltages ranging from 12 volts to 460 volts. The most common, however, is 120 volts.

In the secondary line of the transformer, we find a fuse, which protects the transformer from accidental shorts on the secondary. Following the fuse is an on-off switch, then power rail number 1. The other side of the secondary is connected directly to power rail number 2. These two power rails form the uprights of the ladder. The rungs of the ladder contain the balance of the control circuits. In our example, the top rung contains only a red pilot lamp, which lights when the switch is on.

The second rung of the ladder, reading from left to right, connects first to a set of normally closed relay contacts, labeled 1*CR*2. This means that they are the first contact set of control relay number 2. When we speak of switches and contacts, *normal*

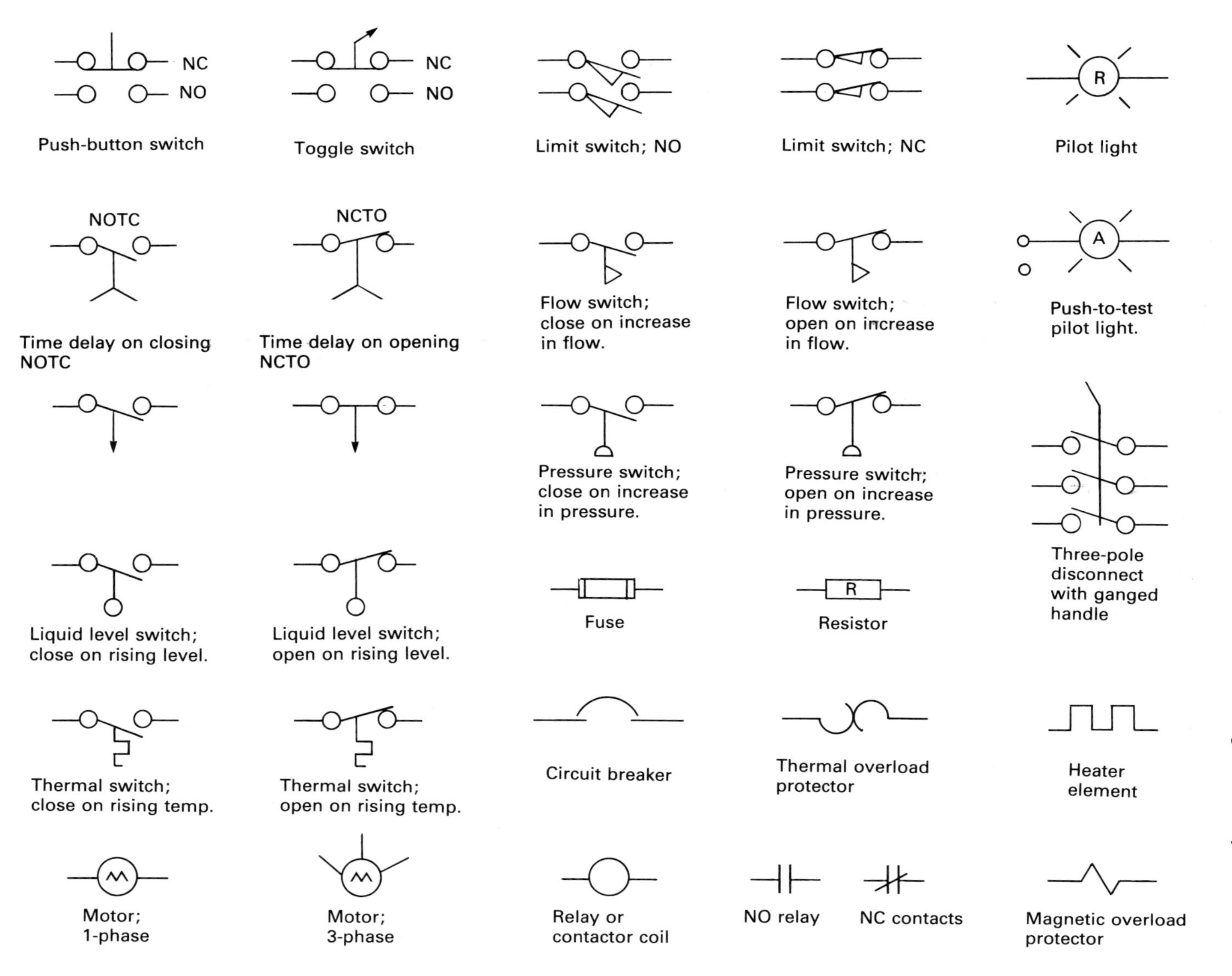


**Figure 23.3** Ladder diagram symbols

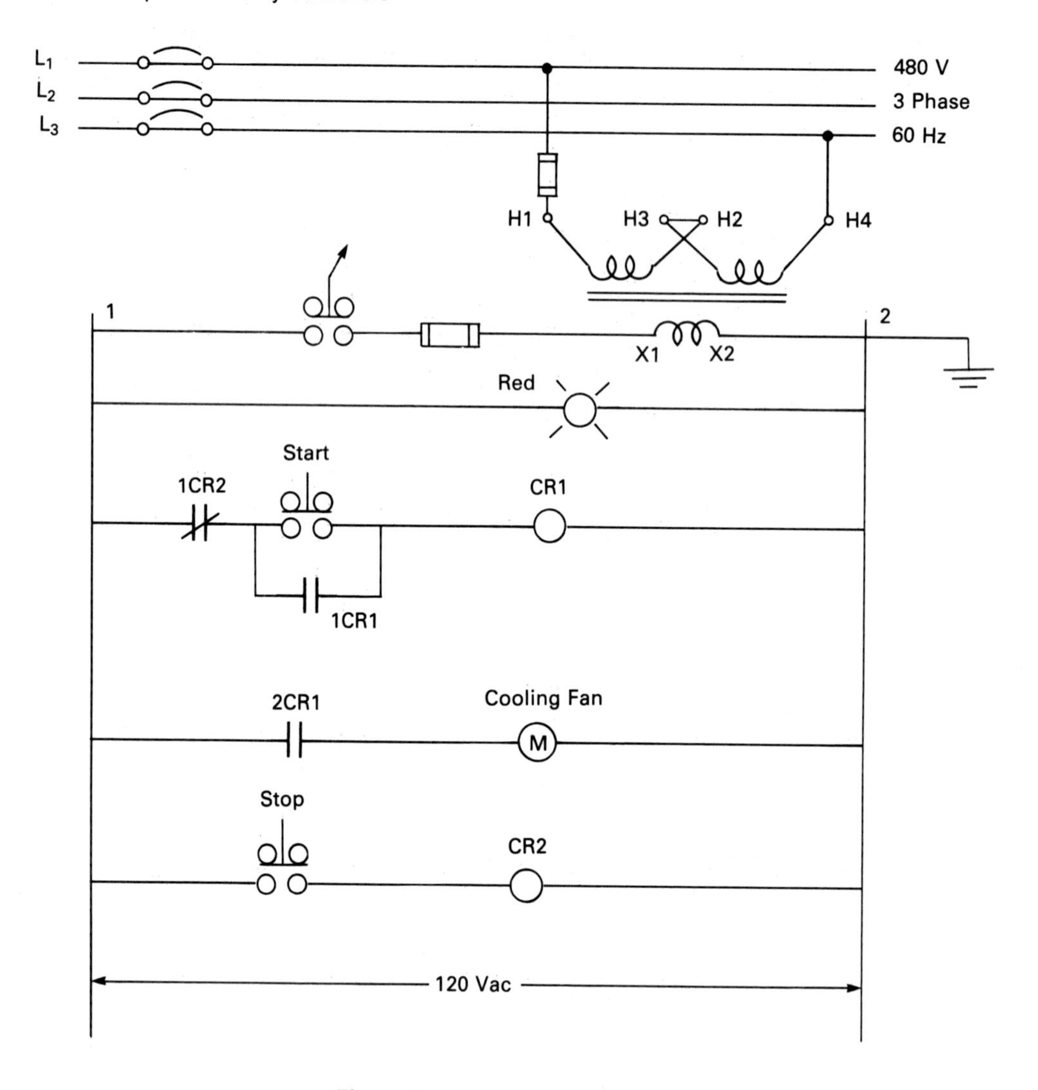


**Figure 23.4** A simple ladder diagram

means not activated. Thus, when control relay number 2 is not activated, these contacts are closed.

Following 1*CR*2, we find the normally open contacts of a momentary-contact push-button switch, labeled *start*. Note that these contacts are shunted by a set of normally open contacts, 1*CR*1, from control relay number 1. Finally, we have control relay *CR*1. When the start button is pressed, contact will be made from rail number

1, through the NC contacts 1*CR*2, through the start button contacts, through relay *CR*1, to rail number 2. The coil of *CR*1 will be actuated. This action will close the NO contacts 1*CR*1, placing a path in parallel with the start switch, holding the relay operated when the switch is released. This *self-latching* operation is quite common in control relay circuits. You may also hear this called a *seal-around* or a *hold-in* circuit.

In the third rung of the ladder, we find a set of NO contacts, 2*CR*1, and a 120 V fan motor, *M*. Thus, when the start button is pressed, the operation of *CR*1 will turn on the 120 V motor.

In the fourth rung, we find a set of NO contacts from a second momentary contact push button, which is labeled *stop*. Next in line is the second control relay, *CR*2. When the stop button is pressed, relay *CR*2 will be actuated. This will open the NC contact set 1*CR*2 on the first rung, releasing relay *CR*1. When *CR*1 releases, contacts 1*CR*1 will open, releasing the self-latch of *CR*1. Contacts 2*CR*1 will also open, turning off the fan motor.

As you can see, it is far easier to read a ladder diagram than it is to describe its operation.

The ladder diagram in Figure 23.5 is a more elaborate ladder diagram in a common application, a motor starter circuit. The field coils of large motors are made up of a few turns of very heavy wire. They offer a very low ohmic resistance to the ac line, allowing excessive amounts of current to flow in the circuits when the motor is switched on. The currents will automatically go down once the inductive field is built up around the coils, but some means must often be found to reduce the surge currents when the motor is started.

The power lines in this illustration are 240-volt three-phase lines. Note the connection of the control transformer for 240-volt operation. Here, the two primaries are wired *in parallel*. Note that the start leads of both primaries are connected together to one phase of the power line, and the end leads of both primaries are connected together to a second phase of the power line. This connects the two primaries in parallel-aiding. If the *H*3-*H*4 connections are inadvertently reversed, the secondary voltage will be zero.

Let's follow the sequence when the start button is pressed. The start button completes the circuit to the high-power control relay *LC*, actuating *LC*. *LC* is a special high-current relay called a *contactor*. When *LC* is actuated, contacts *LC*1, *LC*2, and *LC*3 close, connecting the motor leads to the 240-volt power line. The resistance in each leg of the motor limits the surge current in the field windings to a safe level. Auxiliary low-current contacts *LC*4 also close, actuating the time delay relay, *TDC*. A second set of auxiliary contacts, *LC*5, closes in parallel with the start button, holding *LC* operated when the button is released. Another set of auxiliary contacts, *LC*6, also closes in the path to contactor *AC*, preparing that path for a later operation.

Relay *TDC* is a time delay relay, which will not operate until a preset time has elapsed after the application of power. When the preset time has passed, relay *TDC* will operate. This closes the path to the second contactor, *AC*. Contacts *AC*1, *AC*2,

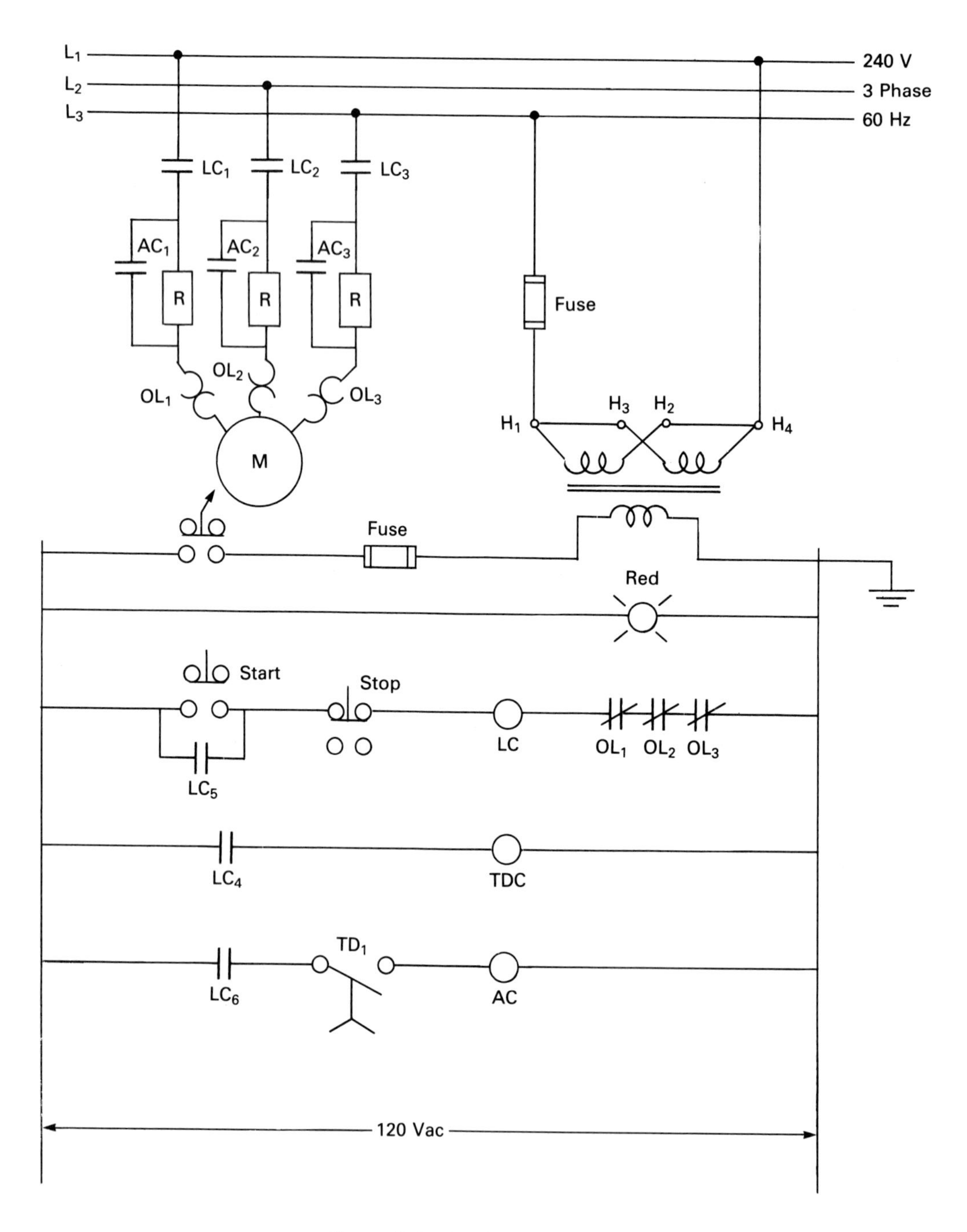


**Figure 23.5** A resistive motor starter

and $AC3$ close, bypassing the resistance and allowing full current to flow to the field coils of the motor.

Pressing the stop button opens the circuit to the line contactor, *LC*. When *LC* releases, the motor is disconnected from the line.

The thermal overload protectors mounted on the motor can also turn off the motor. Each field coil is protected by a thermal overload, and the NC contacts of the overload protectors are connected in series with the coil of *LC*. Thus, if any one of these contacts opens, *LC* will be released and power will be disconnected from the motor.

## 23.3 TIME DELAY RELAYS

The motor starter in Figure 23.5 used a time delay relay, in this case a delay-on-make relay. Time delay relays come in three basic configurations: delay on make, delay on break, and interval.

The delay-on-make relay delays operating for a preset period of time. When the preset time has passed, the relay operates. The delay-before-make relay releases when power is removed from the coil.

The delay-on-release relay operates the moment power is supplied to the coil, but it delays for a preset period of time after power is removed from the coil.

The interval relay operates as soon as power is applied to the coil, and it operates for a preset period of time. Once the preset time is reached, the relay releases.

Figure 23.6 is a time delay relay that provides all three of these options in one package. The unit has six switch-selected ranges, from 1 second to 10 hours. The

**Figure 23.6** A time delay relay *(Courtesy of Automatic Timing and Controls)*

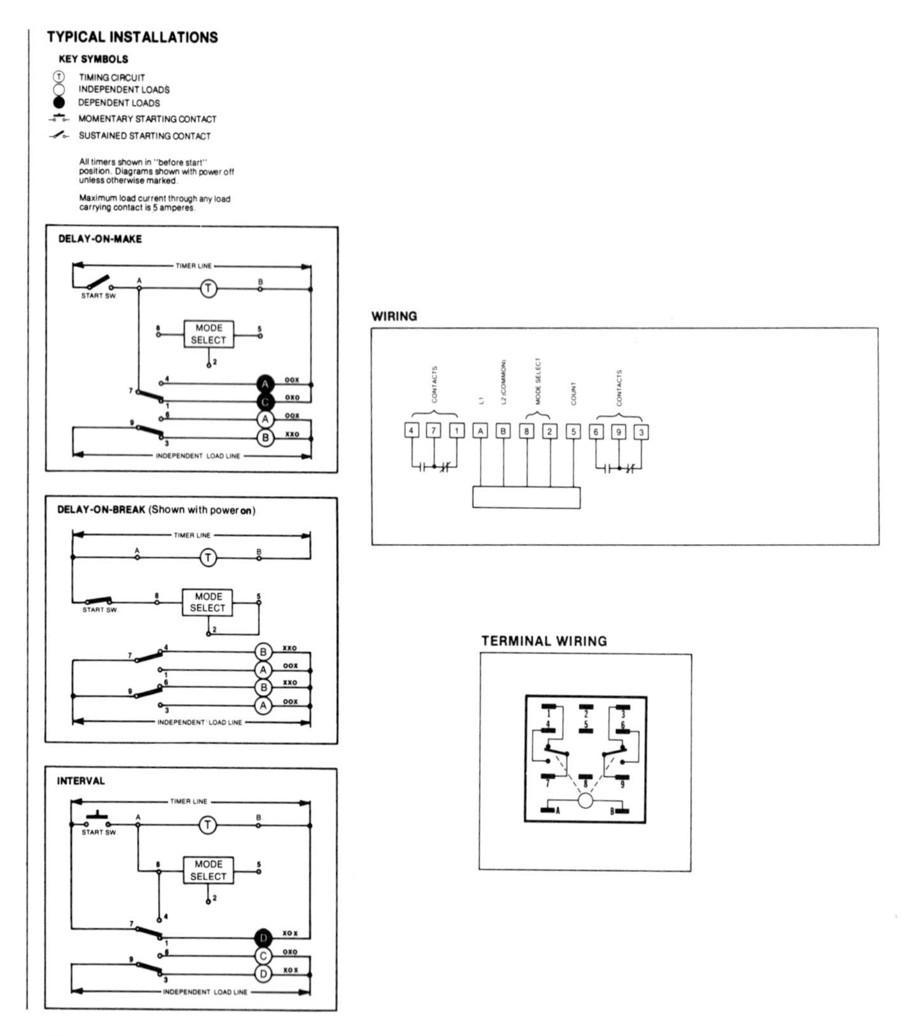


**Figure 23.7** Installation diagram of the time delay relay *(Courtesy of Automatic Timing and Controls)*

numbered dial on the bottom of the relay panel allows fractions of the switch-selected timing period to be selected.

In Figure 23.7, we see the wiring diagrams for the time delay relay of Figure 23.6. From this, we can see that the mode of operation depends upon how the mode select pins are wired. Note that the relay contact set numbered 1, 4, and 7 is *dependent* in the delay-on-make mode and the interval mode. That is because connection to this contact set is dependent on the start switch. Thus, in these modes, the relay has only one set of independent contacts. This limitation is easily overcome, if necessary, by allowing these contacts to operate a separate control relay.

## SUMMARY

In this chapter, we have looked at the use of control relays and relay controllers. We have seen that:

1. A relay is an electromagnetic switch.
2. A full set of industrial controls can be implemented using relay logic.
3. A time delay relay can be configured as delay-on-make or delay-on-break.
4. The standard diagram for relay control systems is the ladder diagram.
5. Ladder diagram symbols are different from electronic schematic symbols because they are largely pictorial rather than symbolic.

## SELF-TEST

1. When power is applied to the coil of a relay, the resulting electromagnetic field attracts the ______________________________ to the pole piece.
2. When the armature moves toward the pole piece, an actuator lever attached to the armature operates a set of ______________________________.
3. Control system relays most often have screw terminal connections or they are ______________________________ units.
4. When relay contacts are contaminated by dirt, they can be cleaned with denatured alcohol and a(n) ______________________________ tool.
5. When relay contacts are burned or welded, the best repair is ____________ ______________________________.
6. Relays are rated according to the ______________________ of their contacts.
7. A heavy-duty relay designed to switch large currents is called a(n) ______ ______________________________.
8. A(n) ______________________________ overload protector provides protection against long-term overload.
9. A(n) ______________________________ overload protector provides protection against short-circuit currents.
10. A set of relay contacts labeled NC will be closed when the relay coil is ______________________________.
11. The label *CR*3 on a ladder diagram refers to ___________________________.
12. A relay marked *TDC* is a(n) ______________________________ relay.
13. A control transformer can be wired to either 240 volt or 480 volt ac because it has two ______________________________.
14. When wired for 240 volt operation, the two primaries of the control transformer are wired in ______________________________.

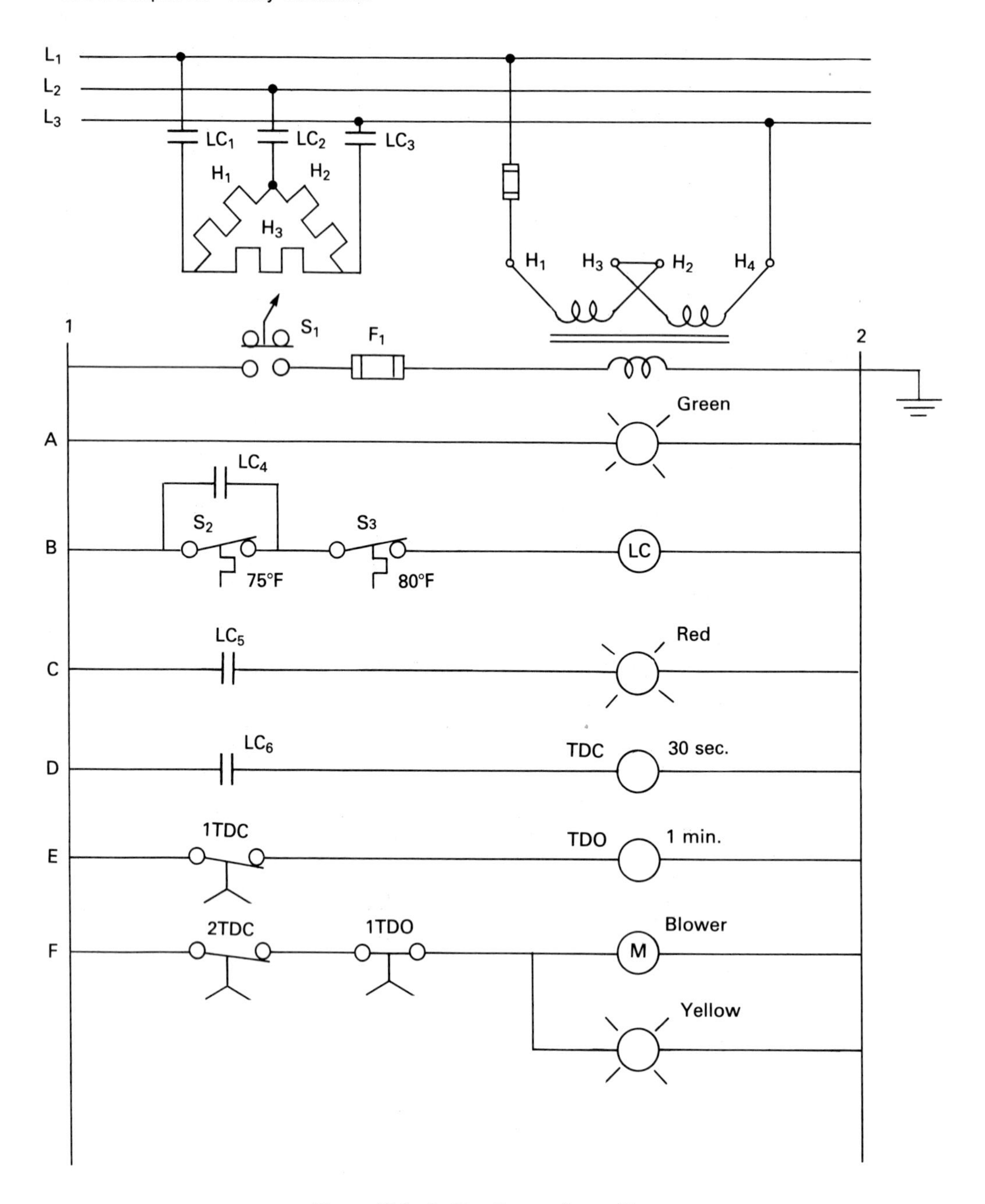


**Figure 23.8** Ladder diagram for problems

## QUESTIONS/PROBLEMS

Use the diagram in Figure 23.8 for the following problems.

1. The heater elements in the diagram are connected in a(n) ______________ ______________ configuration.
2. Switch *S*2 is a(n) ______________________________ controlled switch.
3. Switch *S*3 will open when the temperature rises above ________________ ______________.
4. When the temperature is below ____________________________ degrees, the line contactor will be actuated.
5. Relay *TDC* is a(n) ____________________________ relay.
6. When will relay *TDC* close its contacts?
7. There is an error on line *F* of this diagram. Describe the error and the correct connection.
8. Assume that contact set 1*TDO* is a NO contact set. Describe the operation of this controller.

# 24

# Analog Controllers

Relay controllers provide only on-off control: switch closures and openings. They are not capable of determining the degree of adjustment required, nor of providing proportional, integral, or differential control. In other words, they are essentially undamped systems.

In many cases, however, some form of damping of a system is required. Consider a gantry crane moving a heavy load. As the crane begins to move on its overhead track, the natural inertia of the load provides some degree of proportional control as the motor builds up to speed. As the load nears its destination, it will be traveling at full speed. A relay controller would detect the instant that the load reached its destination and turn off the motor, but because of the inertia of the load, the crane would most likely overshoot its goal. A *proportional control system* is needed here to slow the motor as the crane nears its destination.

In this chapter, we will examine some of the analog controllers that can offer such control.

## OBJECTIVES

You will have successfully completed this chapter when you can:

1. Describe a proportional control system.

2. Describe an integral control system.
3. Describe a derivative control system.
4. Explain how these three systems may be combined to create a PID system.

## 24.1 TYPES OF CONTROL

A *proportional controller* detects the *magnitude*, or size, of the error in the controlled variable and outputs a control signal to the actuator that is proportional to the amount of correction required. Since the actuator control signal becomes smaller as the process nears the control setpoint, the amount of correction being applied is reduced, and the correction process slows. Proportional control reduces overshoot.

An *integral controller* detects the *duration* of an error signal: how long the error has been present. The longer the error has existed, the larger will be the actuator control signal. Integral control helps correct for steady-state error, which is present in all control systems.

A *derivative controller* detects the *rate of change* in an error signal: how quickly the controlled variable changes. When there is a sudden change, the derivative controller outputs a large actuator control signal. When there is a gradual change, the derivative controller outputs a small actuator control signal. Derivative control is *never* used alone. It must always be used in conjunction with proportional control. The derivative controller is used to increase the response speed of a control system when sudden disturbances in the process variable are likely.

## 24.2 TYPES OF CONTROL INPUT

The transducers used in control systems are available with many different types of analog and digital outputs. Obviously, the analog-transducer outputs are the ones used with analog control systems. These may be either current outputs or voltage outputs.

Among the transducer *current outputs*, the most typical is the ISA standard 4–20 mA signal. Other ranges used on occasion are 0–5 mA, 1–5 mA, 10–50 mA, and −20 mA to +20 mA. For new designs, the 4–20 mA ISA range is recommended. Recall that the transmission of control signals via current is less susceptible to electromagnetic interference because the systems are low-impedance systems.

The transducer *voltage outputs* also cover many ranges, with far less of an effort at standardization. Among the ranges of voltage outputs are 0–30 mV, 0–50 mV, 0–100 mV, 0–4 V, 0–5 V, 0–6 V, 0–8 V, 0–10 V, −5 V to +5 V, and −10 V to +10 V.

With this wide range of possible analog control inputs, it is important to match the controller input module with the transducer signal. Probably the best way to do this is to use the ISA 4–20 mA as a standard throughout the plant. When this is not possible, an alternative is to order transducers and controllers from the same manufacturer. Virtually all of the manufacturers in the control field offer engineering assistance to their customers. When you have any doubt, take advantage of this help.

The catalogs offered without charge by manufacturers of control systems and components are good sources of information. Many of these catalogs contain encyclopedic information, far beyond a mere listing of the company's products. They include selection charts, complete descriptions of the theory of operation of components, and often much more.

## 24.3 ANALOG CONTROLLER OUTPUTS

Many of the analog controllers you will find in catalogs are designed to accept an analog input, but their output consists of either on-off control or a set of relay contacts.

On-off control outputs are usually transistor switches. The controller provides the base voltage to an NPN transistor to drive the transistor into saturation.

Figure 24.1 illustrates this type of output as diagrammed in a manufacturer's catalog. The external load must use the same ground reference as the controller, although it may use a different source of $V^+$. The $V^+$ is limited to 30 Vdc, and the current is limited to 50 mA or so.

Using such a controller to operate machinery requires additional components to isolate the controller's low-voltage dc circuitry from the high-voltage, high-current load of the machine. Plug-in relays are often used to convert the on-off output into a set of relay contacts. Later in this chapter, we will see how we can also use solid-state switching for the same task.

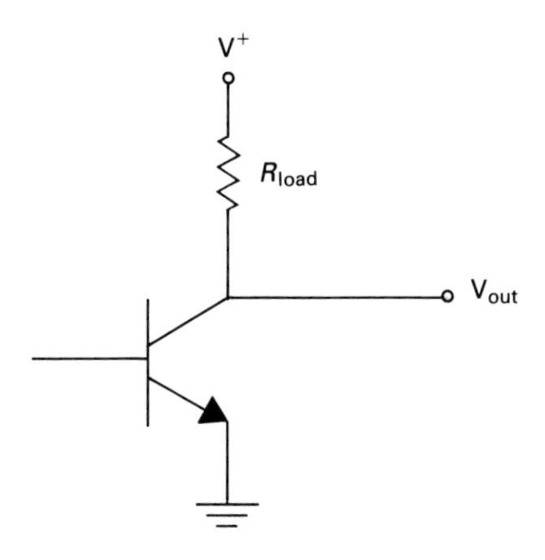


**Figure 24.1** A transistor switch on-off output

Some controllers output a 4–20 mA output signal or a low-voltage 10 mV to 50 mV output signal. This signal is used with proportional actuators, or to provide frequency control to inverters and cycloconverters. *Power regulators* are also designed to accept such inputs and output variable power to resistive ac loads such as strip heaters.

## 24.4 SOLID-STATE INTERFACING

Relay contacts suffer from airborne dirt and tend to arc and burn out over a period of time. Current technology is replacing relays with solid-state control devices, which have none of these problems.

You may think of a transistor as an amplifier, but it also makes an effective switch, as was mentioned in the previous section of this chapter. Figure 24.2 illustrates the fundamentals of switch design using transistors. In this illustration, the input to the transistor is a 5 V signal from the controller. Resistor $R_B$ limits base current to a safe value. Resistor $R_C$ is the load for the circuit. It may be a relay, a lamp, or a motor.

In designing the circuit, we must first determine the current demanded by the load, $I_C$. From this, and the $h_{FE}$ (or beta) of the transistor, we can determine the base current, $I_B$. Once this is determined, we can calculate for the value of $R_B$. We divide by two to assure that the transistor will be saturated under worst-case conditions.

### Example 24.1

Design a transistor switch for a load of 50 mA and a switching voltage of 3.5 V. The $h_{FE}$ of the transistor is stated to be between 60 and 120.

**Solution:** We will use the lower value of $h_{FE}$. This is the minimum current gain of the transistor.

$$I_B = \frac{50 \text{ mA}}{60}$$

$$I_B = 833 \ \mu\text{A}$$

Now we will calculate $R_B$ — allowing 0.7 V for the base-emitter voltage.

$$R_B = \frac{(3.5 \text{ V} - 0.7 \text{ V})}{833 \ \mu\text{A}}$$

$$R_B = 3.36 \text{ k}\Omega$$

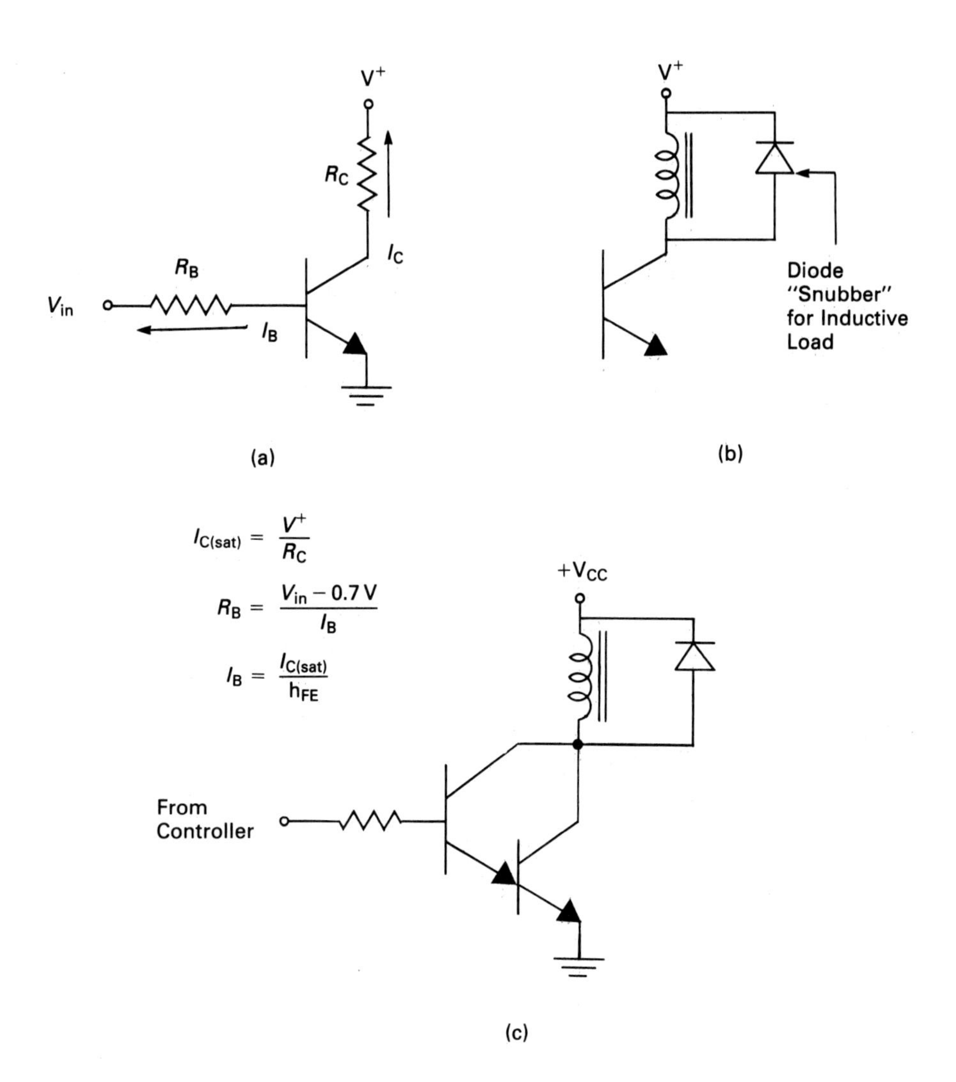


**Figure 24.2** Design procedure for a transistor switch

This is the resistance that would *just* saturate the transistor. To allow for a safety margin:

$$R_B = \frac{3.36\ \text{k}\Omega}{2}$$

$$R_B = 1.18\ \text{k}\Omega$$

The nearest standard value to 1.18 kilohms is 1.2 kilohms, which we will use.

It is unlikely that a service technician will ever be required to design such a switch, but an understanding of the design procedure will be helpful in selecting replacement parts. The design procedure emphasizes that the replacement transistor must have an $h_{FE}$ that is equal to or greater than the $h_{FE}$ of the original part.

Other specifications that must be considered are the breakdown voltage, $BV_{CEO}$, and maximum current, $I_{CO(max)}$. A safety margin of two or more must be allowed in both cases. In other words, if the circuit power is to be 24 V, the $BV_{CEO}$ should be at least 48 V. If the circuit current is to be 50 mA, then $I_{CO(max)}$ should be at least 100 mA.

When the load is inductive, the inductive pulse that is generated when current is cut off to the load can destroy the transistor switch. An additional silicon snubber diode should be connected in parallel with the inductive load to absorb this pulse. The diode is connected in reverse bias, as shown in Figure 24.2b.

The Darlington-pair connection shown in Figure 24.2c is used to increase the current gain of the transistor. For high-current loads, such as stepper motors, the controller's output current may be insufficient to drive a single transistor into saturation. Since the $h_{FE}$ of a Darlington pair is in the order of 10,000 or more, the controller need only supply a very small drive current to the base of the Darlington pair to assure saturation.

Transistor switches are essentially low-current, low-voltage dc devices. They are adequate for driving dc relays, low-voltage dc motors, stepper motors, or indicator lamps, but they are not useful for handling large-current loads, high-voltage loads, or ac loads. For these tasks, the thyristor is used.

Figure 24.3 illustrates the basic structure of a *silicon controlled rectifier,* or *SCR*. Figure 24.3a shows the four-layer makeup of the SCR, which is a PNPN "sandwich" of semiconductor material. In Figure 24.3b, we see the two-transistor equivalent circuit. Note here that the collector of each transistor is connected directly to the base of the other transistor. Thus, if some external event causes $I_{C_1}$ to increase, *that* will cause $I_{B_2}$ to increase, which will, in turn, cause $I_{C_2}$ to increase, which will cause $I_{B_1}$ to increase, and so on. The circuit has its own internal feedback mechanism, which causes it to switch on *and stay on* when it is triggered. This *latching action* is characteristic of the SCR.

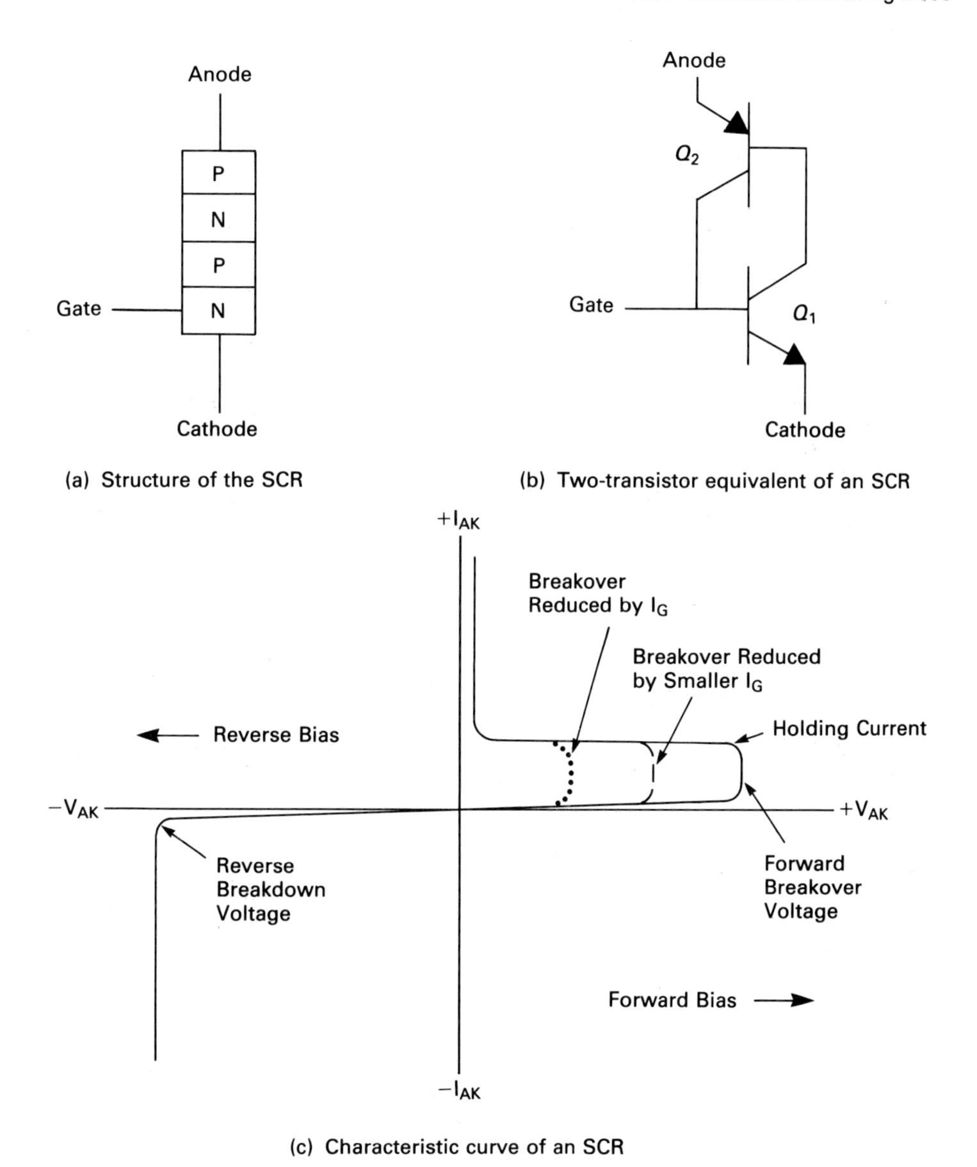


**Figure 24.3** The silicon controlled rectifier

The SCR is triggered by making the gate positive with respect to the cathode. Because of the built-in latching action, once the SCR is turned on, it will continue to conduct until something causes the current in the anode-cathode path to decrease below a specific point.

The characteristic curve in Figure 24.3c shows the basic operating characteristics of the SCR. Note that in the reverse bias connection, the SCR behaves much like an ordinary diode. As $V_{AK}$ increases, only a small reverse leakage current will flow. When the reverse breakdown voltage is reached, the current increases rapidly with the breakdown of the depletion regions within the SCR. Unless the current is limited by external components, the SCR will be destroyed by excessive heat from power dissipation.

In the forward-bias mode, as $V_{AK}$ is increased, only a small leakage current will flow. When $V_{AK}$ reaches the forward breakover voltage, the current begins to increase rapidly. But when the current reaches a critical value, called the *holding current*, the internal feedback of the SCR switches the latch on, and the forward-voltage-drop $V_{AK}$ across the SCR drops to only a few volts.

Once this has occurred, the SCR behaves like a closed switch. As long as the current, $I_{AK}$, remains above the holding current, the SCR will stay on.

The SCR is not, however, used in the breakover mode. What makes it a useful switching device is the gate connection. When gate current is allowed to flow, the SCR's forward breakover voltage is *reduced*. This is shown by the dashed line and dotted line characteristic curves in the graph. Note in the PNPN sandwich that the gate is a P-type semiconductor. In order for gate current to flow, the gate must be made positive with respect to the cathode. Because of the latching action of the SCR, once $I_{AK}$ reaches the holding current level, the gate signal can be removed and the SCR will remain on. That means that we can switch the SCR on with a short pulse from a computer or other control circuit, and it will remain on when the pulse is removed.

When selecting a replacement SCR, two important parameters must be met. The *forward blocking voltage* of an SCR is the largest voltage that can be applied to the SCR that will not exceed the forward breakover voltage. This voltage should be at least twice the circuit voltage. When the source voltage is ac, the forward blocking voltage of the SCR should be at least *three times* the applied rms voltage. The second parameter is the forward current, $I_{AK}$. This should also be allowed a safety factor of at least two.

Let's take a look at a few control circuits so that we can see how the SCR is used to control both ac and dc loads.

In Figure 24.4, we see one of the most used SCR control circuits. During the positive alternation of the ac line, capacitor $C$ is charged through the variable resistance until the charge on the capacitor is sufficient to forward bias the gate of the SCR. When gate current flows, the SCR is switched on, and current flows in the load. As the positive alternation nears zero, current in the load decreases below the holding current of the SCR, and the SCR switches off. The process repeats on the next alternation.

By adjusting the resistance in the gate circuit, we can control the firing angle of the SCR. When the SCR fires (triggers) late in the positive alternation, current can flow in the load for only a brief time before the SCR switches off. When the SCR fires early in the positive alternation, current can flow in the load for nearly the full half-cycle before the SCR switches off. This circuit is used as a lamp dimmer circuit

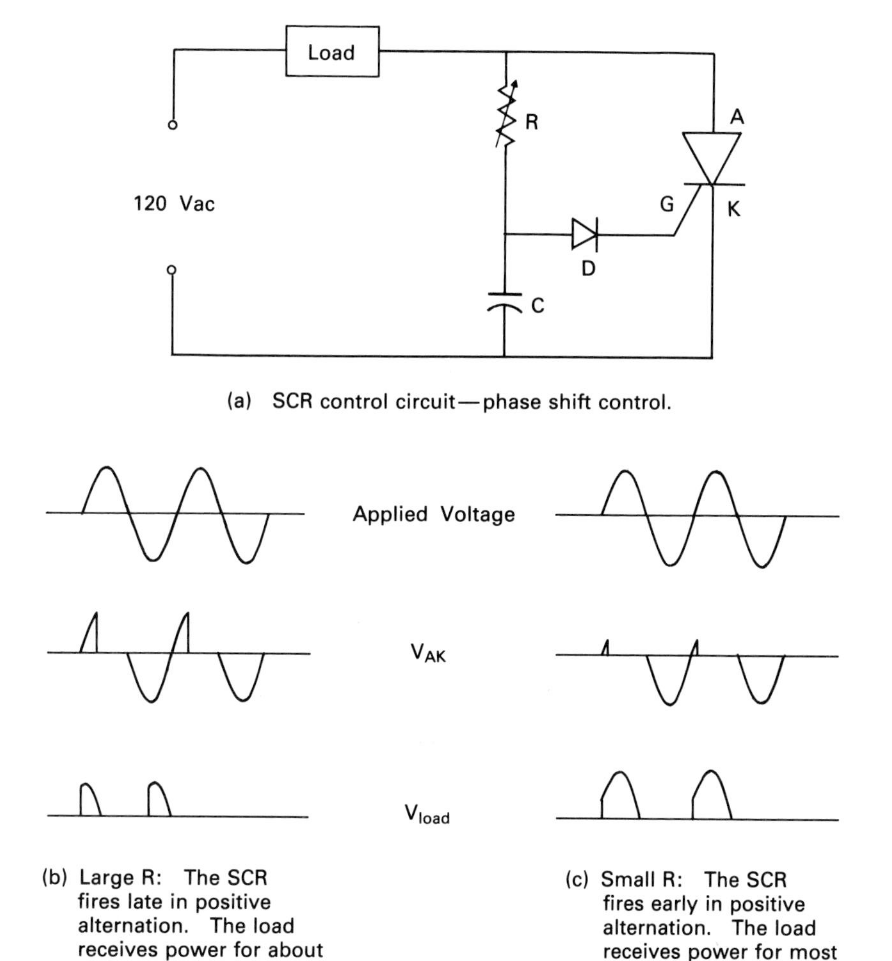


**Figure 24.4** Controlling the dc load

or to control the speed of the brush-type series wound universal motor found in small appliances and tools.

Note diode *D* in the gate circuit. It prevents any voltage from the negative alternation from reaching the gate. The gate-cathode connection of an SCR has a low reverse breakdown voltage, usually near 5V. If the gate becomes reverse biased by the line, the SCR will be destroyed.

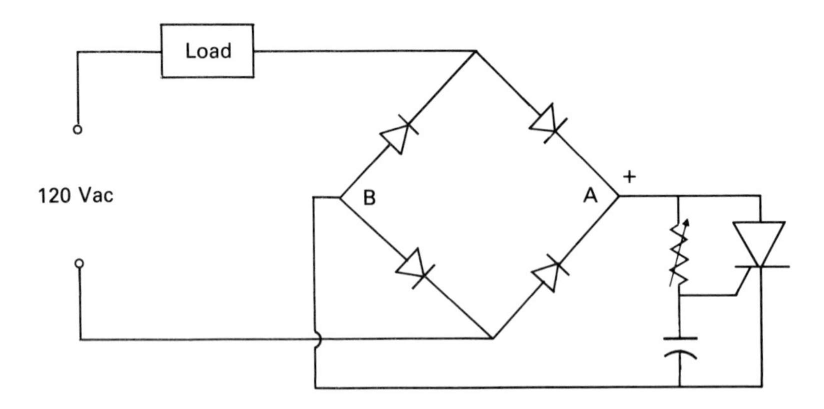


**Figure 24.5** Ac control with an SCR

The SCR acts like a diode when it is switched on, providing us only half-wave control for the circuit shown in Figure 24.4. In the next circuit, we will see how the SCR can control full-wave ac.

In Figure 24.5, we find the SCR in the center of a bridge rectifier. At first glance, we might think that the SCR will be destroyed by forward current when it is switched on because there is no resistance in series with it. However, note that the *load* is in *series* with the ac source voltage for the bridge. The load provides the current limiting.

When the SCR is switched off, there is no current between points *A* and *B*, and thus no current in the load. When the SCR is switched on, there is a short between points *A* and *B*, which allows current to flow in the load. Note that we use the same type of phase shift control for the SCR in this circuit.

In both of these circuits, we have control through only 90 degrees of the ac waveform. If the capacitor has not been charged to a high enough voltage during the first 90 degrees of an ac cycle, it will *never* be charged enough to start gate current because the applied voltage is *falling* during the second 90 degrees. A *unijunction-transistor* trigger circuit, however, can provide us with a full range of control.

The device shown in Figure 24.6 is a unijunction transistor, or *UJT.* The structure of the UJT, shown in Figure 24.6a, coupled with its doping levels, causes the UJT to act as an ideal *relaxation oscillator.*

Figure 24.6b shows the internal equivalent circuit of the UJT. The two internal resistors, $R_{B_1}$ and $R_{B_2}$, form the *interbase resistance,* $R_B$. $R_B$ is typically around 10 kilohms. These two resistors act as a *voltage divider,* reverse biasing diode *D.* The ratio of $R_{B_1}$ to $R_B$ is called the *intrinsic standoff ratio,* which controls the voltage at the junction, *the intrinsic standoff voltage.* The intrinsic standoff ratio ranges from around 0.5 to 0.7.

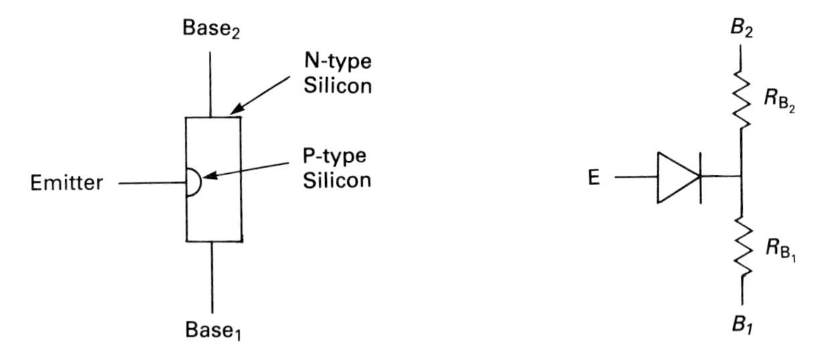


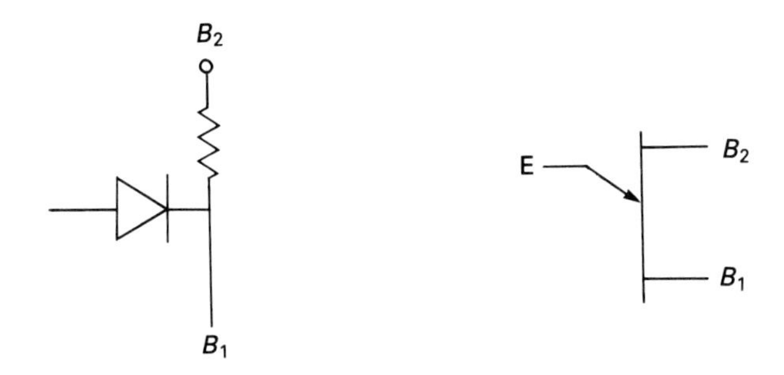


**Figure 24.6** The unijunction transistor (UJT)

When the voltage at the emitter becomes slightly larger than the intrinsic standoff voltage, the emitter diode becomes forward biased, and current will flow from the emitter through $R_{B_1}$. Because of the doping levels in the UJT, when current begins to flow in the emitter circuit, the resistance of $R_{B_1}$ drops suddenly to nearly zero ohms.

The circuit of Figure 24.7 is that of a simple relaxation oscillator designed around the UJT. When power is first turned on, capacitor *C* begins to charge through resistor *R*. Resistors $R_1$ and $R_2$, along with the UJT's internal resistances, set up the intrinsic standoff voltage within the UJT. (Because $R_1$ and $R_2$ are small compared with the interbase resistance, they have only a small effect on the intrinsic standoff voltage.) When the charge on *C* becomes large enough to forward bias the emitter diode, resistance $R_{B_1}$ (inside the UJT) suddenly drops to only a few ohms. This low resistance is in parallel with *C*, and discharges it.

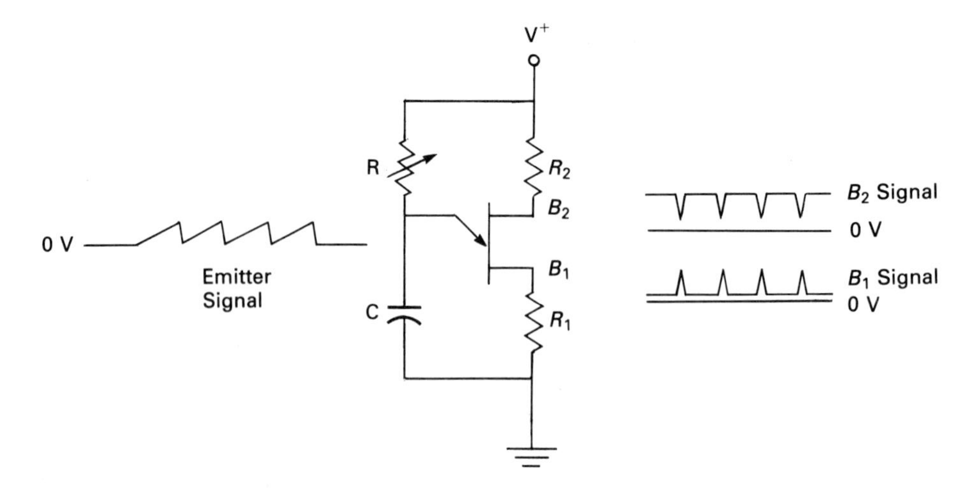


**Figure 24.7** A UJT relaxation oscillator

When $C$ is almost completely discharged, emitter current will stop, and $R_{B_1}$ will increase to its normal value. The process will then repeat. There are three different output waveforms available from this circuit, as shown in the figure. Of these, the one of most interest at this point is the series of *positive pulses* available at $B_1$. These positive pulses are useful for triggering SCRs, Figure 24.8.

In this circuit, the SCR is being used to control an ac load. When it is triggered, the SCR will once again place a short circuit across the bridge rectifier circuit. Resistor $R_1$ and Zener diode $D_5$ provide a safe operating level for the UJT. The Zener diode keeps the operating voltage for the UJT constant except for the brief periods when the supply voltage is near zero, or when the SCR is turned on.

The waveforms show the effects of a long RC charging time, Figure 24.8b, and a short RC charging time, Figure 24.8c. Note that we now have control over nearly 180 degrees of any cycle.

When replacing a UJT, the most important parameters are the intrinsic standoff ratio, $\eta$, and the interbase resistance, $R_B$. If these are not the same, the device will need recalibration. Of course, you must always be sure that the circuit voltages are not greater than the maximum ratings for the device. As always, a safety factor of two or more should be used.

The TRIAC offers a simpler means of control for ac circuits. Figure 24.9 shows the basic idea behind the TRIAC. It acts like two SCRs mounted in parallel with the anode of each connected to the cathode of the other. Note that the characteristic curve is the same in both directions. The three leads of the TRIAC are labeled main terminal 1, main terminal 2, and gate.

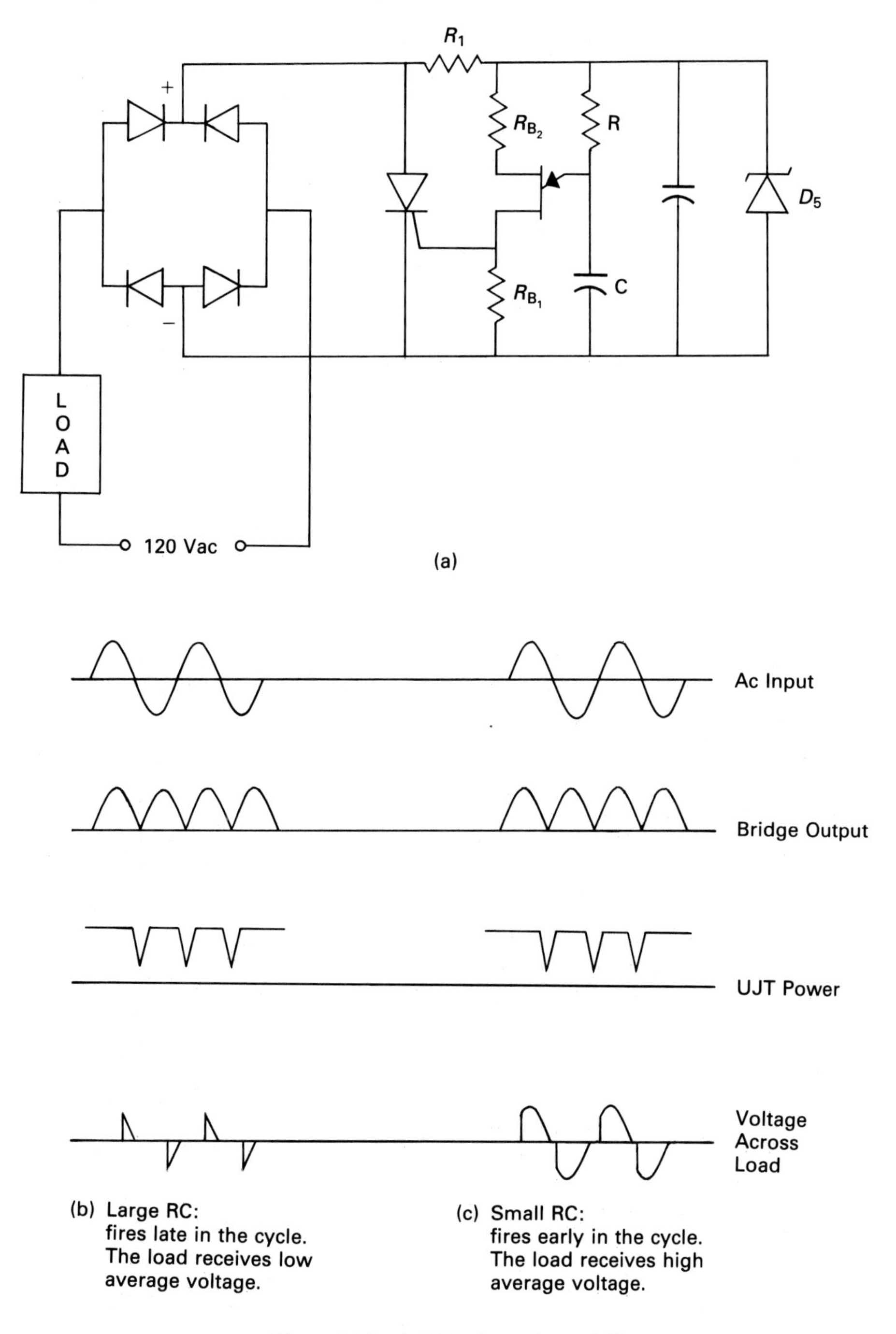


**Figure 24.8** A UJT trigger for an SCR

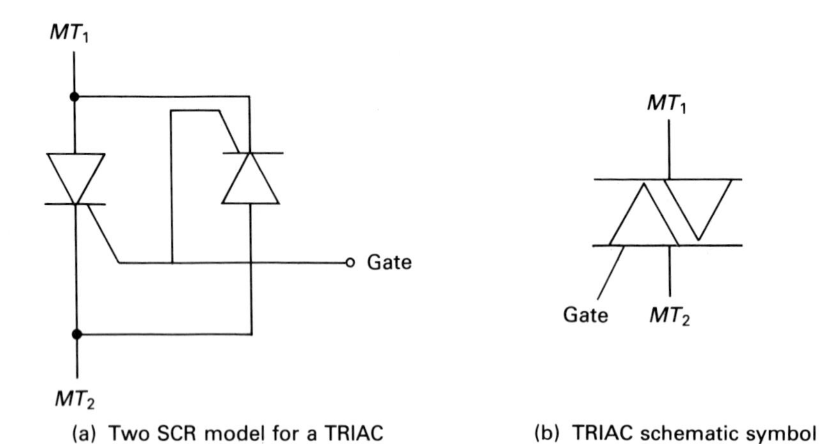


**Figure 24.9** The TRIAC

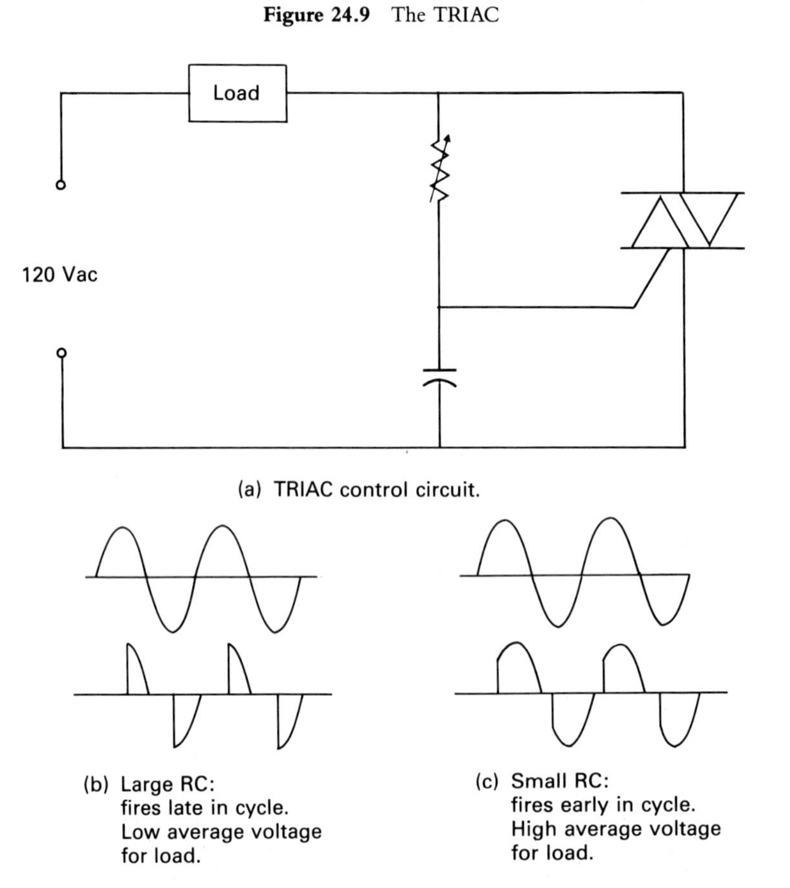


**Figure 24.10** Controlling ac with a TRIAC

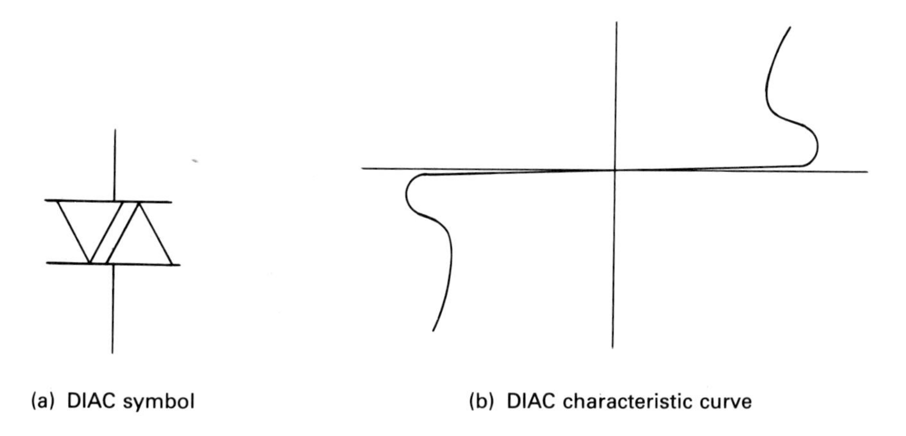


**Figure 24.11** The DIAC

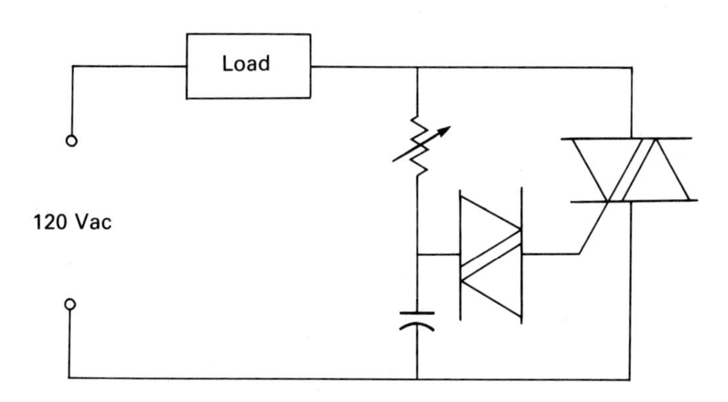


**Figure 24.12** TRIAC/DIAC control

The circuit in Figure 24.10 is exactly like the circuit in Figure 24.4 except that it controls an ac load. The *RC* network on the gate controls the firing angle of the TRIAC. The difficulty with this circuit is that the gate of a TRIAC is forward biased at about 0.7 V. That means that the range of control for a given value of *R* and *C* is quite limited.

The DIAC in Figure 24.11 is designed to improve the range of control of the TRIAC controller circuit. The DIAC is a PNPN device *without a gate*. The characteristic curve in Figure 24.11 shows that the DIAC conducts only when the applied voltage exceeds the breakover voltage of the DIAC. Typical breakover voltages range from 24 to 35 V, depending on TRIAC type.

The circuit shown in Figure 24.12 shows how the DIAC is used to extend the range of control for a TRIAC. The DIAC illustrated has a breakover voltage of 28 V. Therefore, the capacitor must now charge to 28 V before the gate becomes forward biased. This circuit provides a broad range of control. It is very similar to the commercially available light dimmers you find on sale in a hardware store.

When replacing a DIAC, it is important to match the breakover voltage of the replacement DIAC with the original DIAC. Failure to do this will cause the circuit to operate at a different firing angle, and require recalibration of the system.

## SUMMARY

In this chapter, we have looked at the analog controller. We have seen that:

1. Analog controllers are capable of proportional control: outputting a correction signal that is proportional to the error signal.
2. Analog controllers are capable of integral control: outputting a correction signal that is proportional to the duration of the error signal.
3. Analog controllers are capable of derivative control: outputting a control signal that is proportional to the rate of change of the error signal.
4. Most analog controllers simply output a relay contact closure or a transistor on-off signal.
5. Solid-state switching can replace the relay in many applications.
6. Thyristors — SCRs, TRIACs, and DIACs — provide a good means of controlling power to a load, with the advantages of solid-state switching.

## SELF-TEST

1. A controller that combines proportional, integral, and derivative control is called a(n) ______________________ controller.
2. Relay controllers require ______________________ input signals.
3. Relay controllers are only capable of ______________________ output control.
4. A proportional controller will output a(n) ______________________ signal for a large error signal than for a small one.
5. An integral control signal will have a large output voltage (or current) when the error signal has been present for a(n) ______________________ time.

6. A derivative controller will output a large output signal when the error signal changes ______________________.
7. The on-off output of many controllers is a switched-on transistor. This output must be low voltage and ______________________ current.
8. An SCR is triggered on by making the gate ______________________ with respect to the cathode.
9. As long as $I_{AK}$ is greater than the SCR's ______________________ current, the SCR will remain switched on without need for continuing gate current.
10. Causing gate current to flow in an SCR will reduce the ______________ voltage of the SCR.
11. By triggering an SCR with a UJT relaxation oscillator, we can attain control over ______________________ degrees of any alternation with an SCR.
12. The TRIAC is used for control of ______________________ circuits.
13. The DIAC is used to ______________________ the range of control for a TRIAC circuit.
14. When the ______________________ diode of a DIAC is forward biased, the internal resistance, $R_{B_1}$, decreases to nearly zero.

## QUESTIONS/PROBLEMS

1. You are repairing a transistor switching circuit which has been damaged by a voltage surge. Your replacement transistor has an $h_{FE}$ of 150. The original transistor had an $h_{FE}$ of 100. Will the circuit work with the replacement? Explain your answer.
2. You must replace an SCR which was controlling a 440 Vac motor. What is the minimum forward blocking voltage for a replacement SCR?
3. The UJT in an SCR firing circuit has been replaced. The original had an interbase resistance of 10 kilohms and an intrinsic standoff ratio of 0.6. The replacement has the same interbase resistance, but the intrinsic standoff ratio is 0.5. Explain what effect this will have on the operation of the circuit, and what change(s) you may have to make to correct for the difference.

# 25

# Digital Controllers

By far the greatest number of control systems being installed today are *digital controllers*. Digital controllers are less susceptible to unwanted oscillations than analog controllers, and can be made more immune to the EMI noise that can disturb analog-control accuracy. Digital controllers can also be designed to behave like relay controllers, but with a fraction of the power consumption and heat, and without the problems of contact contamination and arcing that cause problems with relays.

In this chapter, we will examine some of the digital controllers that are in common use today.

## OBJECTIVES

You will have successfully completed this chapter when you can:

1. Explain the difference between analog and digital systems.
2. Describe the TTL and the CMOS logic families.
3. Explain the differences between a general-purpose process controller and a dedicated process controller.

## 25.1 DIGITAL LOGIC

As the Earth orbits the sun, the light we receive at any point on the surface varies from the full darkness of midnight to the blaze of the noonday sun. That light is *analog* in nature. It is capable of having any brightness between the two extremes.

Enter a darkened auditorium and turn on the ceiling lights. The auditorium becomes instantly brightly lit. That light is *binary* in nature. It has two levels: on or off. There is no level between these extremes.

Digital electronics uses binary signals: signals with two voltage levels, usually called *high* and *low*, or *1* and *0*. Using these two voltage levels and simple transistor switches called *gates*, digital systems are capable of extremely complicated decision making. Remember, the digital computer is made up of little more than an arrangement of digital logic gates.

There are two *families* of logic in common use for control systems: *transistor-transistor logic* (*TTL* or $T^2L$) and complementary metal oxide semiconductor logic (*CMOS*). Each of these families has several *subfamilies*, which offer special advantages in some cases. While an AND gate in a TTL circuit performs the same *logic* function in any one of the subfamilies of TTL logic or in any of the CMOS logic subfamilies, the characteristics of the ICs, and even the operating voltages of the ICs, are different enough that substitution of one for another often creates operating problems in the system. We will illustrate that point so that you can see why it happens.

## 25.2 TTL LOGIC

The first widespread use of digital logic ICs came about with the development of transistor-transistor logic ICs. Earlier logic families were difficult to work with and required special consideration in circuit design.

There are two different series of standard TTL logic: the 7400 series and the 5400 series. The two are pin and function compatible and can be freely intermixed in most circuits. The 7400 series is the commercial or industrial series, while the 5400 series is the military series. The differences between them are in terms of temperature and power supply voltage tolerance.

The commercial 7400 series of TTL ICs operates in the temperature range of 0 degrees C to 70 degrees C. Power supply voltages must be within +/−5% of 5 Vdc. The military-version 5400 series operates in the temperature range of −55 degrees C to +125 degrees C. The power supply voltages must be within 10% of 5 Vdc. From this, we can see that a 5400 TTL IC may be substituted for a 7400 TTL IC at any time. Substituting a 7400 series for a 5400 series requires some care, but unless conditions are extreme, it should cause no problems.

Newer designs have been created in the 7400 format to meet special needs: lower power consumption or higher speeds of operation. These include the 74L00 series low-power chips, the 74H00 series high-speed chips, the 74S00 series, which uses Schottky transistors for even higher speeds, and the 74LS00 series low-power Schottky chips. All of the TTL chips have identical input voltage requirements and output voltage characteristics, but they differ widely in their ability to sink or source current.

Recent developments in IC design have yielded the 74F00, *fast* series, the 74AS00, *advanced Schottky* series, and the 74ALS00, *advanced low-power Schottky* series of TTL chips, as well.

The output high voltage for any TTL chip, $V_{OH}$, is guaranteed to be greater than 2.4 V. The output low voltage is guaranteed to be less than 0.4 V. On the input side, the TTL family recognizes any voltage greater than 2.0 V to be a valid "high" input, and any voltage smaller than 0.8 V to be a valid "low" input. Input levels between 2.0 V and 0.8 V are considered to be *bad levels.* When troubleshooting an erratic circuit, look for bad-level signals on the inputs of your ICs. Substituting a 74LSxx for a 74xx IC can be one cause of this, since the *LS* series has a fanout of only one standard TTL input. Bad levels cause erratic and intermittent operation.

All of these *subfamily* members are pin compatible with the standard 7400 series. In other words, the pinouts of the 74LS08 and the 7408 chips are the same. As we saw from the previous paragraph, however, that does not make them direct substitutes for each other in a circuit.

The following chart illustrates the differences found within the TTL subfamilies.

**TTL COMPARISON CHART**

| TYPE | $T_{PLH}$ | $P_D$ |
|---|---|---|
| 7400 | 10 nS | 40 mW |
| 74L00 | 33 nS | 4 mW |
| 74H00 | 6 nS | 90 mW |
| 74LS00 | 10 nS | 8 mW |
| 74S00 | 3 nS | 75 mW |
| 74F00 | 2.5 nS | 16 mW |
| 74AS00 | 4 nS | 30 mW |
| 74ALS00 | 1.5 nS | 4 mW |

$T_{PLH}$ is the *propagation delay* of the gate when it changes from a low output to a high output. $P_D$ is the power dissipation of a quad two-input NAND gate. The figures given are typical; for specific figures for any IC, consult the manufacturer's data manual.

The 7400 series and the 74LS00 series presently enjoy the most popularity. Great care should be used when substituting one subfamily for another because the differences in fanout and propagation delay may well cause a mysterious malfunction in the device you are repairing. A TTL gate, for example, can fan out to 50 *LS* gates, but an *LS* gate has a fanout of only *one* TTL load. The designer selected the logic family according to the system needs — leave it that way and you will have far fewer problems.

## 25.3 CMOS LOGIC

Newer designs are moving away from TTL logic toward the newest CMOS technology. CMOS, or *complementary metal-oxide semiconductor,* has undergone a revolution in IC design in the past few years. Early CMOS chips were used primarily in places where low power consumption was an important factor and the long propagation delays were not objectionable.

The advent of lap-top portable computers, however, was made possible only by the latest designs in CMOS technology, which rival the 74LS00 chips in speed but consume only a fraction of the power. Power consumption is important, of course, in battery-operated equipment, but it also offers advantages in line-operated equipment. Low power consumption in a digital system does not make significant savings in your electric bill, but the equipment *does* run cooler. Less heat in an equipment cabinet means that the equipment will be more reliable, last longer, and may not need forced-air cooling fans.

The transistors in CMOS chips are all enhancement-type MOSFETs. The name *CMOS* is derived from the configuration of the output transistors, which are connected as a complementary pair.

Figure 25.1 shows the complementary output transistors of a CMOS gate. Note that $Q_1$ is a P-channel MOSFET and $Q_2$ is an N-channel MOSFET. Recall from your earlier training that an enhancement MOSFET acts like an open switch until a bias voltage is applied. The bias voltage that turns on a P-channel MOSFET is negative, and the bias voltage that turns on an N-channel MOSFET is positive. Since the gates are connected together, they cannot be turned on at the same time.

The advantage of MOSFET logic circuits is that they consume very little current. Unlike the bipolar transistors of TTL logic, the input of a MOSFET is *insulated* and no current flows in the input circuit. And since only one of the two output transistors is on at any given time, there is no current path in the output side of the gate, either. Practically the only current that flows is leakage current.

Low power consumption means that many more MOSFET logic gates can be put on a single silicon chip without the problems of overheating. On the downside, MOSFET insulated gates are very susceptible to damage from static electricity when they are out of the circuit. The technician's tools, soldering iron, and workbench must be grounded. Indeed, the *technician* must be grounded! A special wrist strap and lead wire connects the technician to ground. Several megohms of resistance prevent dangerous currents from flowing in the ground strap — it is only necessary to drain off static charge.

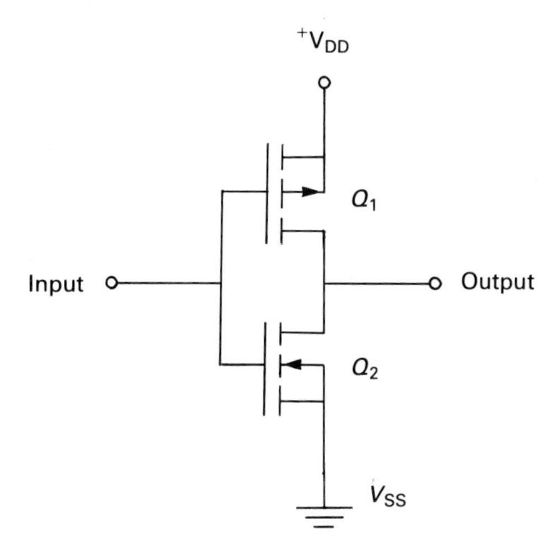


**Figure 25.1** A CMOS output section

Functionally, a CMOS AND gate is identical to a TTL AND gate; electrically, they are quite different. We will look at those differences, and the differences between the CMOS subfamilies, so that you can make intelligent choices in replacing parts and in new designs.

The earliest CMOS gates carried the family designation 4000 series. These included the 4000B series, which had buffered outputs, and the 4000A series, which was the early design. The 4000 series operates with any power supply voltage between 3 Vdc and 15 Vdc. The seldom-used 74C00 series is pin-compatible with the 7400 series TTL gates, although input voltage requirements and fanout are not suitable for direct substitution in a TTL circuit. Neither of these is used much in new designs today.

The new CMOS subfamilies are the 74HC00 and the 74HCT00. The chips are pin-compatible with each other, but they are not direct substitutes for each other. The 74HC series is used in all CMOS circuit designs. The 74HCT series is used in hybrid designs along with TTL logic. Power supply voltage requirements are the same for both: they will operate with any power supply voltage between 2 Vdc and 6 Vdc. The *designs center* power supply voltage for both, however, is 5 Vdc.

Because of the design of the output section of a CMOS chip, the output voltages are very nearly the same as the power rail voltages. For all family members, $V_{OH}$ is 0.05 V less than $V^+$, and $V_{OL}$ is +0.05 V. On the input side, there are some differences, which are shown in the following chart.

## CMOS COMPARISON TABLE*

| TYPE | $T_{PLH}$ | $P_D$ | $V_{IH}$ | $V_{IL}$ |
|---|---|---|---|---|
| 4000B | 50 nS | 10 nW | >3.5 V | <1.5 V |
| 74C00 | 100 nS | 10 nW | >3.5 V | <1.5 V |
| 74HC00 | 15 nS | 100 μW | >3.5 V | <1.0 V |
| 74HCT00 | 15 nS | 100 μW | >2.0 V | <0.8 V |

**BASED ON $V_{DD}$ = 5 Vdc*

From this chart, we can see the differences between the main types of CMOS logic chips. They can be readily combined in the same circuit, but note that only the 74HCT subfamily has a $V_{IH}$ that can interface directly with a TTL chip. The 4000B and the 74HC00 require a higher voltage for a logic one input than a TTL chip will provide.

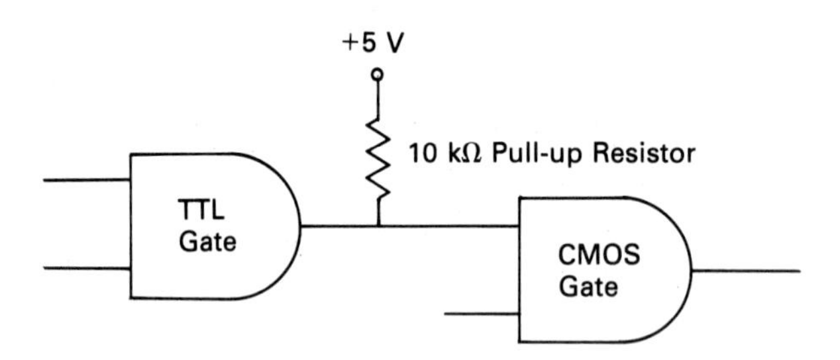


**Figure 25.2** Interfacing CMOS and TTL

Figure 25.2 shows how the interface between a TTL and a CMOS gate is done. The 10 kΩ *pull-up* resistor will increase the $V_{OH}$ of the TTL gate to the higher input voltage required by the CMOS gate. The only consideration needed when interfacing a CMOS output to a TTL input is fanout. Generally, a CMOS gate will fan out to only one TTL gate input.

As usual, consult a data manual before making any substitutions.

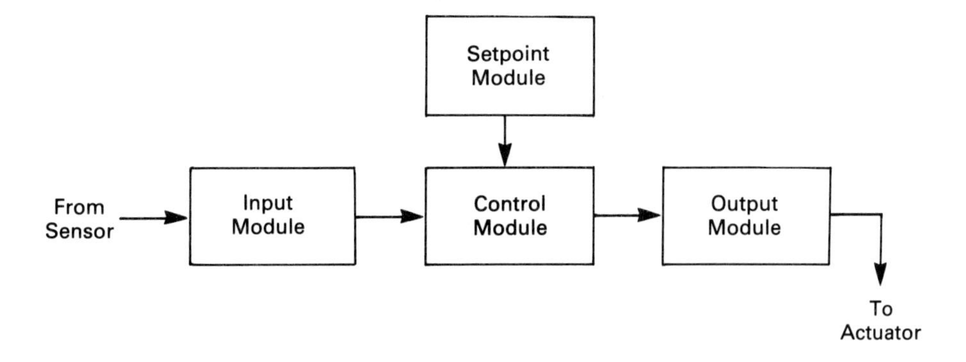


**Figure 25.3** Block diagram of a typical process controller

## 25.4 DIGITAL PROCESS CONTROLLERS

Digital process controllers have begun to dominate the industrial control market. They offer flexibility, reliability, and low cost.

A typical digital process controller has four basic modules in its design:

1. The *input module* provides operating voltages for the input transducers, and an analog-to-digital converter to convert the input signal into a binary value for the controller's digital electronics to work with. Input modules are available for dc or ac voltage inputs, for current loop inputs, for rate-of-flow variable-frequency inputs, and with cold-junction compensation for thermocouple inputs.
2. The *setpoint module* accepts input from the operator. Setpoint inputs may be BCD thumbwheel switches, numeric keypads, or potentiometer settings. These setpoints are also converted into a binary value for the controller.
3. The *control module* processes the binary signals from the input and setpoint modules. The control module is the decision maker of the system. Control modules are available with adjustable PID control, or with on-off control.
4. The *output module* drives the actuator, based on the decisions made in the control module. Output to the actuator may be a relay contact closure, a TRIAC on-off control, or a phase-angle TRIAC PID output.

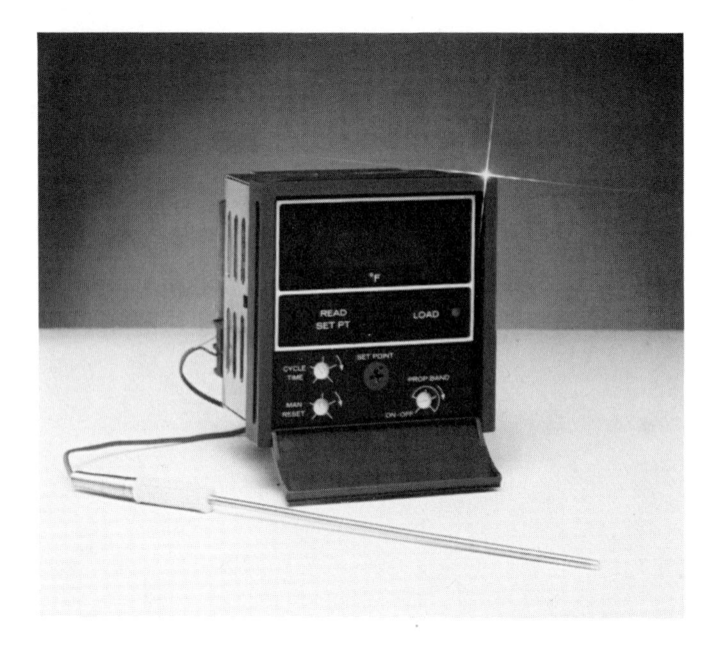


**Figure 25.4** A digital process controller *(Courtesy of OMEGA ENGINEERING, An OMEGA Technologies Company)*

Figure 25.4 is a typical digital process controller. The main settings for this model are on the rear panel to prevent tampering by untrained or unauthorized personnel. This unit is available with input modules for type *J* or type *K* thermocouples, for input from an RTD, rate-of-flow indicators, two different levels of dc voltage, ac line voltage, or pressure and strain gage bridges. It offers outputs for PID control or on-off TRIAC control. The display may be *scaled* to read in common engineering units so that the system operator can tell at a glance how the system is responding to control.

## 25.5 DEDICATED DIGITAL CONTROLLERS

Some manufacturers produce controllers that are dedicated to a particular transducer system and designed for a specific control task. This approach provides a perfect match between system components, but it has the minor disadvantage of using several different types of controllers in a given system.

The Astrosystems *dial-a-limit* controller shown in Figure 25.5 is an example of a dedicated controller. The transducer (A) in this illustration is an absolute shaft position encoder, which was discussed in Chapter 3. The master control unit (B) accepts the position encoder signal and presents it to the attached setpoint modules (C). The output of the setpoint modules is used to operate the control relays (D), shown in

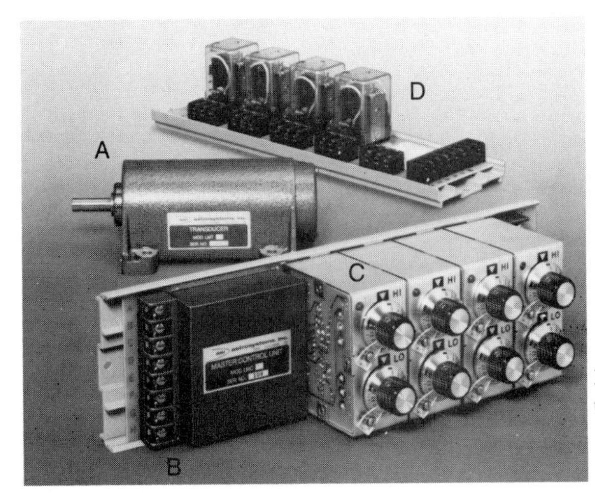


**Figure 25.5** A dedicated digital controller *(Courtesy of Astrosystems, Inc.)*

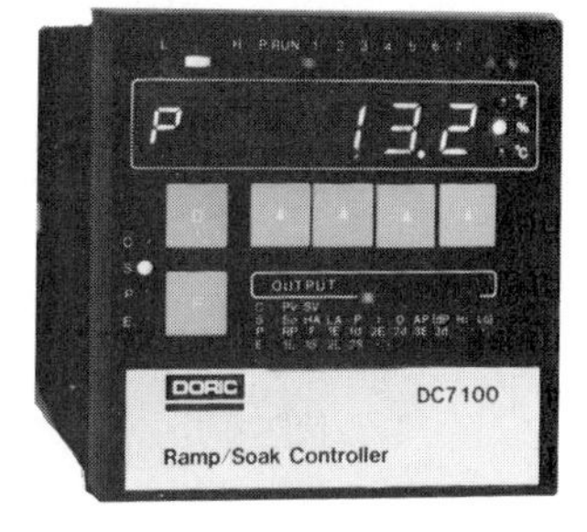


**Figure 25.6** A ramp/soak heat controller *(Courtesy of Emerson Electric Co., Doric Scientific Division)*

the illustration. The illustrated unit can detect four different shaft positions and provide a relay operation for each.

The Doric *ramp/soak* controller shown in Figure 25.6 is specifically designed to control process temperature. The input side has switch-selectable options from among eight different thermocouple types, two different RTD ranges, or 10 mV to 50 mV ranges. The addition of a precision resistor can provide sensitivity to an ISA 4 mA to 20 mA transducer.

The output of the controller may be selected from among relay control, 0–24 V drive for a solid-state relay, or an ISA 4 mA to 20 mA current drive. PID control is built in. The controller can be programmed for ramp up rate, soak time, and ramp down time, and can control as many as three different processes simultaneously.

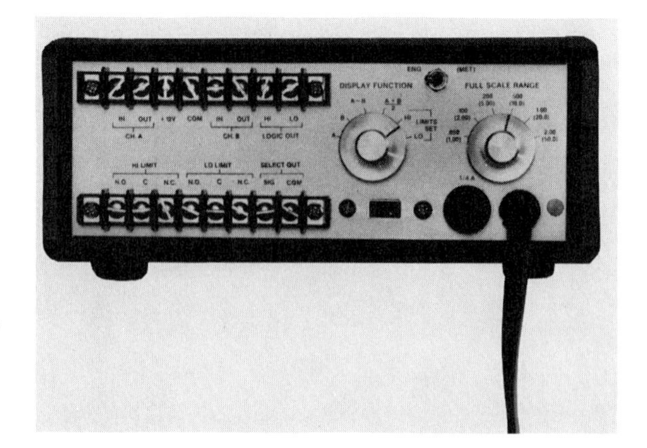


**Figure 25.7a** The Electro Mike controller — rear panel view *(Courtesy of Electro Corp.)*

**Figure 25.7b** The Electro Mike controller — front panel view *(Courtesy of Electro Corp.)*

The Electro Mike controller shown in Figures 25.7a and b is designed to interface with the Electro Mike transducers shown in Chapter 3. From the rear panel detail in Figure 25.7a, we can see that the controller accepts inputs from two sensors. The center rotary switch selects the function being displayed on the front-panel seven-segment display. The end rotary switch selects the sensitivity of the system. The unit may be used with one or two sensors to measure and control the position of materials.

The front panel provides controls for the calibration of both input sensors and for upper and lower setpoints. The outputs may be relay closures or a logic level output, provided directly from the rear panel.

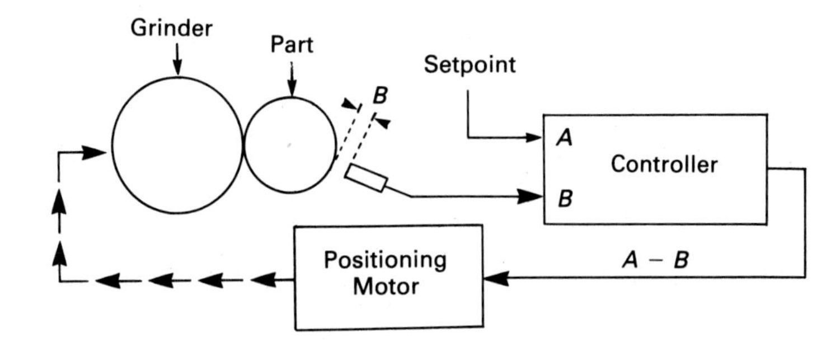


(a) Controlling part diameter

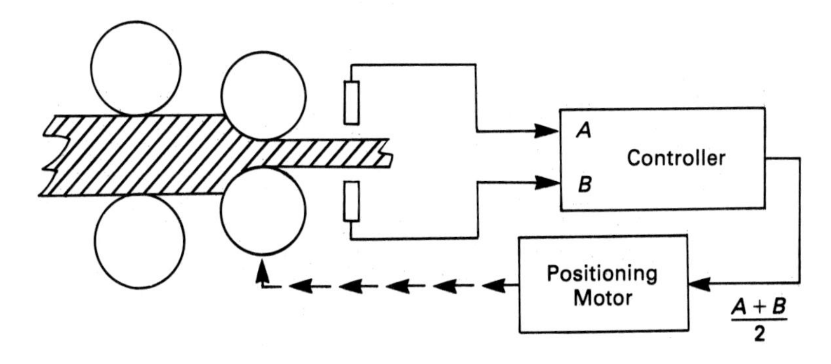


(b) Controlling material thickness

**Figure 25.8** Electro Mike applications

Figure 25.8 shows some typical Electro Mike applications. In Figure 25.8a, a single detector measures the diameter of the part being ground. The rear panel switch is set to the "$A - B$" position, and the output controls a positioning motor for the grinder.

In Figure 25.8b, two sensors are being used to control the thickness of a material as it passes through a rolling mill. The average thickness is found by calculating the average distance each sensor is from the material:

$$\frac{A + B}{2}$$

## SUMMARY

The digital controller and the integrated-circuit families that have led to its popularity were described in this chapter. We have seen that:

1. TTL and CMOS logic families have several subfamilies.
2. Care must be used when substituting among the various subfamilies within either group of logic families.
3. Digital controllers in this chapter can be classified in two categories: generalized process controllers and dedicated or specialized controllers.

## SELF-TEST

1. TTL logic is based on the ____________________ transistor.
2. CMOS logic is based on the ____________________ transistor.
3. All TTL logic families operate with a(n) ____________________ volt power supply.
4. The 74xx series of logic chips requires a power supply voltage that is regulated to within ____________________ percent.
5. The 4000 series of CMOS logic chips will operate with any power supply voltage between _______________ and _______________ volts.
6. CMOS logic ICs generally use ____________________ power than TTL logic ICs.
7. While troubleshooting a TTL digital controller, you measure the power supply voltage and discover that it is 4.85 volts. Is this voltage within the tolerance of the logic circuit?
8. While troubleshooting a TTL logic circuit, you measure a gate input signal with an oscilloscope. The signal measures 1.8 volts. Is this signal within the tolerance of the logic circuit?
9. While troubleshooting a CMOS logic circuit that uses 74HC technology, you measure a low input signal at 0.7 volts. Is this signal level within the tolerance of the logic circuits?
10. When a TTL IC is being used to drive the input of a CMOS IC, a(n) ____________________ resistor must be used to increase the voltage output of the TTL IC for a logic high output.

## QUESTIONS/PROBLEMS

1. You are servicing a digital controller which uses a 15 MHz clock rate. A 74HC02 NOR gate is bad, but your parts department is out of stock on this part. The stock clerk suggests a 74C02, which has the same pinout diagram. Will this part work in the circuit? Explain your answer.
2. A single 74LS00 NAND gate is driving eight other LS TTL gates. Can a 7400 NAND gate be substituted for it? Assume that power supply dissipation is not a problem. Explain your answer.
3. A circuit design used a 4004B inverter to drive a standard TTL input. The circuit operates in some cases, but not in others. What is the most likely problem?

# 26

# The Microprocessor as a Controller

In 1971, the Intel corporation introduced the 4004 microprocessor IC. More than any other development, this event revolutionized the control industry. Indeed, it has colored the lives of all of us, and continues to do so daily.

The original 4004 microprocessor was developed with industrial control circuits in mind. It was quickly followed by the 4040, an enhanced version of the 4004, then by the 8008 eight-bit microprocessor, and then by the 8080 and 8085 microprocessors. Motorola released the 6800 microprocessor at about the same time the 8080 was released, and the world has not been the same since.

In the excitement over personal computers in the home and office, the original purpose of the microprocessor may seem to have been lost in the shuffle. In this chapter, we will see that that is far from the case.

## OBJECTIVES

You will have successfully completed this chapter when you can:

1. Define *microprocessor* and explain some of its advantages as a controller.
2. Explain how a microprocessor is interfaced to the outside world.

## 26.1 THE MICROPROCESSOR

A *microprocessor* is a single IC that contains most of the logic required for a small computer. The following items are contained within the IC or as peripheral chips:

- The timing and control circuitry that keeps a bewildering array of operations in the proper sequence.
- An arithmetic logic unit capable of performing thousands of operations in a second.
- Program memory, which holds a sequence of instructions for the microprocessor to carry out.
- Input/output circuitry, which enables the microprocessor to interface with the "outside world."

Some microprocessors contain all of the parts for a small computer within the single IC. We often see these ICs without recognizing them. They serve as controls for our automobile ignition systems, our microwave ovens, even as channel selectors in our TV sets.

It is beyond the scope of this book to include a full description of microprocessors and microprocessor programming. However, *interfacing* microprocessors is an important part of industrial control. Because most schools and courses concentrate (correctly) on a single microprocessor, the technician often does not understand how circuits are interfaced. In this chapter, we will look at the most popular microprocessors and how they are interfaced.

## 26.2 THE PRINCIPLES OF MICROPROCESSOR INTERFACING

There are basically two different ways that a microprocessor can be interfaced, although both techniques are not available in all microprocessors.

In *memory-mapped* interfacing, the microprocessor reads from or writes to a *memory location*. Instead of memory, however, the microprocessor is reading from an *input port* or writing to an *output port*.

In *register I/O*, the microprocessor reads directly from an input port or writes directly to an output port. Register I/O does not use memory space for I/O, but it has a special addressing scheme for I/O only.

An I/O port requires special consideration in design. It must have *an address* that is unique from any other device in the system, and it must be connected to the *control bus* of the microprocessor so that it can respond to signals at the correct time.

An *input port* must be coupled to the system bus through *three-state buffers* so that the port and the microprocessor are not in contention for the bus at the same time. The input port may also have *latches* to store the input data until the microprocessor is ready for it.

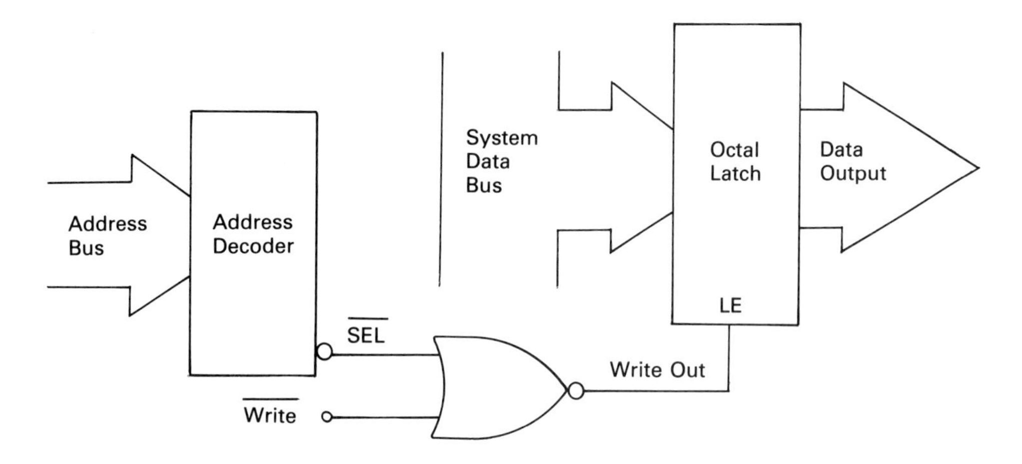


**Figure 26.1** The basic input port

Figure 26.1 is a general input port for a microprocessor system. It contains all of the components required for an input port for either memory-mapped or register input. The *address decoder* decodes the address of the port and outputs a low (SEL*) only when the correct pattern is on the address bus. SEL* is combined with READ* by the 7402 NOR gate. Only when SEL* and READ* are *both* present will the 7402 output a high (INPUT READ). INPUT READ enables the three-state buffers and the microprocessor to read the data from the bus.

An *output port* most often will have latches, as well. The microprocessor presents data to the output port for only a brief period of time — 500 nanoseconds or less. If the data is not latched, it will most likely be lost.

Note that the signal names used in these diagrams and throughout this chapter are *generic*, except when they refer to specific *control bus signals* from a particular microprocessor. In other words, SEL* is not used by every manufacturer to identify that particular signal. The terms are specific to this text. The convention of placing an asterisk (*) after a signal name signifies that the signal is *active low.*

Figure 26.2 shows the design of a general output port. The address decoder outputs SEL* when the port address is on the address bus. SEL* is combined with WRITE* by the 7402 gate. The output of the 7402 is a high pulse (WRITE OUT) only when the correct address is on the decoder and the WRITE* pulse is present at the same time.

The specifics may be different for different microprocessors, but the basic concepts are always the same. In the next several sections of this chapter, we will look at specifics for the most popular microprocessors used for control today.

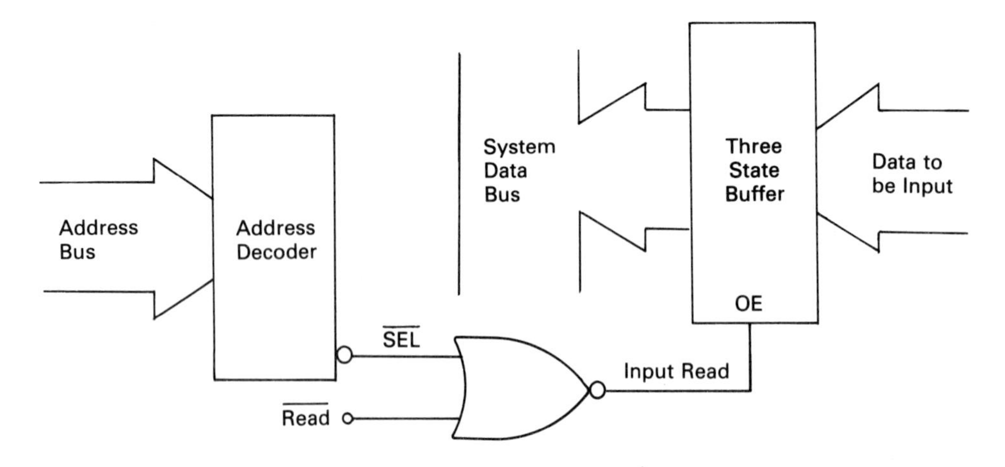


**Figure 26.2** The basic output port

## 26.3 THE 8080 AND 8085 MICROPROCESSORS

The 8080 was the first eight-bit microprocessor. Although its capabilities are limited by today's standards, it is still a powerful device and it is still in use in many older control systems. It offers both register and memory-mapped I/O. The 8085 microprocessor simplifies the wiring of the system because many of the control bus signals that required special decoding circuits for the 8080 system are generated within the microprocessor itself for the 8085. The 8085 microprocessor will execute all of the software written for the earlier 8080.

The 8080/8085 instruction set includes two I/O commands, IN <B2> and OUT <B2> for controlling register input and output. <B2> represents the second byte of a two-byte instruction. It is the *port address*, not the data to be input or output. This eight-bit address is capable of selecting 256 different input ports and 256 different output ports. Output data must first be stored in the *accumulator* or *A* register of the microprocessor. Input data will be read from the input port *into* the accumulator of the 8080/8085.

Figure 26.3 illustrates an input port for an 8080/8085 microprocessor system. The operating sequence is shown in the timing diagrams.

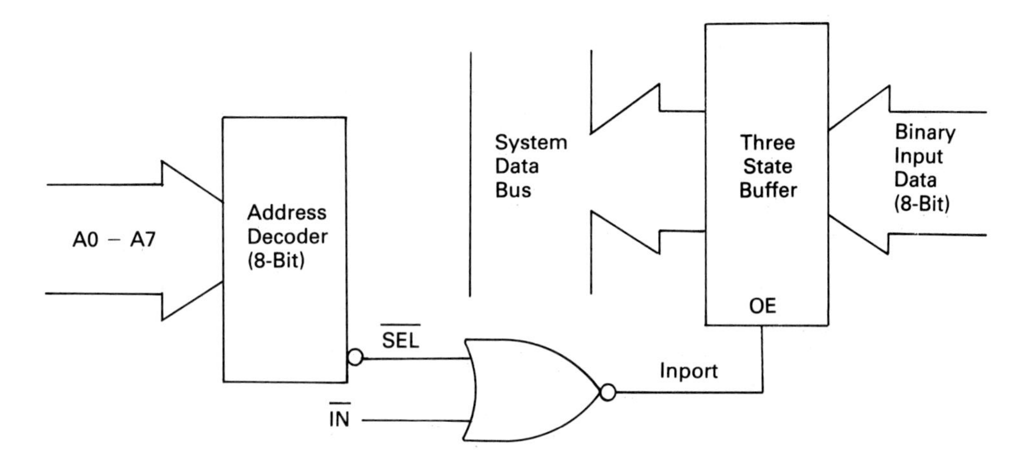


**Figure 26.3a** An 8080/8085 register input port

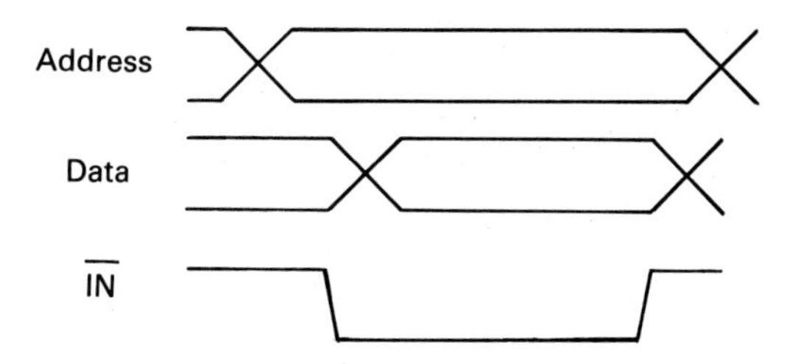


**Figure 26.3b** 8085 outport timing

1. The microprocessor places the input port's eight-bit address on *both* the high-order and the low-order halves of the address bus. This address is decoded by the address decoder to yield SEL*.
2. When the port address is stable on the bus, the microprocessor outputs a low pulse on the control bus line IN*.
3. SEL* and IN* are combined into one signal by the 7402, yielding a single high-going pulse, INPORT.
4. INPORT enables the eight-bit three-state buffer, so that the input data is placed on the data bus.
5. The microprocessor reads the data from the data bus into its accumulator.

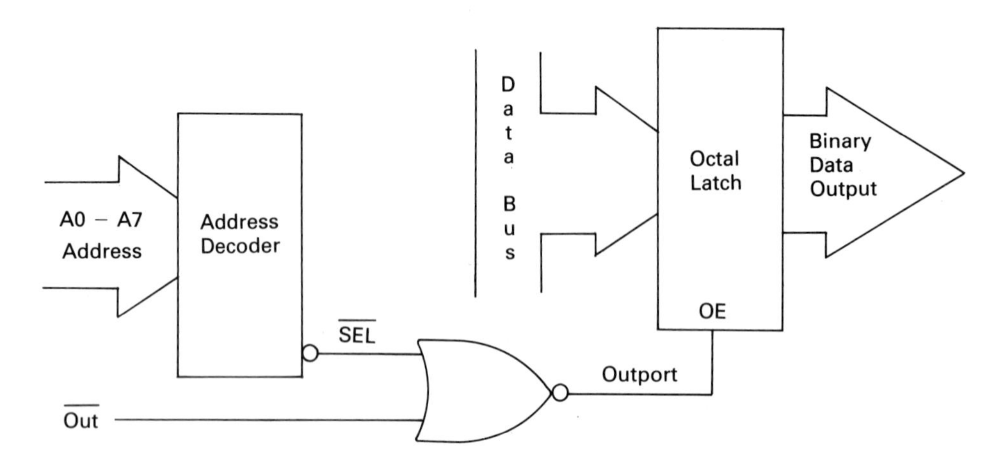


**Figure 26.4a** An 8080/8085 register output port

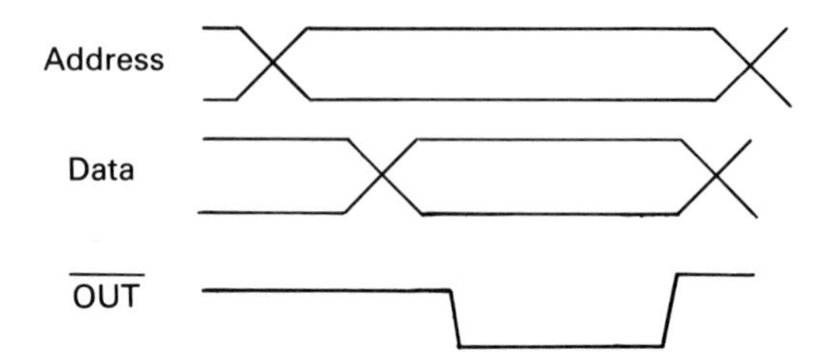


**Figure 26.4b** 8085 outport timing

Figure 26.4 is the diagram of an 8080/8085 register output port. The data to be output is first stored in the 8080/8085 accumulator register. When the OUT <B2> instruction is executed, the sequence of operations is:

1. The microprocessor places the output port's eight-bit address on *both* the high-order and the low-order halves of the address bus. This address is decoded by the address decoder to yield SEL*.
2. The microprocessor places the contents of the accumulator on the data bus.
3. When the port address and the data are both stable on the buses, the microprocessor outputs a low pulse on the control bus line OUT*.

4. SEL* and OUT* are combined into one signal by the 7402, yielding a single high-going pulse, OUTPORT.
5. OUTPORT clocks the eight-bit latches which form the interface between the microprocessor and the outside world. The data from the data bus is clocked into the latches and is now available for the outside-world device to use.

For memory-mapped I/O, a section of memory space must be set aside strictly for I/O use. The microprocessor then "reads from memory" to perform an input operation, or "writes to memory" to perform an output operation. Since memory-mapped I/O uses an area of system memory rather than dedicated address space as in register I/O, the address decoder must be 16 bits wide.

For register I/O with the 8080/8085, data must transfer through the accumulator. In other words, all input data comes into the accumulator, and all output data must first be stored in the accumulator. For memory-mapped I/O, however, *any general-purpose register* may serve as the transfer point.

Figure 26.5 shows a memory-mapped input port for the 8080/8085. Note that the address decoder now is 16 bits wide. The sequence for an input operation is:

1. The microprocessor places the input port's 16-bit address on the address bus. This address is decoded by the address decoder to yield SEL*.

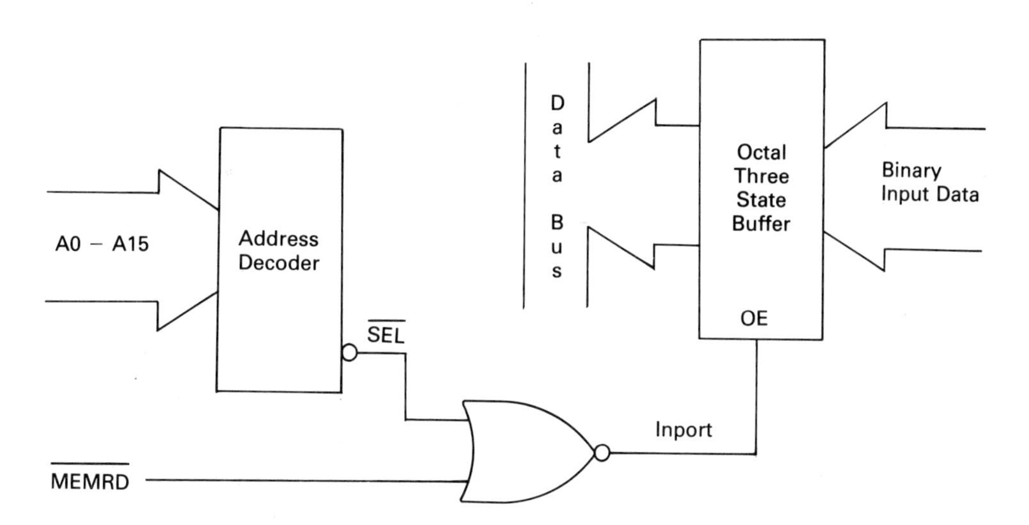


**Figure 26.5** An 8080/8085 memory-mapped input port

2. When the port address is stable on the bus, the microprocessor outputs a low pulse on the control bus line MEMRD*.
3. SEL* and MEMRD* are combined into one signal by the 7402, yielding a single high-going pulse, INPORT.
4. INPORT enables the eight-bit three-state buffer, so that the input data is placed on the data bus.
5. The microprocessor reads the data from the data bus into one of its internal registers.

Figure 26.6 illustrates a memory-mapped output port for an 8080/8085 system. The sequence of operations is:

1. The microprocessor places the output port's 16-bit address on the address bus. This address is decoded by the address decoder to yield SEL*. At about the same time, the microprocessor places data from one of the internal registers on the data bus.
2. When the port address is stable on the address bus, the microprocessor outputs a low pulse on the control bus line MEMWR*.
3. SEL* and MEMWR* are combined into one signal by the 7402, yielding a single high-going pulse, OUTPORT.
4. OUTPORT clocks the eight-bit latches which form the interface between the microprocessor and the outside world. The data from the data bus is clocked into the latches and is now available for the outside-world device to use.

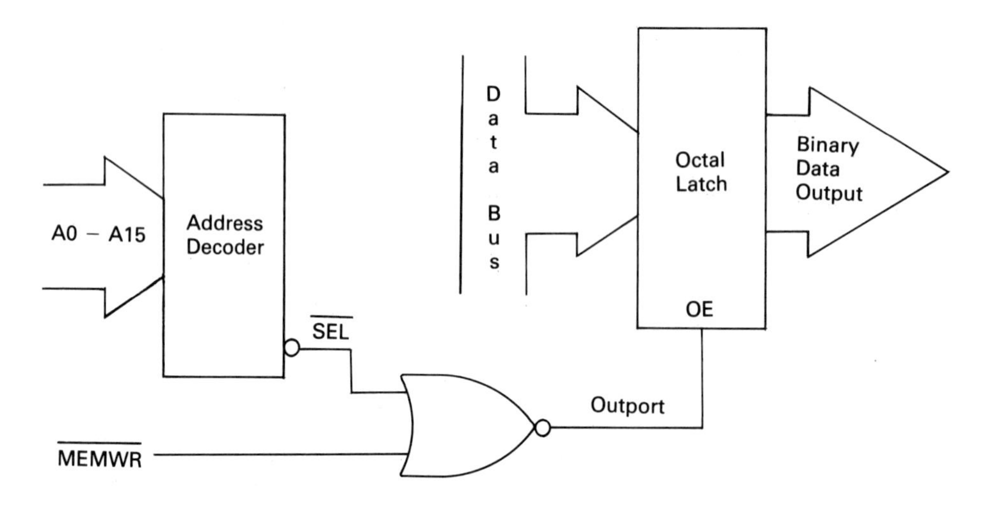


**Figure 26.6** An 8080/8085 memory-mapped output port

The instruction used for the memory-mapped I/O operation determines which of the 8080/8085's internal registers will serve as the I/O transfer point.

The main advantage of memory-mapped I/O is *speed*. Both input and output take three fewer clock cycles for memory-mapped operation than for register I/O. This in itself may not seem like a great advantage, but memory-mapped *input* also allows arithmetic or logic operations to be performed directly on the input data byte during the input operations, which can add an advantage of several more clock cycles. Such advantages can be of significance in some *real-time* operations.

## 26.4 THE Z80 MICROPROCESSOR

The Z80 microprocessor is very similar to the 8080/8085 microprocessor. It is produced by the ZILOG Corporation, which was formed by the original developers of the 8080 microprocessor. The Z80 is software compatible with the 8080 microprocessor, but it has a greatly expanded instruction set and has much greater capabilities than either the 8080 or the 8085. The Z80 was popular in personal computers and industrial controllers. While the 16-bit microprocessor has displaced the Z80 in personal computers, it is still the favorite microprocessor for control applications.

The Z80 microprocessor can perform register I/O from *any general-purpose register.* This greatly enhances the I/O instruction set for the Z80. Special I/O instructions also allow the Z80 to input or output entire *blocks* of data, which can speed up the reading of data from a number of sensors.

Figure 26.7 is a Z80 input port for register I/O. The sequence of operations for register input varies with the specific instruction. The following sequence is general:

1. The microprocessor places the port address on the low-order half of the address bus. This is decoded by the address decoder into SEL*.
2. The microprocessor places a low on the signals IOREQ* and RD*. These are combined by the OR gate into IN*.
3. IN* and SEL* are combined into INPORT* by a second OR gate.
4. INPORT* enables the input data onto the data bus through the three-state buffers.
5. The microprocessor reads the input data into the proper internal register.

Note that the Z80 uses the low-order address bus to address I/O ports. One clever feature of the microprocessor is that it can perform *both input and output at the same time.* Depending on the instruction used for I/O, either the accumulator data or the data from the *B* register will appear on the high-order half of the address bus during register I/O. This feature can be used for simultaneous input and output, or to extend the number of I/O ports beyond the normal 256 available with ordinary register I/O.

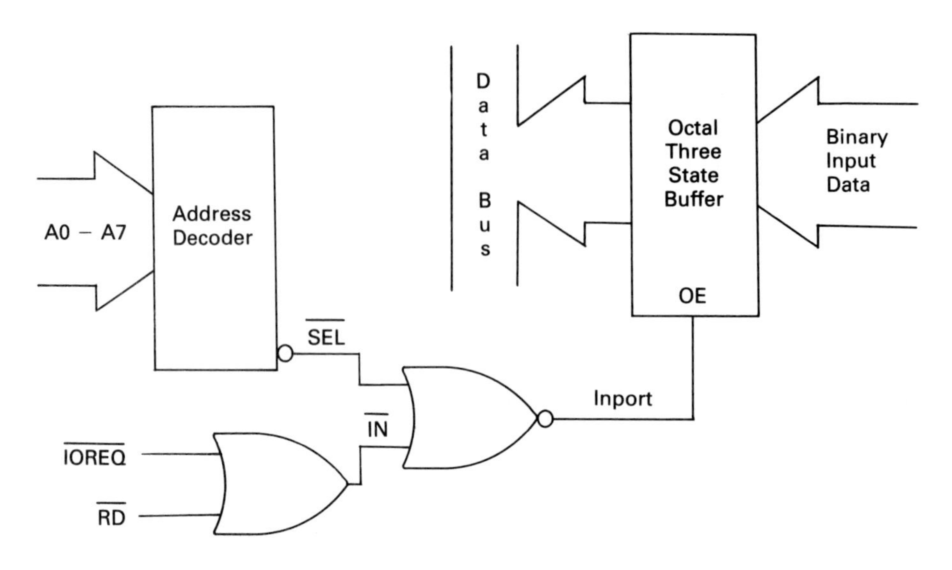


**Figure 26.7** A Z80 input port

Figure 26.8 is a Z80 output port. The sequence of operations is:

1. The microprocessor places the port address on the low-order half of the address bus. This is decoded by the address decoder into SEL*.
2. The microprocessor also places the data to be output onto the data bus during the same clock cycle.
3. The microprocessor places a low on the signals IOREQ* and WR*. These are combined by the OR gate into OUT*.
4. OUT* and SEL* are combined into OUTPORT* by a second OR gate.
5. OUTPORT* clocks the output data into the eight-bit data latches. The data is now ready for use by the external device.

## 26.5 THE MC6800 MICROPROCESSOR

The Motorola MC6800 microprocessor is another microprocessor that has found wide use in control systems. Unlike the 8080/8085 and Z80 microprocessors, the MC6800 is *memory oriented*, rather than register oriented. It contains no general-purpose registers. It is designed to perform all of its operations on memory through two internal accumulators.

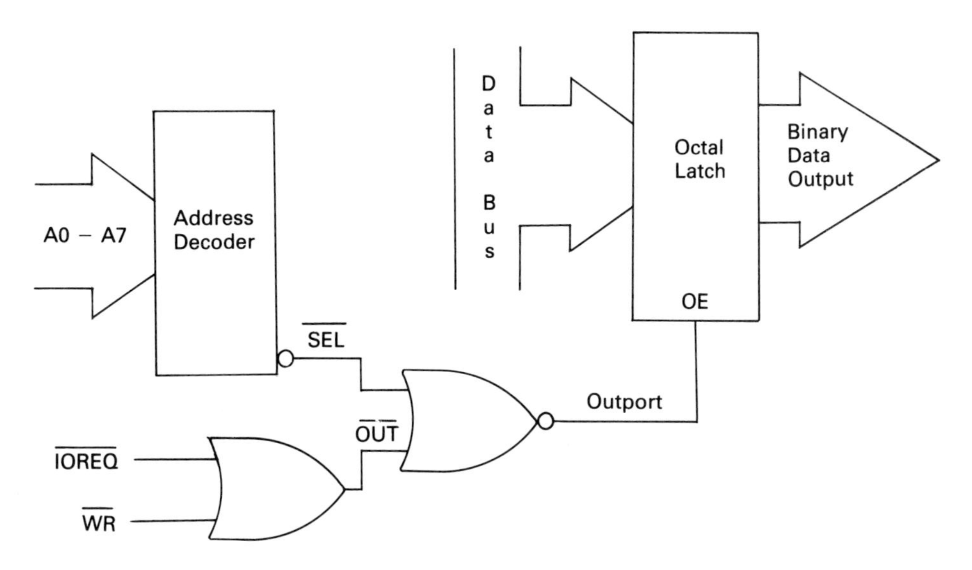


**Figure 26.8** A Z80 output port

This design scheme means that *all MC6800 I/O is done using memory-mapped techniques.* The 6800-series microprocessors include the system clock in their read/write operations, as well as the normal control bus signals. Another new signal on the control bus is *VMA, valid memory address.* VMA is high only when the address on the address bus is a valid memory address. During internal processor operations, the address bus may contain spurious highs and lows. VMA is used to prevent these signals from enabling any memory or memory-mapped ports.

Figure 26.9 shows the basics of a 6800 input port. The sequence of operation is:

1. The microprocessor places the port address on the address bus. This is decoded by the address decoder into SEL1.
2. When the address has had time to become stable, the microprocessor places a high on VMA and a high on the R/W control line. VMA is ANDed with the write signal and with phase two of the system clock into SEL2.
3. SEL1 and SEL2 are ANDed to yield SEL, which enables the three-state buffers. The buffers pass the input data to the data bus.
4. On the falling edge of the phase-two clock, the microprocessor reads the input data.

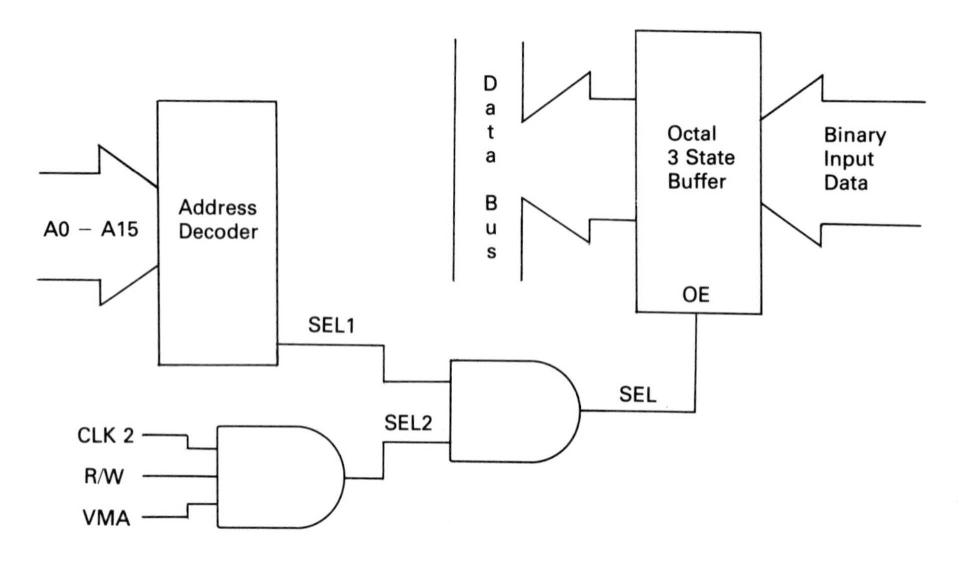


**Figure 26.9** A 6800 input port

The 6800 output port follows a similar approach.

The 6800 memory-mapped output port is shown in Figure 26.10. The sequence of operation is:

1. The microprocessor places the port address on the address bus. This is decoded by the address decoder into SEL1.
2. When the address has had time to become stable, the microprocessor places a high on VMA and a low on the R/W control line. The low on the R/W control line is inverted and ANDed with VMA and phase two of the system clock into SEL2.
3. SEL1 and SEL2 are ANDed to yield SEL, which enables the output latches.
4. The microprocessor places the output data on the data bus, and it passes into the already enabled latches. On the falling edge of the phase-two system clock, the data is stored in the latches for use by the external device.

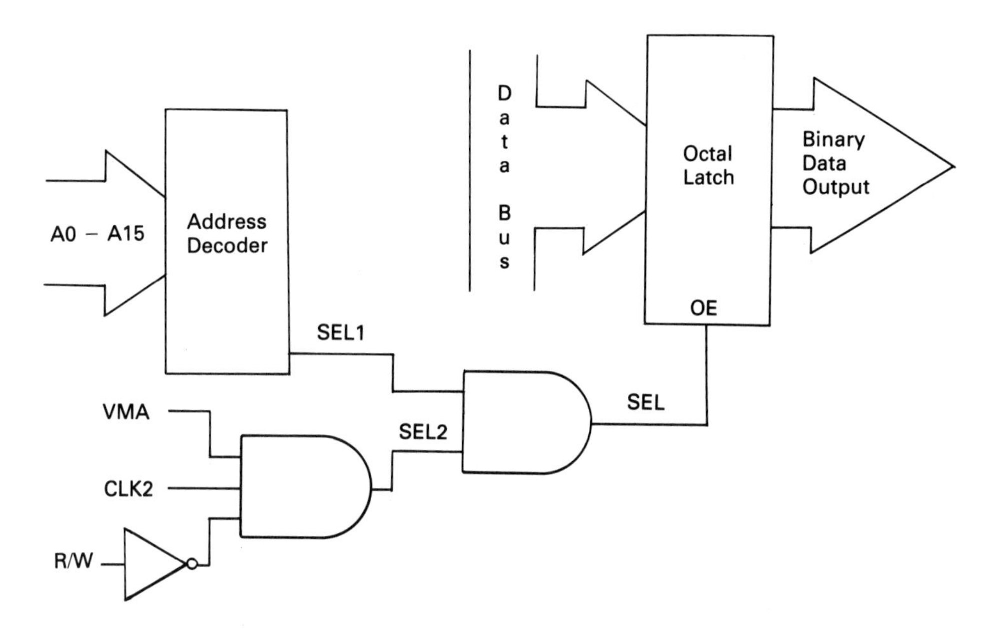


**Figure 26.10** A 6800 output port

## 26.6 THE 6502 MICROPROCESSOR

The 6502 microprocessor is finding limited use in control circuits. It is included here because it is used in a wide variety of personal computers, which are being pressed into service as controllers. The 6502 read and write cycles are similar to the 6800 cycles, but they lack the VMA signal. As with the 6800, only memory-mapped I/O is possible with a 6502 system.

Figure 26.11 shows a 6502 input port. The sequence of operations is:

1. The microprocessor places the port address on the address bus. This is decoded by the address decoder into SEL1.
2. At the same time, the microprocessor places a high on the R/W control line. This is ANDed with phase two of the system clock into SEL2.
3. SEL1 and SEL2 are ANDed to yield SEL, which enables the three-state buffers. The buffers pass the input data to the data bus.
4. On the falling edge of the phase-two clock, the microprocessor reads the input data.

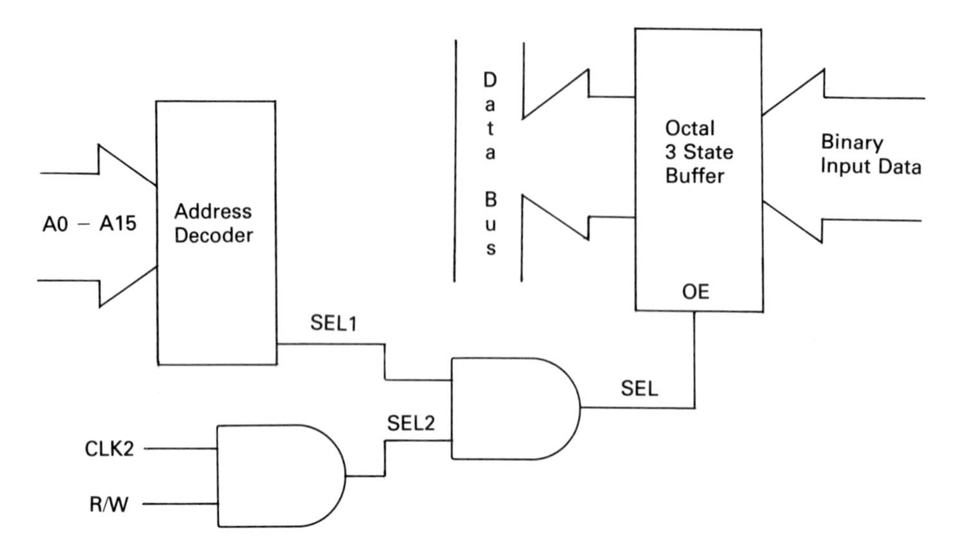


**Figure 26.11** A 6502 input port

In Figure 26.12, we can see the basics of a 6502 output port. The sequence of operations is:

1. The microprocessor places the port address on the address bus. This is decoded by the address decoder into SEL1.
2. When the address has had time to become stable, the microprocessor places a low on the R/W control line. The low on the R/W control line is inverted and ANDed with phase two of the system clock into SEL2.
3. SEL1 and SEL2 are ANDed to yield SEL, which enables the output latches.
4. The microprocessor places the output data on the data bus, and it passes into the already enabled latches. On the falling edge of the phase-two system clock, the data is stored in the latches for use by the external device.

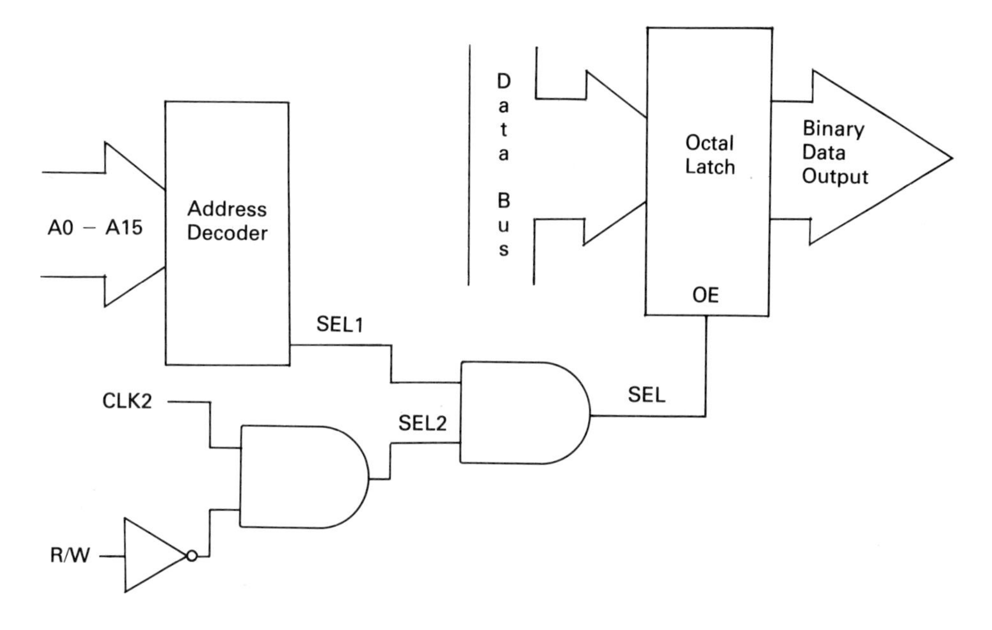


**Figure 26.12** A 6502 output port

## 26.7 THE 8088 AND 8086 MICROPROCESSORS

The INTEL 8088 and 8086 16-bit microprocessors are the microprocessors selected by IBM for their line of personal computers. The improvements over the eight-bit microprocessors include direct addressing of one megabyte of memory space and a powerful instruction set. Enhanced versions of these microprocessors include the 80286 and 80386. These later generation microprocessors are similar to their cousins, but include access to larger areas of memory and much enhanced instruction sets. These microprocessors are capable of much higher speeds of operation and much faster processing times than their eight-bit predecessors.

They are also finding their way into control circuits, either on dedicated control boards or through the coupling of IBM personal computers and their clones (imitators) into control applications. Because of the popularity of the IBM PC, we must include these powerful microprocessors in our work.

Like the other microprocessors by INTEL, the 8088/8086 microprocessors are *register oriented*: they contain many user-programmable internal data registers. Thus, these microprocessors also have *register I/O*. Eight-bit data can be input to or output from either the high-order half of the accumulator (AH) or the low-order half of the accumulator (AL). However, there are two distinct addressing modes for I/O with these microprocessors.

The *fixed-port* mode is used when no more than 256 ports will be needed for either input or output. In this mode, the instruction format is the same as with the 8080/8085 microprocessors: IN <B2> or OUT <B2>, where <B2> is the port number. The address decoder for this mode must decode an eight-bit address.

When more than 256 ports are needed, the *variable-port* mode must be used. In this mode, a sixteen-bit port address is first stored in the microprocessor's internal sixteen-bit register, DX. Once this is done, the instruction used is IN AL,DX or OUT AL,DX (to input to the low-order accumulator). The address decoder for this mode must decode a sixteen-bit address. As many as 65,536 different I/O ports are available in this mode.

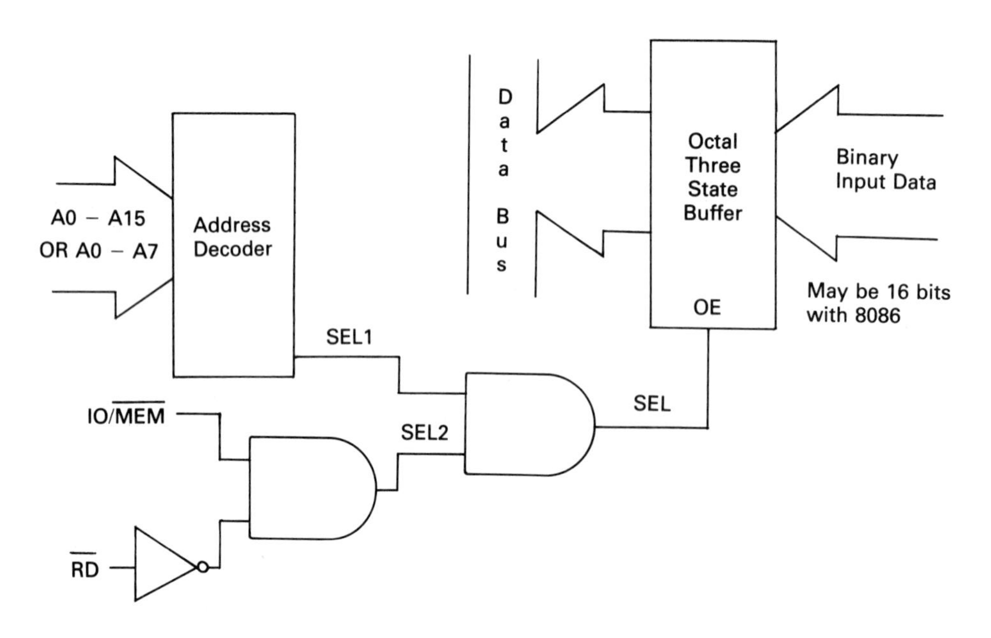


**Figure 26.13** An 8088 input port

In Figure 26.13, we can see the hardware essentials for an 8088 input port. The sequence of operation is:

1. The microprocessor places the port address on the address bus. This is decoded by the address decoder into SEL.
2. The microprocessor places a high on the control-bus lead IO/MEM*.
3. The microprocessor places a low on the control-bus signal RD*. This signal is inverted and ANDed with SEL and IO to enable the three-state buffers, which place the input data on the data bus.
4. The microprocessor reads the data from the data bus into the AL or AH register.

The only variable in this idea is the address decoder, which must decode an eight-bit address for a fixed port and a 16-bit address for a variable port.

In Figure 26.14, we see an 8088 output port. The sequence of operations is:

1. The microprocessor places the port address on the address bus. This is decoded by the address decoder into SEL.
2. The microprocessor places a high on the control-bus lead IO/MEM*.

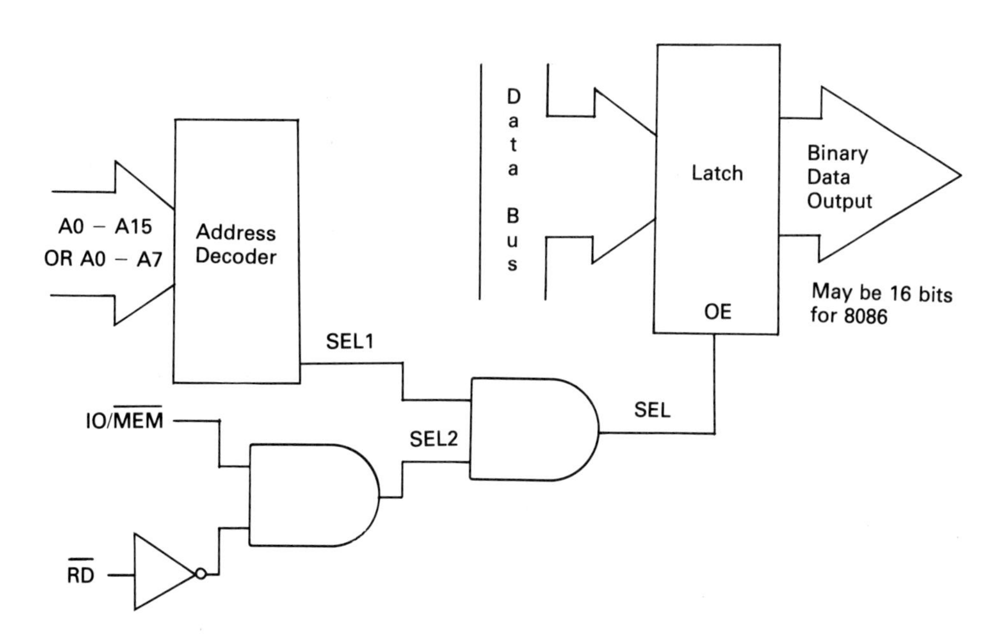


**Figure 26.14** An 8088 output port

3. The microprocessor places a low on the control-bus signal WR*. This signal is inverted and ANDed with SEL and IO to enable the eight-bit latches.
4. The microprocessor places the data from the accumulator on the data bus, where it passes into the latches. On the rising edge of WR*, the data is latched and is available for use by the external device.

Memory-mapped I/O is also available. It entails decoding a 20-bit address, rather than the eight-bit or 16-bit address for the register I/O ports. Other than that, the main difference between the memory-mapped I/O port and the register I/O port is that the IO/MEM* control-bus lead is made low by the microprocessor to signal that a memory operation is taking place. To use this signal in the ports shown in Figures 21.12 and 21.13, it must be inverted.

## 26.8 TROUBLESHOOTING I/O PORTS

Considering the wide variety of input and output port methods, troubleshooting any I/O port is fairly simple. It does require some knowledge of machine language programming and an understanding of the control-bus signals in use for that particular port. You need not, however, be a master programmer to succeed. The steps are as follows:

1. Write a short machine-language routine that will access the port as a part of an infinite loop. In other words, to troubleshoot an input port, write a routine that will input data from the port over and over again.
2. Use a dual-trace oscilloscope to monitor the signals on the control bus. For an input port, the IO READ signal must be active *at the same time as the output of the address decoder.* The IO READ signal varies according to the microprocessor. For the 8080/8085, it is IN* or OUT*. For the 6800, you must pay attention to VMA, the R/W control line, the system clock, and so on.
3. If all of the control-bus signals are good, check the output of the various combinational logic circuits.

A logic probe will not provide all of the information needed to service an I/O port: timing of the signals is as important as the presence of the signals. The control-bus signal you see with a logic probe may not be related to your port at all. This is especially true with memory-mapped systems.

## 26.9 PARALLEL I/O CHIPS

All of the manufacturers of microprocessors also manufacture parallel I/O chips, which are designed to simplify interfacing. These chips connect directly to the various control-bus lines and provide all of the timing and control functions internally. The only external support they require is usually an abbreviated form of address decoding.

Because of the complexity of the many different parallel I/O chips, a detailed discussion is beyond the scope of this text. Refer to the manufacturer's literature or to your microprocessor textbook for details about specific parallel I/O chips.

## 26.10 A/D CONVERTER CHIPS

Much of the data collected by sensors is analog data. Since our microprocessors are digital, this analog data must be converted to digital form before the microprocessor can work with it. Today, many manufacturers supply analog-to-digital converter chips (A/D or ADC) to ease this task. We will concentrate on just one such IC so that you can get an idea of how it works. The basics are the same for any ADC, although the method of conversion and the data acquisition times may differ.

Figure 26.15 is the pin-out diagram of the ADC 0804 analog-to-digital converter. This IC connects to the data bus through built-in three-state buffers, eliminating the need for a separate buffer IC. A low on the chip-select input will enable the chip. A low on the WR* pin at the same time will command the chip to begin converting the analog voltage at the differential inputs into a digital quantity. It uses the successive approximation technique for conversion.

When the conversion is complete, the ADC 0804 signals the microprocessor by placing a low on its INTR* pin. In response to this signal, the microprocessor once again makes CS* low and sends an RD* pulse. This enables the converted eight-bit data through the three-state buffers to the data bus.

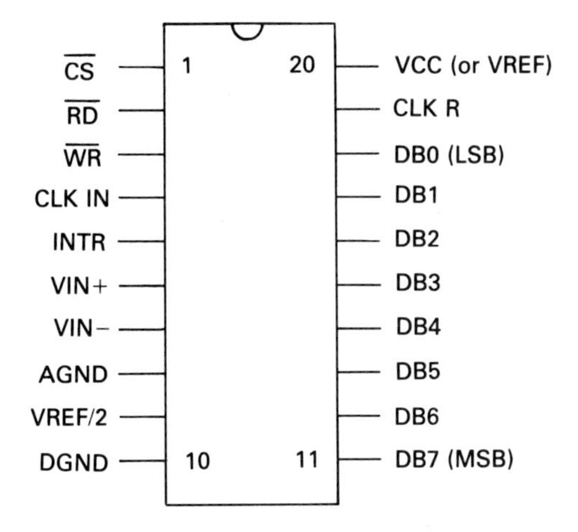


**Figure 26.15** The ADC 0804

The ADC 0804 is designed specifically to be interfaced to a microprocessor through an I/O port. The CS* signal for the IC can come from an address decoder, and the port read/write signals can be developed directly from the control bus.

There are several variations on ADC chips for microprocessor use. For more resolution (sensitivity to smaller differences in $V_{in}$), ten- and twelve-bit ADC converters are available. For use with an eight-bit microprocessor, the data register must be read *twice* to get full resolution. Another variation allows as many as sixteen different voltages to be measured with one ADC chip. Each different $V_{in}$ has its own input pin, which is selected by different port addresses.

## 26.11 INTERFACING WITH HIGH-POWER DEVICES

Using a microprocessor system to control industrial actuators requires the system to find a safe way to interface the low voltage dc power system of the microprocessor to the high-voltage, high-current industrial power system. Any single component is subject to failure — we don't want a component failure to allow the industrial power to reach the microprocessor system and destroy it.

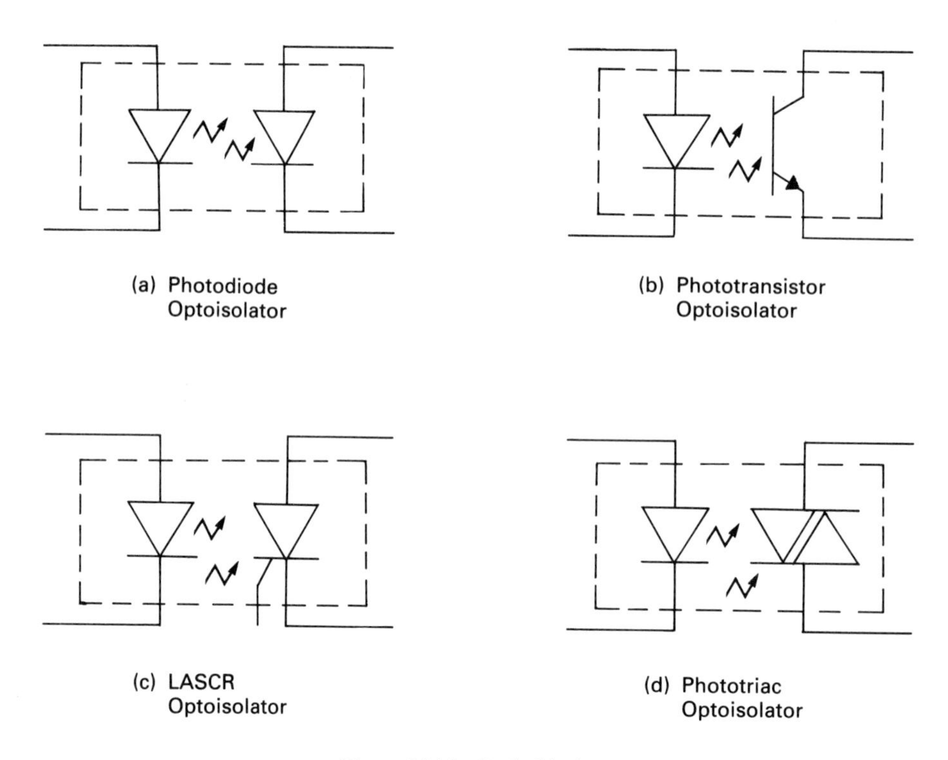


**Figure 26.16** Optical isolators

Figure 26.16 illustrates various types of *optical isolators.* An optical isolator consists of a light source, usually an infrared LED, a light path, and a photosensitive switch. The light path is clear epoxy or glass, either of which will pass the light readily and offer high insulation value. Voltage isolation is typically 5000 volts or greater. That means that as long as the voltage difference between the computer power system and the industrial power system is less than 5000 volts, there will be no electrical connection between them.

In Figure 26.17, we see the application of an optical isolator to control an ac motor. An output low from the microprocessor output port lights the LED. The light from the LED turns on the photosensitive TRIAC, turning on the motor. Note that the optical isolator's TRIAC is used to control a second TRIAC. This is because most optical isolator TRIACs are low-current devices and cannot handle the full current of any but the smallest of ac motors. A single eight-bit output port can control as many as eight motors in this manner.

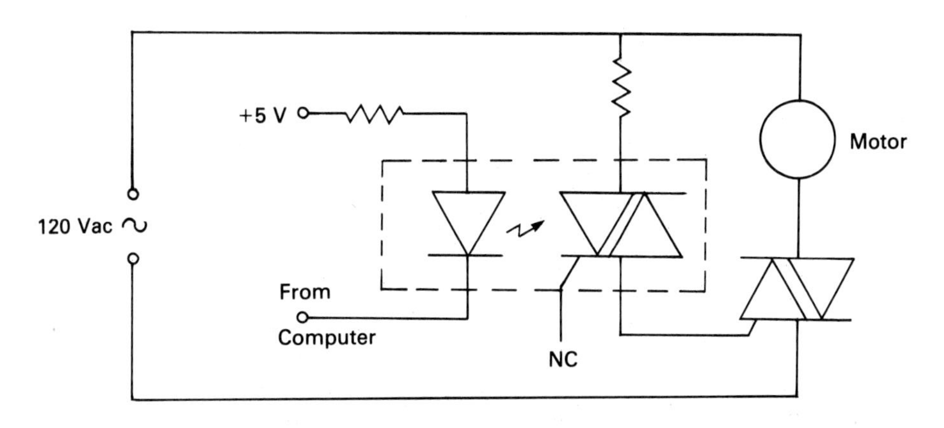


**Figure 26.17** Controlling an ac motor

The phototransistor output optical isolator can be used to control low-voltage dc motors directly, as can the LASCR (light-activated SCR) type. The difference is that once the LASCR is triggered on, it stays on, as you might expect from an SCR. To control higher voltage motors, these devices can be used to operate larger external semiconductors or relays.

## SUMMARY

In this chapter, we have examined input and output ports for the most popular microprocessors on the market today. We have seen that:

1. An I/O port must have an address decoder so that one specific port may be selected.
2. For an input port, the data bus must be isolated from the outside world devices by a three-state buffer.
3. For an output port, the data must usually be captured in latches so that the external device can have time to work with the data.
4. When inputting analog data, an analog-to-digital converter IC is used.
5. When controlling high-voltage, high-current devices with a microprocessor, optical isolators should be used to protect the microprocessor system from the higher voltages.

## SELF-TEST

1. In ______________________________ I/O, the microprocessor treats I/O as though it were memory.
2. In ______________________________ I/O, a separate addressing scheme is used for I/O ports.
3. An input port must have ______________________________ to isolate the data bus from the outside world.
4. Most output ports contain ______________________________ to store the output data until the external device can use it.
5. I/O ports contain ______________________________ so that the microprocessor can access one specific port.
6. Analog data must be converted into ______________________________ form for use by the microprocessor.
7. ______________________________ isolators are used to enable a low-voltage microprocessor system to control high-voltage, high-current devices without risk of damaging the microprocessor system.

# 27

# The Computer as a Controller

In the last chapter, we saw that the microprocessor can be easily connected to circuits for controlling actuators. What makes the microprocessor so valuable is that it is *programmable*. Circuits using discrete logic must be designed for a specific task. If the task is changed, then the circuit design must also be changed. When a microprocessor is used, however, all that must be changed is the *program*.

It is this programmability that makes microprocessor-based control systems economical. When a process changes, only the program must be changed. In a system using discrete logic circuits, the whole system must be changed, new logic circuits designed, new pc boards etched. For each process in an operation, a new controller must be designed. With a microprocessor-based system, the same type of controller may be used for each process.

In this chapter, we will look at the programmable controller and at the personal computer as a controller.

## OBJECTIVES

You will have successfully completed this chapter when you can:

1. Describe the use of programmable controllers.
2. Describe the use of personal computers in control applications.

## 27.1 THE PROGRAMMABLE CONTROLLER

A *programmable controller* is a microcomputer that is dedicated solely to control applications. Its operating system is designed specifically for control work. It is designed to make programming and using the controller simple from the point of view of floor operating personnel.

The programmable controller shown in Figure 27.1 is the Director One programmable logic controller by Struthers-Dunn, Inc. This unit can handle up to 128 discrete I/O points. Input and output modules are available for all standard variations of I/O.

The unit is programmed for a sequence of steps which are entered as mnemonics from a hand-held Diagnostic Programmer module. Timer, counter, shift register, math, and logic functions are available for as many as 1,970 program steps.

The Director 4002 system shown in Figure 27.2 accepts three-digit BCD data from the input modules shown on the left of the photo. The processor can be programmed for PID control, special math functions, and conditional programming.

Both units can be coupled to personal computers through an RS232 serial data port. The computer can be programmed to log data, or to send control programs to the controllers.

These two units are typical of the many programmable controllers available today. They range in I/O capability from as few as 20 I/O points to the 8192 I/O points of the 565 unit. Programming generally follows relay ladder logic — if you can read a ladder diagram, you can read the controller's program. In the more sophisticated units, PID function blocks may be added to the ladder logic for a finer degree of control.

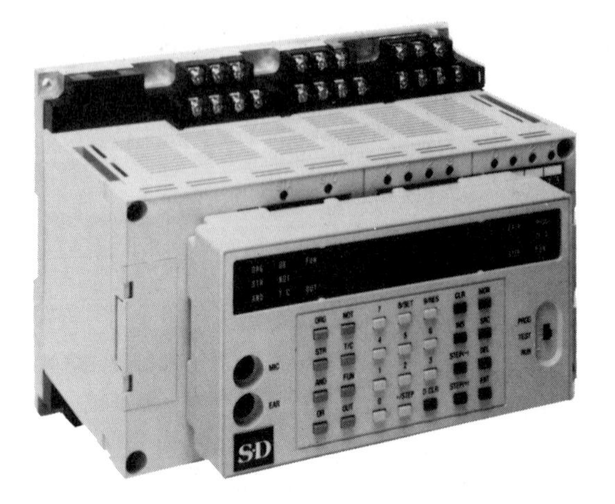


**Figure 27.1** A programmable controller *(Courtesy of Uticor Technology, Inc.)*

**Figure 27.2** A larger programmable controller *(Courtesy of Uticor Technology, Inc.)*

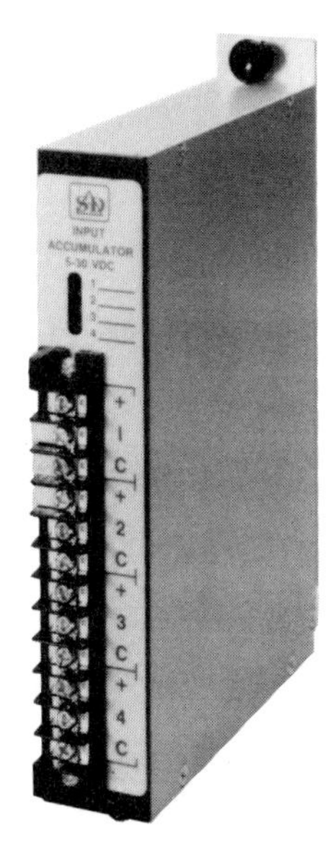


**Figure 27.3** A microprocessor counter *(Courtesy of Uticor Technology, Inc.)*

## 27.2 OTHER MICROPROCESSOR SYSTEMS

In addition to the fully programmable controllers we saw in the previous section, microprocessors are also being used in specialized controllers.

The Accumulator Module in Figure 27.3 is designed to accept incoming pulses at rates up to 10 KHz. It contains four two-digit BCD counters which can be used independently or cascaded in combinations up to eight digits.

While there is nothing particularly unique in a counter that can perform all of these functions, the use of the microprocessor reduces the cost and complexity of the system. Since microprocessors are relatively inexpensive today, they are being used in more and more simple control applications, which once were the exclusive ground of discrete logic. One company even used one to count the number of balls played in a pinball game!

The stepper motor control module shown in Figure 27.4 is a microprocessor-based module which provides a serial pulse train that is interfaced to a translator for driving the stepper motor. All data required for control of the stepper motor is provided by the user program. Required data are the maximum step rate, number and direction of steps, ramp up, and ramp down times.

The motion controller is designed to be used in conjunction with the Director series of programmable controllers shown in Figures 27.1, 27.2, and 27.3.

## 27.3 THE PERSONAL COMPUTER AS A CONTROLLER

Even the personal computer has found its way to the shop floor as a controller — a major advance from playing video games.

The personal computer can be converted for control work in two different ways: as an add-on external system or as a plug-in card. External add-ons usually interface

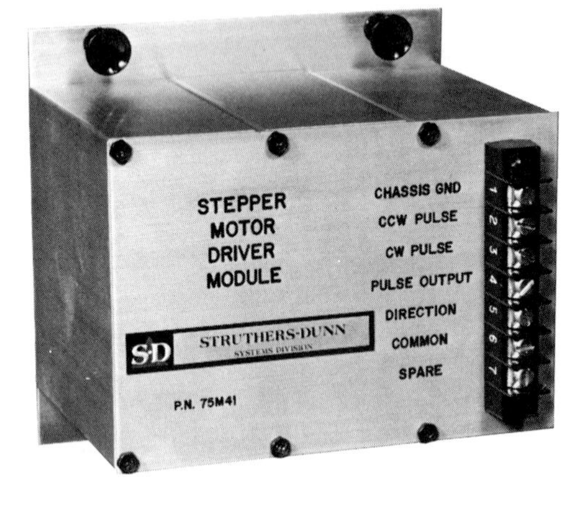


**Figure 27.4** A microprocessor motion controller *(Courtesy of Uticor Technology, Inc.)*

**Figure 27.5** Add-on controller modules *(Courtesy of Uticor Technology, Inc.)*

with the computer through the RS232 port — the same port you use for your modem connection.

The add-on controller shown in Figure 27.5 is a typical add-on. It has its own housing and power supply. The number of I/O points in this system depends on the number of cards: a single input board will normally handle twelve separate inputs. The boards are designed for specific input types — strain gages, RTDs, and thermocouples each require their own type of board. The boards provide signal conditioning and A/D conversions and output the data to the personal computer in serial form on the RS232 link. The computer program must then make decisions based on the data and output appropriate control signals.

Some plug-in controller cards are designed for use with the IBM PC, XT, AT, or one of the clone computers, or with the Apple computer models.

These cards typically perform A/D conversion for as many as sixteen separate inputs. Signal processing is not included. Once again, the computer program must make the decisions and output control signals. Similar plug-in cards are available for output signals. These are limited in their current capacity, and should be used to drive low-voltage relays.

## SUMMARY

In this chapter, we have seen a few of the commercial controllers that are designed around microprocessors. We have seen that:

1. A programmable controller is a microcomputer designed specifically for control applications.
2. Most programmable controllers are programmed using relay ladder logic. Additional features, such as mathematical functions, may be added.
3. Microprocessors are also being used to simplify the design of special-purpose controllers.
4. Add-on units and plug-in cards are being used to convert the personal computer for work as a controller.

## SELF-TEST

1. Programmable controllers have the advantage of being ________________ ________________, eliminating the need to design a custom controller for each application.
2. Programmable controllers are usually programmed with ______________.
3. Special-purpose controllers use microprocessors to simplify the design and ____________________________ the parts count.
4. When a personal computer is used as a controller, plug-in or add-on circuits are used for ____________________________.
5. When a personal computer is used as a controller, the decision making must be done by the ____________________________.

## Troubleshooting Controllers

Troubleshooting a controller usually involves troubleshooting the interface circuits: the input and output portions of the system. As always, a careful visual inspection and a check of the power supplies can often save hours of frustrating work. Check the power supplies of a digital or microprocessor controller with an oscilloscope as well as a DVM. Some "fuzz" riding on the dc level is normal in these systems, but watch for pulses on the power supply lines that are greater than 5% of the power line voltages. These intermittent pulses are caused by switching transients when counters and other logic devices change states. They can introduce an unwelcome "glitch" into the system. Failure of a bypass capacitor allows these pulses to reach levels large enough to confuse the system's logic.

Most of the system's inputs involve sensing contact closures or A/D conversion. Troubleshoot by simulation.

Substitute a short jumper wire or a dual-ended alligator clip lead for a contact closure that is not being sensed. If the controller responds properly, the problem is external. If the controller does not respond properly, the problem is most likely in the input transistor or optical isolator.

Test the A/D converter by supplying it with a known voltage from a potentiometer voltage divider off the system power supply. If the controller does not respond properly, check for a *conversion complete* signal from the A/D converter and for a response from the system's read pulse with a dual-trace oscilloscope or a logic analyzer.

One of the most common failures on the output side of a controller is failure of a relay. Inspect the contacts for evidence of arcing or contact welding. Except in the mildest cases of this, do not attempt to clean the contacts. Replace them, and replace the snubber circuit that was designed to prevent such failure.

If the relay fails to operate, check for coil voltage with a DVM and, if operating voltage is present, check for coil continuity with an ohmmeter. When the operating voltage is missing, the switching transistor is the most likely suspect. When you must replace the switching transistor, check the snubbing diode as well. It is most likely open.

If the system output is a voltage, test it under simulated load conditions. The system may output the proper voltage with no load connected, but fail under full-load conditions. The most likely cause of this is a failed or failing output or driver transistor. Use a similar full-load simulation for SCR and TRIAC outputs.

Troubleshooting any system requires patience, logic, experience, and luck. Do not hesitate to rely on the experience of the manufacturers of your controllers. All of these companies maintain staffs of experienced service engineers who will advise you by phone or mail. They can often be reached by a toll-free phone call. They are as much a part of your troubleshooting arsenal as are your DVM and oscilloscope.

# Section VI

# Robotics

In the past few years the word *robot* has undergone a change in meaning in the public consciousness. No longer the stuff of science fiction, the word *robot* today brings to mind a number of different images.

Assembly line workers see the robot as an enemy, a mechanical competitor for their jobs and the cause of unemployment. Some production supervisors see the robot as a tireless worker, never absent from the job, never demanding wage increases, and not unionized. The general public sees the robot as a many-jointed machine that moves heavy loads around at a blinding pace. Only our children still see the robot as a friendly tin guy, wandering around shopping malls.

In fact, except for the children's view, a robot is all of these things and more. In this section of the text, we will look at the concept of the robot in more detail. We will define what a robot is, what it is not, and we will see what makes it work.

# 28

# What Is a Robot?

Is robotics the threat to job security that labor believes? Is it the answer to production woes? Can the implementation of robotics in today's factories solve problems in the United States and in the world? Perhaps, but these questions must be left to time for an answer.

The fact is, the robot is a machine; more specifically, an *automatic machine.* In this chapter, we will define more precisely what a robot is and what it is not.

## OBJECTIVES

You will have successfully completed this chapter when you can:

1. Describe exactly what a robot is.
2. Define the term *degree of freedom.*
3. Explain the difference between rectilinear robots, cylindrical robots, spherical robots, and fully articulated robots.

## 28.1 A DEFINITION OF A ROBOT

The Robot Institute of America (RIA) is a branch of the Society of Manufacturing Engineers (SME). The RIA definition of a robot was accepted by the eleventh International Symposium of Industrial Robotics in 1981. That definition is:

> A robot is a reprogrammable multifunctional manipulator designed to move material, parts, tools, or other specialized devices through variable programmed motions for the performance of a variety of tasks.

Let's analyze the definition:

- A robot is a reprogrammable machine. That means that its task can be changed by changing a program — a set of instructions. Teach it how to unload a pallet and it will unload pallets. Teach it how to drill a hole and it will drill a hole.
- A robot is multifunctional. That means that it can do more than one job. A numerically controlled lathe can be reprogrammed, but it will always be a lathe. A robot can be a parts mover today and a painter tomorrow. Although it is unlikely that such a change would be made, it *could* be made. This fits in with the idea that a robot can perform a variety of tasks.

Let's see what a robot is *not*. The "robot" that sometimes wanders your shopping mall is a remote-controlled machine, controlled by a concealed operator with an RC transmitter and a microphone. It's fun, but it is not a robot.

The "robot arm" of the space shuttle is not a robot either, for much the same reason. NASA calls it a "manned manipulator arm" because it needs an operator to steer it and command its actions.

## 28.2 THE ROBOT ASSEMBLY

A robot is an assembly of several major sections. It is made up of the arm, the end effector tooling, the power supply, the controller, and peripheral tooling.

The *arm* provides the necessary motion to move the tool or part into the proper position for an operation. The sole function of the arm is to provide motion for the end tooling.

The *end effector* is the tool that performs the actual work. We tend to think of a robot arm ending in a gripper mechanism, but most robots in industry have more specialized end-of-arm tooling. Because robots themselves are so flexible, there is an almost infinite variety of end effectors, special tools, tool holders, or manipulator assemblies designed specifically for attachment to the robot arm itself. Each end effector is designed for the specific task that the robot is required to perform.

The *power supply* provides the necessary power for moving the robot arm through its range of motion. Most industrial robots use electric motors or stepper motors for their motion. The only disadvantage is the increasing cost of electrical power. Some arms use hydraulic or pneumatic power. The power source for hydraulic or pneumatic arms is generally much larger than an electric supply. The cylinders and pistons for these arms are larger than their electrical equivalents, as well.

The *controller* provides the instructions that tell the arm what to do. Robot controllers are almost always computers. Relay ladder logic cannot operate with a sufficient degree of complexity or at high enough speeds for robotics use.

*Peripheral tooling* is not a part of the robot itself. It consists of tools and jigs designed to hold a part or a tool in the correct position for the robot. At times it is necessary to rethink the way a part is made so that it can be adapted to robotic assembly and handling. As robot vision systems are improved, the need for such additional tooling will be reduced.

## 28.3 DEGREES OF FREEDOM

Different types of robotic arms have different axes of motion — these are called *degrees of freedom.* Among the many ways of classifying robots, degrees of freedom is one of the most important.

Figure 28.1 illustrates the different degrees of freedom available to a fixed-base fully articulated robotic arm. Motion of the entire arm about the fixed base is called *waist motion* or *arm sweep.* The next movement above the waist is the *shoulder* or *vertical motion.* The third movement is called *elbow extension.* Virtually all fully articulated robotic arms have these three degrees of freedom.

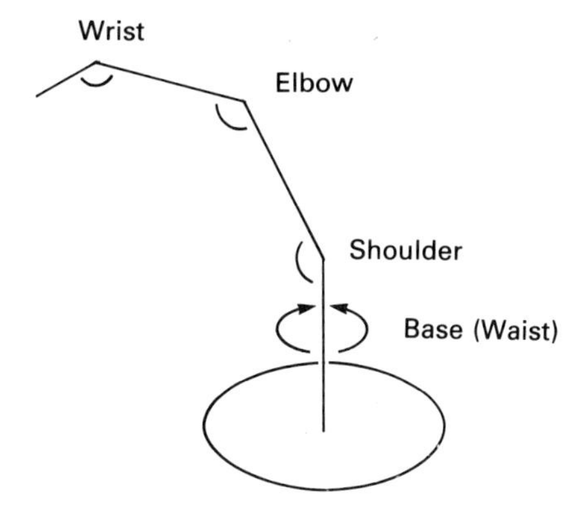


**Figure 28.1** Degrees of freedom

The movement of the wrist is less universal. Three different types of wrist motion are possible: pitch, yaw, and roll. Because of differences in need and the cost of including all three, many robotic arms have only one or two types of wrist motion. *Pitch* is an up-and-down motion, in the same plane as the motion of the elbow or shoulder. *Yaw* is side-to-side motion, at right angles to the motion of the shoulder and elbow. *Roll* is a rotation of the wrist about the axis of the forearm, the motion you use when you tighten a screw.

Many robotic arms for industry have *none* of these wrist motions included in their initial design. Rather, pitch, yaw, and roll are left to the *end-of-arm tooling*, or *end effector.*

## 28.4 ROBOTIC ARM ARTICULATION

The area within which a robotic arm can work is called its *work envelope.* The end effector can reach every point within the work envelope. The shape of the work envelope depends on the number of degrees of freedom and the type of *articulation* that the arm has.

The robot arm shown in Figure 28.2 has *rectilinear articulation.* It can move its end effector in only three directions: up and down, left to right along a track, and front to back. The work envelope of this robot is box shaped. Rectilinear robots have a smaller range of motion than the other types, but they are the easiest to program because of the simple rectangular coordinate system they use.

The *cylindrical robot* in Figure 28.3 has a greater range of motion than the rectilinear robot because the arm can swing around its base in *circular* or *polar* motion. The up-and-down and front-to-rear motion of this robot is rectilinear. The work envelope is a cylinder with a core at the center that cannot be reached.

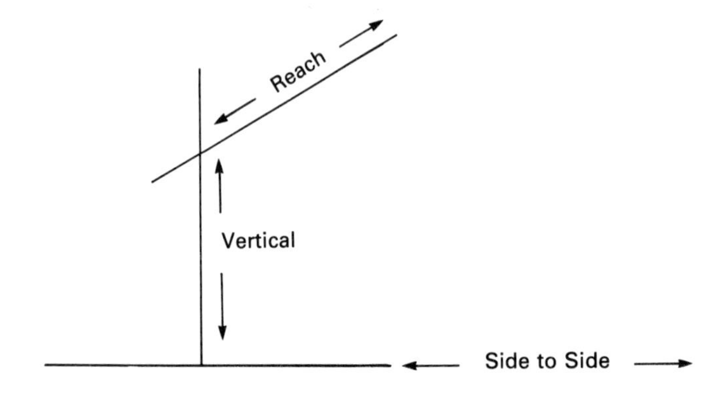


**Figure 28.2** The rectilinear robot

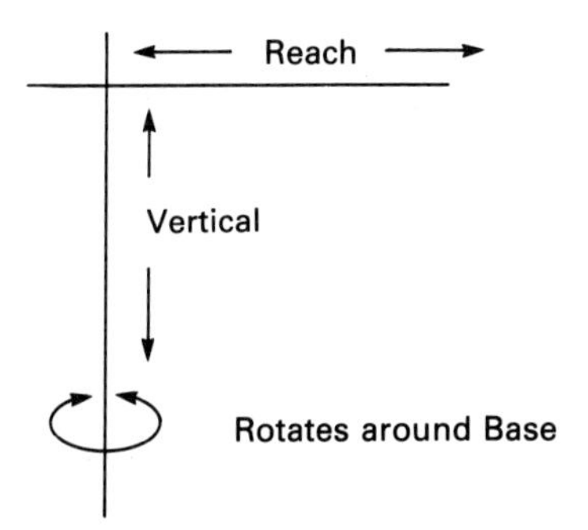


**Figure 28.3** The cylindrical robot

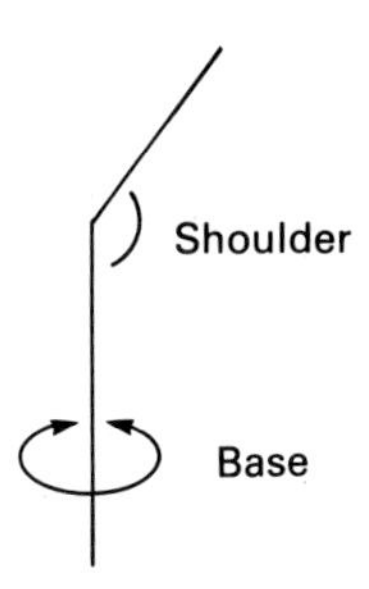


**Figure 28.4** The spherical robot

The *spherical robot* shown in Figure 28.4 has polar articulation at the waist and the shoulder, but uses rectilinear motion for reach. The work envelope is roughly spherical, minus a pie-shaped wedge.

The *fully articulated robot* of Figure 28.5 uses polar articulation for all degrees of freedom. This is the most flexible scheme of articulation. It is also the most difficult to program. The fully articulated arm is the most popular arm in industry because of its extended ranges of motion. Its work envelope is usually a pie-shaped crescent in the horizontal plane and an irregular spheroid in the vertical plane.

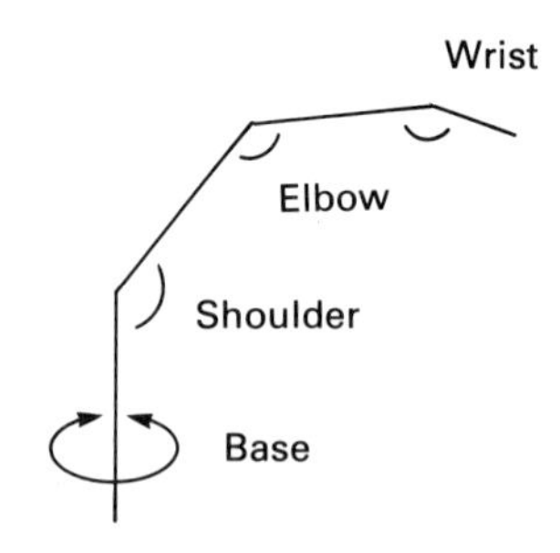


**Figure 28.5** The fully articulated robot

## 28.5 TYPES OF CONTROL

In addition to the range of motion a robotic arm can have, an important consideration in robot selection is the type of control available for the robot. There are essentially two types of control: servo-positioning and nonservo-positioning control.

The *servo-positioned* robot is equipped with feedback sensors so that the controller knows the exact position of the end effector at all times. This is an important feature in *continuous-path operation*, such as spray painting and welding, where every motion of the tool is important to a correct result.

The *nonservo-positioned* robot lacks the positional feedback and the fine degree of control found in the servo-positioned robot. Each degree of freedom has a home position, which serves as a point of reference for all motion. The home position is sometimes detected by a limit switch. These robots are often powered by stepper motors. The controller assumes that when a certain number of steps have been taken, the arm is in a particular position. If any joint slips, the controller has no means of detecting the error. Nonservo-positioned robots are used in pick-and-place operations such as loading or unloading parts pallets and assembly lines.

## SUMMARY

In this chapter, we have begun to look at the robot. We have seen that:

1. A robot is a reprogrammable machine.
2. A robot is a multifunctional machine.

3. The robot assembly consists of an arm, an end effector, a power supply, a controller, and peripheral tooling.
4. The robot's axes of motion are called degrees of freedom.
5. The area within which a robotic arm's end effector can work is called the work envelope.
6. In a servo-positioned robot, the controller knows where the end effector is at all times.
7. In a nonservo-positioned robot, the controller is not aware of the position of the end effector.

## SELF-TEST

1. A CNC milling machine can be reprogrammed to manufacture a different part. Is this machine a robot?
2. The ______________________________ provides the motion to move the robot's tool into position to perform a task.
3. The _________________________ is the portion of a robot that performs the actual work.
4. The energy for robotic motion comes from the robot ________________.
5. Special jigs and tooling to hold parts in the proper position for a robot falls into the category of __________________________ .
6. The number of different ranges of motion available to a robotic arm is called ____________________________.
7. Up-and-down motion of the wrist is called ________________________.
8. Side-to-side motion of the wrist is called __________________________.
9. Rotary motion of the wrist is called ______________________________.
10. The axes of motion of a rectilinear robot are all at __________________ angles to each other.
11. A cylindrical robot features ___________________________ rotation about the base.
12. A spherical robot features polar movement at waist and shoulder, but the reach is ____________________________.
13. A fully articulated robot features _________________________ movement at all axes of rotation.
14. The ____________________________ robot is easiest to program.
15. With a(n) __________________________ robot, the controller knows at all times exactly where the end effector is.
16. What type of robot would you use for a painting operation? For a pick-and-place palletizing operation?

# 29

# Robotic Drives

Controlling the activities of a robot involves controlling the operation of the robot's *drive system*. While the *controller* is almost invariably a computer, the drive system that the computer must control depends largely on the design of the robotic arm — which, in turn, depends upon the function of the arm.

For example, an arm designed for point-to-point movement of heavy materials may well have a hydraulic drive system, while an arm designed for continuous-path welding operations will most likely have an electric motor drive. In this chapter, we will look at the main drive systems used in today's industrial robots.

## OBJECTIVES

You will have successfully completed this chapter when you can:

1. Describe the operation of a robot electric drive system.
2. Describe the operation of a robot hydraulic drive system.
3. Describe the operation of a robot pneumatic drive system.

## 29.1 ELECTRIC DRIVE SYSTEMS

Electric drive systems for robots are known for their quiet operation and the relatively small size of their power supplies. The growth of robotics in the past few years has led to major advances in electric motor design.

The stepper motor is used mainly in small "training robots." Some of these robots have sufficient accuracy to be used in very light industrial and laboratory tasks, but for the most part the stepper motor robot is a rarity in industry.

Although it is used to power the pumps or compressors for hydraulic or pneumatic robots, the ac motor is not widely used in powering electric robots. Ac motors are not as easily controlled for either speed or position as are dc or stepper motors.

The dc motor is the backbone of industrial electrically powered robots. Torque, acceleration, position, and speed are easily controlled with a new type of dc motor.

The latest innovation in dc motors is the *brushless permanent magnet excited dc motor.* At first glance, this motor looks something like a PM stepper motor. Closer inspection may give the motor the appearance of a synchronous ac motor. In design, it is a sort of "inside out" permanent magnet dc motor.

Figure 29.1 illustrates the idea behind the brushless dc motor. The rotor is a ferrite or ceramic permanent magnet, and the *stator* contains the coils that are normally found in the armature of the brush-type dc motor. Coupled to the rotor shaft is a magnetic

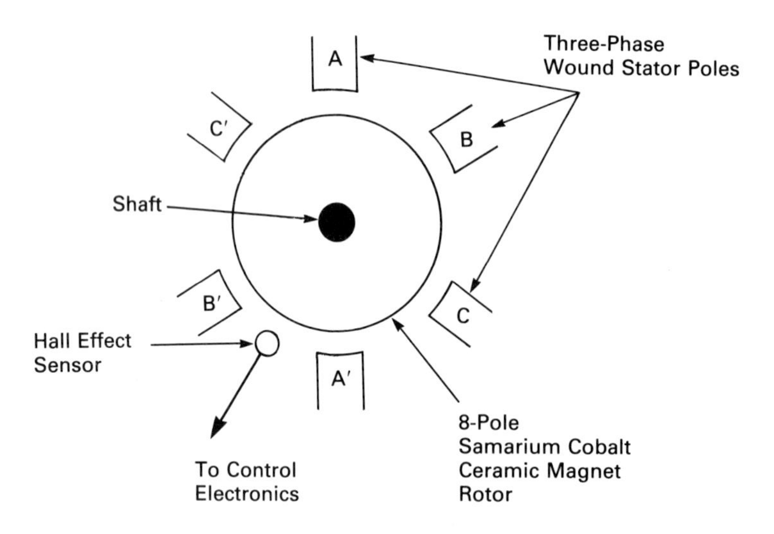


**Figure 29.1** A brushless dc motor

position sensor, which replaces the brush and commutator arrangement seen on a wound rotor dc motor. A commutation signal is sent back to an electronic controller, signaling it to change field windings, thus creating a rotating magnetic field similar to that found in an ac motor.

The control electronics can control the speed, acceleration, and torque of this new motor by varying the current to the field windings or the rate of rotation of the magnetic field. The control electronics can be commanded at any time to stop changing windings, providing the holding torque of a PM stepper motor.

The permanent magnet excited dc motor is available in horsepower ratings ranging from 0.25 hp to 10 hp. Advances in magnetics and motor design will no doubt increase the horsepower ratings over time.

Meanwhile, the brush-type dc motors, which were discussed in Chapter 19, are used for larger robots.

## 29.2 HYDRAULIC DRIVE SYSTEMS

Hydraulically powered robots are used to handle the heavy loads in today's industrial robot applications. Such robot designs utilize the mechanical advantage that can be gained with fluid power. In addition to the linear actuators and vane-type fluid motors shown in Chapter 22, rotary actuators and hydraulic gear motors are widely used in industrial robotics.

The rotary actuator shown in Figure 29.2 operates much like the vane motor shown in Chapter 22. When fluid under pressure enters port *A* in the illustration,

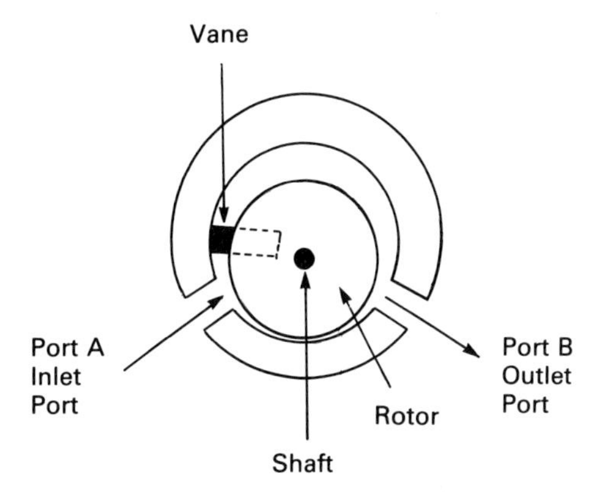


**Figure 29.2** A rotary actuator

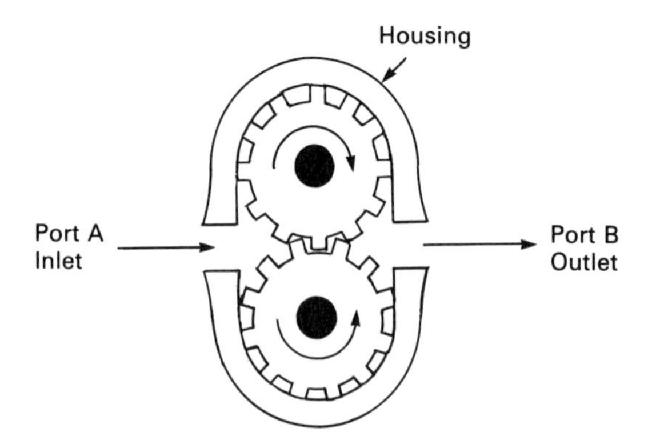


**Figure 29.3** A hydraulic gear motor

the vane is forced to rotate in a clockwise direction. Once the vane has moved past port *B*, the motion is stopped. Rotation is less than 360 degrees. Such an actuator may be used to rotate the elbow of a robot, or to close its gripper. Motion can be reversed by exchanging the input and output ports of the actuator through the use of direction control valves.

The hydraulic gear motor shown in Figure 29.3 is widely used to provide full rotary power in robotics applications. Fluid entering the motor at port *A* forces the two gears to turn as it flows *around the outside of the gears.* Depending on pressure and rate of flow, large amounts of torque can be developed by such a motor. Once again, direction reversal is accomplished by exchanging the two ports.

Great care must be used when servicing hydraulic systems. The hydraulic fluid must be kept meticulously clean. The smallest speck of dirt in the fluid can cause severe damage to actuators and pumps.

Another area of care is safety. A pinhole leak in a hydraulic hose releases an invisible stream of hydraulic fluid under pressure. Such a stream is quite capable of slicing off an arm or a finger before you are aware that you have been injured. ***Do not slide your hand along a hydraulic hose looking for a leak.***

## 29.3 PNEUMATIC DRIVE SYSTEMS

Pneumatic drive robots are less popular than hydraulic systems, but they may be found in several applications. Because the drive fluid of a pneumatic system is air, excess is simply bled off into the atmosphere, rather than piped back to a reservoir as is done in hydraulics. This reduces the cost of the system itself, since only the feed pipes and tubings are needed. In addition, since air is cheaper than hydraulic oil, operating costs are also lower.

Because the air in a pneumatic system has a degree of compressibility, a certain amount of sag is expected. Pneumatic systems are less accurate than hydraulic systems. They also are less capable of handling the heavy weights that are handled by hydraulic robots. They are used mainly in point-to-point operation with medium loads.

The same constraints for cleanliness and safety that apply to hydraulic systems also apply to pneumatic systems.

## SUMMARY

In this chapter, we have taken a brief look at the power systems used today in robotics. We have seen that:

1. Electrically powered robots mainly use the brushless permanent magnet excited dc motor.
2. Hydraulic robots are used to handle heavy loads and where high-speed operation is required.
3. Pneumatic robot systems are less powerful and less accurate than hydraulic robot systems.

## SELF-TEST

1. The dc stepper motor is used mainly in ______________________ robots.
2. The ______________________ is the primary dc motor used in modern-day electrically driven robots.
3. When a robot must handle a large load, a(n) ______________________ robot is selected.
4. Pneumatic robots are less accurate than hydraulic or electrically powered robots because of the ______________________ of the air in the system.
5. Hydraulic robots use ______________________ for rotations less than 360°.

# 30

# Robotics Applications

Robotics! In the minds of most people, this word brings visions of a fantastic future filled with laboring metal slaves. Advocates predict a paradise with all labor performed by these metal monsters, while opponents envision a future of unemployment and social unrest. In other words, robotics is a typical technological advancement. As in most technological advancements, proponents of robotics claim that the applications are limited only by the imaginations of future users. As in most technological advancements, opponents predict dire consequences from the application of the science.

In this chapter, we will look at a few of the many applications that have been found for robots in today's industrial world.

## OBJECTIVES

You will have successfully completed this chapter when you can:

1. Describe some of the applications of robotics in industry today.

## 30.1 ROBOTICS IN THE FINISHING DEPARTMENT

The typical industrial robot is best used today in specialized areas. Robotic arms are best used in areas that are too hazardous, too tedious, or too strenuous for human workers. One of the most hazardous areas in any large facility is the paint department. Modern finishing materials are usually hazardous on two fronts: they are highly volatile, often explosive materials, and they are hazardous to the health of the spray finisher.

Robots and automatic technology can reduce or eliminate both of these hazards. Hydraulic robots operate spark-free, reducing or eliminating the danger of explosion. Similarly, the robot has neither skin nor lungs and need not be protected from the dangerous health consequences of working with finishing materials. As an added bonus, a properly programmed robotic painter is so efficient in the application of finishing materials that it will usually pay for itself within two years in paint savings alone.

Figures 30.1a, b, and c show the application of the primer coat of paint to a Chrysler Corporation car. In Figure 30.1a, the entire car body is dipped into a primer undercoat material. In Figure 30.1b, the car body is sprayed with a coat of primer by a fully articulated robotic arm. The car body and the paint are electrically charged to assure adherence of the paint to the car body. Note that the control and power supply portions of the arm are protected from overspray by plastic covers. In Figure 30.1c, we can see a rectilinearly controlled sprayer applying a third coat to the car body. While these sprayers cannot be classified as true robots, they are a good example of automation at work.

In Figure 30.2, we have a clear view of the laser beam that is used to control the turbo-type paint sprayers. The laser beam follows the contour of the vehicle body, and the turbo sprayers' position is controlled by sensors. This assures that the sprayers will maintain a uniform distance from the finished surface, guaranteeing a uniform layer of paint.

All finishing operations do not involve paint. Many parts must be treated with corrosion inhibitors which are hazardous to human health. Figure 30.3 shows one such operation. Note here that the end-of-arm tooling has provisions for the treatment of three different parts.

The fully articulated robotic arm shown in Figure 30.4 is applying a uniform bead of sealant around the transmission hump of a truck body. The fumes from such sealants are frequently harmful to human applicators, requiring elaborate safety precautions.

## 30.2 WELDING OPERATIONS

Welding was one of the first operations to be automated by robotics. The actinic light from welding operations is extremely hazardous to the eyes, and the "sputter" from welding is an ever present danger to the skin.

**Figure 30.1a** Immersion priming *(Courtesy of Chrysler Corporation)*

**Figure 30.1b** Electrostatic spraying *(Courtesy of Chrysler Corporation)*

**Figure 30.1c** Electrostatic spraying *(Courtesy of Chrysler Corporation)*

**Figure 30.2** Laser-controlled painting robots *(Courtesy of Chrysler Corporation)*

**Figure 30.3** Robot dipping parts in an anticorrosive *(Courtesy of ASEA ROBOTICS Inc.)*

**Figure 30.4** Applying sealant *(Courtesy of Cincinnati Milacron)* Safety equipment may have been removed or opened to clearly illustrate the product and must be replaced prior to operation.

The robotic arm selected for welding operations is usually a fully articulated arm that can assume positions in a few seconds that may take human welders several minutes to reach. The arm must be capable of high repeatability and accuracy.

Figure 30.5 shows some of the welding operations that take place in an automobile plant. Note that several welds can take place at one time on a single unit. This is far too dangerous for human welders to attempt.

## 30.3 ASSEMBLY ROBOTS

Lighter robots are now beginning to appear on the assembly line, a task only recently deemed too complex for machines to handle. Many of the advances in robotic assembly have resulted from recent research into vision systems for robotic controllers.

The assembly station in Figure 30.6 shows a pair of fully articulated robots that can assemble complicated gear boxes. The vision-control system shown in the background can locate and identify the properly shaped gear for each assembly step and direct the arms in the assembly process.

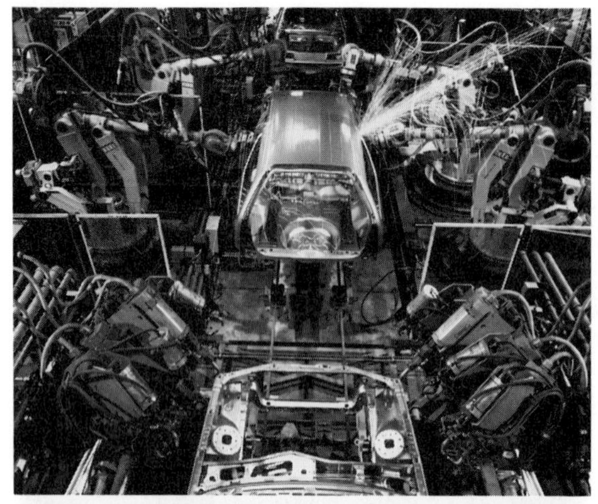

**Figure 30.5** Welding in an automotive plant

a. *(Courtesy of Ford Motor Company)*

b. *(Courtesy of Ford Motor Company)*

c. *(Courtesy of Ford Motor Company)*

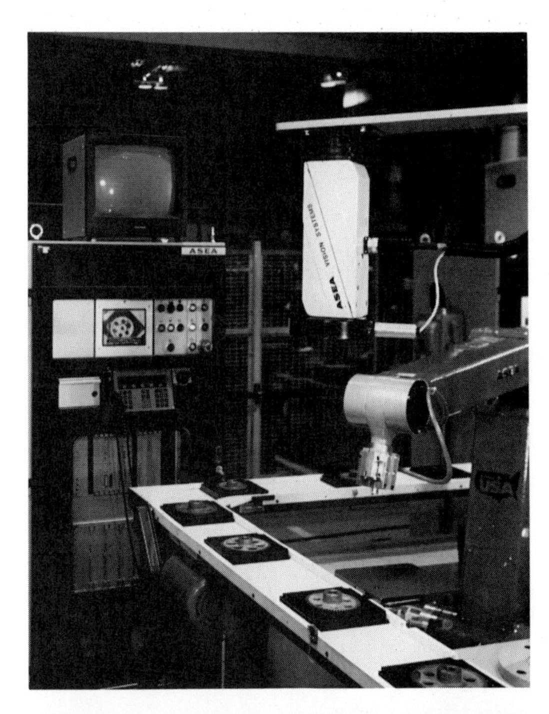

**Figure 30.6** (top and bottom) A robotic gear assembly system *(Courtesy of ASEA ROBOTICS Inc.)*

The assembly station shown in Figure 30.7 uses special tooling to hold a transmission housing in the proper position for assembly. While this is less expensive in the short run, the handling jig for this unit does not have the flexibility of the vision system.

The track-mounted spherical robot in Figure 30.8 is shown assembling a terminal plate for an electric motor. Such applications require great precision, and serve

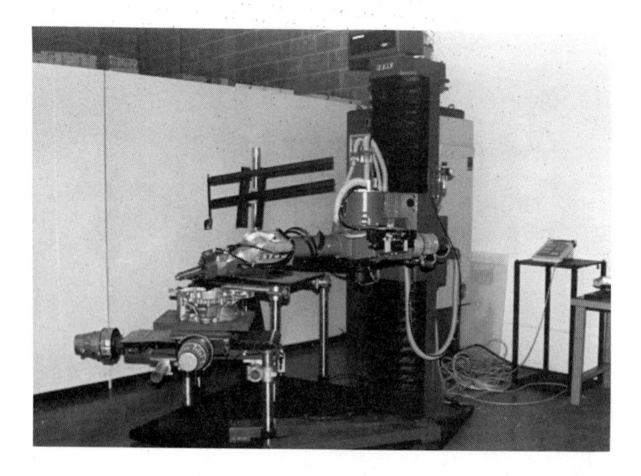

**Figure 30.7** Assembling a transmission housing *(Courtesy of Reis Machines Inc.)*

**Figure 30.8** Assembling an electric motor *(Courtesy of ASEA ROBOTICS Inc.)*

to illustrate the growing abilities of small robotic arms. The overhead track provides an additional degree of freedom for the robot.

The fully articulated arm in the work station in Figure 30.9 is shown installing the gasket around a windshield. Note the suction cups and vacuum lines which allow this arm to transfer the windshield from a conveyer to the special holding jig, and from the jig back to the conveyer. The jig itself is flexible, allowing the arm to work with windshields of different sizes and shapes.

**Figure 30.9** Installing a windshield gasket *(Courtesy of ASEA ROBOTICS Inc.)*

## 30.4 MATERIALS HANDLING

Materials-handling robots include simple pick-and-place robots used for palletizing and depalletizing materials and much more sophisticated robots used for machine tending.

The die casting machine shown in Figure 30.10 is capable of making extremely large cast metal parts. The *extractor* shown removing such a casting from the machine is an example of a heavy-duty materials-handling robot. The castings can be removed from the machine while they are still too hot for human handling, thus increasing the machine's efficiency.

The fully articulated arm in Figure 30.11 removes a part from a conveyer belt, positions it accurately for the boring machine, then places the part back on the conveyer for the next operation. While this part may be handled by a human, the hydraulically powered boring operation is hazardous to human operators. Other machines, such as the punch press, are even more dangerous to operate.

The arm shown in Figure 30.12 is working in the shipping department of a printing plant. It feeds stacks of magazines to specialized machines that tie them into bundles, then moves the bundles to a pallet for outgoing shipments.

**Figure 30.10** Unloading a die casting machine *(Courtesy of Reis Machines Inc.)*

**Figure 30.11** Machine tending *(Courtesy of ASEA ROBOTICS Inc.)*

**Figure 30.12** Palletizing *(Courtesy of Cincinnati Milacron)* Safety equipment may have been removed or opened to clearly illustrate the product and must be replaced prior to operation.

## SUMMARY

In this chapter, we have seen but a few of the many applications being found for robotics in modern industry. The applications truly seem without limit. In one place, a stainless-steel robot arm, its electronics covered with a washable neoprene boot, feeds fifty-pound boxes of meat from a pallet into a microwave oven for precooking. The end product will be soup. In another place, a small robotic arm inserts a floppy disk into a drive, closes the drive door, and presses a key to test the durability of the drive mechanism. Other robots nearby are assembling integrated circuits into printed circuit boards, while still others connect finished boards into automatic test equipment.

Will the result of this automation be lowered prices and prosperity or unemployment and recession? Only time will tell.

## SELF-TEST

1. The best uses for robotic arms today are in areas that are ______________ ______________.
2. Robots used for spray painting are hydraulically powered because they eliminate a(n) ______________ hazard.
3. Robotic arms used for welding are usually of the ______________ type.
4. Vision systems add extra ______________ to robotic assembly units.

# Appendix A

# Answers to Self-Tests and Questions/Problems

## CHAPTER 1

### SELF-TEST

1. a process variable (also called the dynamic variable)
2. transducer
3. control setpoint
4. summing point (also called comparator)
5. error
6. controller
7. actuator
8. undamped
9. critical

### QUESTIONS/PROBLEMS

1. Some possible answers: a refrigerator or freezer, a microwave with temperature control, the agc system in a radio or television set.

2. The control loop will include a temperature sensor, a signal processing section, a summing junction to compare the measurement with the setpoint, a controller, and a heater. The system will require overdamping to prevent overshoot.

## CHAPTER 2

### SELF-TEST

1. expand
2. more
3. accuracy and repeatability
4. different
5. Seebeck
6. voltage
7. thermocouple
8. isothermal
9. reference
10. ice bath
11. software
12. electronic ice
13. positive
14. RTD
15. platinum
16. 154.65 ohms at 100°C.
    108.45 ohms at 0°C.
17. negative
18. thermistor
19. 6000 ohms, 10,000 ohms
20. 2.98V

### QUESTIONS/PROBLEMS

1. −2.95V referenced to the $V^+$ pin, 3.85V referenced to ground.

## CHAPTER 3

### SELF-TEST

1. position
2. push button
3. lever
4. contact bounce

5. sequence
6. closed switch
7. cutoff
8. switching speed
9. Schmitt trigger
10. magnetic reed switch
11. changes
12. presence of a workpiece
13. eddy currents
14. decreases
15. absolute position detector
16. resistive position sensor (potentiometer)
17. capacitive and inductive sensors
18. ac voltages
19. angular
20. incremental
21. complexity

## CHAPTER 4

### SELF-TEST

1. strain
2. micrometer
3. increase
4. weight
5. weight
6. cemented or otherwise fastened
7. less than 1 ohm
8. more
9. more
10. 200%

### QUESTIONS/PROBLEMS

1. 121.76 ohms
2. The resistance will increase 4%.

## CHAPTER 5

### SELF-TEST

1. pressure
2. switch
3. transducer
4. expand
5. spring
6. straighten
7. voltage
8. bridge
9. stainless steel

### QUESTIONS/PROBLEMS

1. a pressure switch
2. 12 mA

## CHAPTER 6

### SELF-TEST

1. float switch
2. predetermined level
3. magnetic switches
4. conductivity
5. infinite
6. decrease
7. nonconductive liquids
8. increase
9. increases
10. frequency

### QUESTIONS/PROBLEMS

1. Select float-type level sensors. They are highly reliable and have a long, trouble-free life.
2. The capacitance will be less with shorter sensor rods, whether the tank is full or empty.
3. The thinner oil will offer less resistance to the flow of air. The pressure will be less.

## CHAPTER 7

### SELF-TEST

1. differential pressure
2. velocity
3. velocity
4. mass
5. moving parts
6. obstructions
7. rate of fluid flow
8. positive displacement
9. electromagnetic
10. frequency

### QUESTIONS/PROBLEMS

1. Select a positive displacement flowmeter because it measures the exact quantity of fluid through the system.
2. Select a doppler flowmeter which does not introduce any obstructions in the pipe.

## CHAPTER 8

### SELF-TEST

1. 32
2. zero
3. preset
4. industrial
5. time base
6. electromagnetic interference or radio frequency interference

### QUESTIONS/PROBLEMS

1. Ready-made industrial counters are more rugged and more reliable than a shop-built custom counter.
2. The count will be too high because the time base allows count pulses to enter the counter for a longer period of time.

## CHAPTER 9

### SELF-TEST

1. frequency of the supply voltage
2. limit switches
3. drum
4. integrated circuit
5. monostable
6. 4.55 mS
7. 3.47 kHz
8. 34.70 kHz
9. 256 $\mu$S
10. digital

### QUESTIONS/PROBLEMS

1. 909 kilohms
2. Divide the 2 MHz clock by 2000 for a 1 ms time base.
3. The crystal frequency may be 100 Hz high or low. This will cause the time base to vary by +/−50 ns.

## CHAPTER 10

### SELF-TEST

1. $V^+$
2. +6.8V
3. The optical isolator keeps the line voltage away from the low voltage parts of the circuit.
4. eddy currents
5. a voltage
6. bridge
7. 1.25V

### QUESTIONS/PROBLEMS

1. 3.818V
2. At 20°C (293.2K) the sensor voltage would be 3.87V. At 30°C (303.2K) the sensor voltage would be 3.77V. Resistors $R_1$ and $R_2$ must be changed to provide 3.87V at the top of the adjustment potentiometer and 3.77V at the bottom of the potentiometer. The potentiometer has 1 kilohm of resistance. The 100 mV across it indicates a current of 100 $\mu$A. $R_1$ must drop 2.93V. $R_1$ should be a 29.3 kilohm resistor. $R_2$ must drop 3.77V. $R_2$ must be a 37.7 kilohm

resistor. This is probably best accomplished by adjusting 50 kilohm potentiometers to the proper values.

## CHAPTER 11

### SELF-TEST

1. calibration
2. lead resistance
3. zero
4. negative
5. $R_C$
6. 3 wire
7. near balance
8. four wire
9. constant current
10. loading

### QUESTIONS/PROBLEMS

1. 479 $\mu$V
2. 200 mV

## CHAPTER 12

### SELF-TEST

1. source resistance
2. negative feedback
3. load
4. all frequencies
5. $R_1$
6. very high
7. voltage follower
8. one

### QUESTIONS/PROBLEMS

1. $A_V = 10$, LG = 20,000, $R_I' = 40{,}000$ Megohms, 3.75 mOhms
2. Due to the large $R_I'$, $V_{in} = 100$ mV. $V_{out} = -1$V
3. 500 kilohms
4. 502.51 ohms
5. buffer the signal with a voltage follower

## CHAPTER 13

### SELF-TEST

1. subtraction
2. cancel the effect of the " + 1" in the gain calculation of the noninverting amplifier.
3. low impedance inputs
4. instrumentation amplifier
5. component parts

### QUESTIONS/PROBLEMS

1. 2.14 kilohms

## CHAPTER 14

### SELF-TEST

1. half power point or −3dB point
2. 0.707
3. steepness
4. steepness
5. load
6. buffering
7. active
8. 40 dB per decade
9. not
10. notch

### QUESTIONS/PROBLEMS

1. 10.61 kHz
2. The cutoff frequency will increase to 106.1 kHz
3. 20% of 15 kilohms is 3 kilohms. Thus, one resistor would be 18 kilohms, the other would be 12 kilohms. The cutoff frequency will be changed only slightly, to 10.83 kHz

## CHAPTER 15

### SELF-TEST

1. ground
2. summing
3. virtual ground
4. summing or adding
5. multiplication

6. division
7. scaling
8. integration
9. differentiation
10. integrator's
11. differentiator's

## QUESTIONS/PROBLEMS

1. $I_1 = 100\ \mu A$, $I_2 = 166.67\ \mu A$, $I_3 = 300\ \mu A$, $I_F = 566.67\ \mu A$, $V_{out} = 8.5$ V.
2. $-V_0 = V_1 + V_2 - V_3$
3. $R_F = 10$ kilohms, $R_1 = 10$ kilohms, $R_2 = 5$ kilohms, $R_3 = 20$ kilohms. Let $V_1 = V_2 = V_3 = 2V$. Then $I_1 = 200\ \mu A$, $I_2 = 400\ \mu A$, $I_3 = 100\ \mu A$, $I_F = 700\ \mu A$, and $V_{out} = 7V$. $2 + 4 + 1 = 7$
4. Let $R_1 = R_2 = R_3 = R_4 = 20$ kilohms. Let $R_F = 5$ kilohms. Assume $V_1 = 1V$, $V_2 = 2V$, $V_3 = 3V$, and $V_4 = 4V$. Then $I_1 = 50\ \mu A$, $I_2 = 100\ \mu A$, $I_3 = 150\ \mu A$, and $I_4 = 200\ \mu A$. $I_F = 500\ \mu A$, and $V_{RF} = V_0 = 2.5V$. The sum of the four voltages is 10V. The average of the four voltages is $10V/4 = 2.5$
5. R = 1 megohm, C = .033 $\mu f$
6. .1 $\mu f$

# CHAPTER 16

## SELF-TEST

1. electromagnetic interference (EMI)
2. induce
3. currents
4. cancel
5. canceled
6. shielded
7. ground
8. currents
9. low
10. 4, 20
11. 1 mA, 5 mA
12. fiber optic
13. digital
14. EMI, RFI

## QUESTIONS/PROBLEMS

1. The low impedance of the current-source model would be less susceptible to interference. Shielded wire installed in conduit will also help reduce noise pickup.
2. The 4 mA to 20 mA standard would be more noise-proof than the 1 mA to 5 mA standard. Induced currents would be the same for either, but would be a smaller percentage of the 4 mA to 20 mA option.

# CHAPTER 17

## SELF-TEST

1. the voltage to the lens-ended lamp
2. zero crossing detector
3. reducing the effect of slowly varying input signals on the counter
4. full
5. These capacitors filter out the noise generated by the motor brushes.

## QUESTIONS/PROBLEMS

1. Reduced lamp voltage will reduce the brightness of the lamp. That will reduce the sensitivity of the system.
2. The photo Darlington will saturate and the +5V will be connected to the noninverting input.
3. The capacitor in the feedback loop reduces the gain of the amplifier for high frequencies. Increasing the capacitor will reduce the high frequency response more. The effect will be to lower the rate at which the counter can count.

# CHAPTER 18

## SELF-TEST

1. cartridge
2. lubricate with graphite or silicon when the heater is installed.
3. high
4. band
5. tubular
6. tubular
7. hot spots from air gaps beneath the heater
8. cast-in
9. strip
10. cooling tubes

## QUESTIONS/PROBLEMS

1. 55.56 $W/in^2$
2. 79.58 $W/inch^2$

# CHAPTER 19

## SELF-TEST

1. force
2. one foot pound
3. power
4. horsepower
5. watts
6. torque
7. armature
8. field
9. field
10. commutator
11. hardened carbon
12. base
13. starting torque
14. wound field
15. shunt wound
16. series-wound
17. compound
18. constant
19. high
20. shunt-wound
21. starting
22. face plate
23. below
24. above
25. resistance

### QUESTIONS/PROBLEMS

1. The series wound motor will be best for the job because of its high starting torque.
2. 1.8 ohms
3. Only armature voltage control will provide full torque at reduced speed.

## CHAPTER 20

### SELF-TEST

1. dc
2. 7.5
3. windings
4. soft iron
5. detent
6. detent
7. pull out
8. coils
9. reduced
10. reduced
11. resistance
12. slew
13. ramping

### QUESTIONS/PROBLEMS

1. 48
2. 200 times
3. 3.75
4. 7.5
5. 150 Hz

## CHAPTER 21

### SELF-TEST

1. zero
2. zero
3. reactive
4. capacitance
5. cost of electricity
6. more smooth

7. synchronous
8. slip
9. centrifugal
10. frequency

### QUESTIONS/PROBLEMS

1. 410.82 VAR, 435 W
2. 1200 rpm
3. 4.17%

## CHAPTER 22

### SELF-TEST

1. compressor
2. pump
3. pressure control
4. cylinder
5. check
6. direction control
7. pilot
8. solenoid

### QUESTIONS/PROBLEMS

1. 66.67 PSI. Load for equilibrium is 666.67 pounds.
2. The cylinder must have an area of 20 square inches.
3. 23.87 PSI

## CHAPTER 23

### SELF-TEST

1. armature
2. switch contacts
3. plug in
4. burnishing
5. replacement
6. current carrying capacity
7. contactor
8. thermal
9. magnetic

10. not actuated
11. control relay number 3
12. time delay on closing
13. primaries
14. series

## QUESTIONS/PROBLEMS

1. delta
2. manually
3. 85°
4. 80
5. time delay
6. 30 seconds after the line contactors operate.
7. 1TDO should be NO
8. Neither the blower nor the yellow pilot lamp could operate.

# CHAPTER 24

## SELF-TEST

1. PID
2. undamped
3. ON/OFF
4. larger
5. long
6. suddenly
7. low
8. positive
9. holding
10. forward breakover
11. nearly 180
12. ac
13. increase
14. four layer

## QUESTIONS/PROBLEMS

1. Yes. The higher $h_{FE}$ will have no effect on circuit operation.
2. The peak voltage is 622V. The SCR should block at least twice that voltage, or 1244V.

3. The replacement UJT will fire at a lower emitter voltage than the original. The emitter resistor will have to be increased to restore system operation.

## CHAPTER 25

### SELF-TEST

1. bipolar
2. MOSFET
3. 5
4. +/−5%
5. 3 and 15
6. less
7. yes
8. no
9. yes
10. pull up

### QUESTIONS/PROBLEMS

1. No, the propagation delay of the 74C02 is too long for the IC to operate at 15 MHz.
2. Yes.
3. The 4004B may not be able to supply enough drive for a TTL input.

## CHAPTER 26

### SELF-TEST

1. memory-mapped
2. register
3. three-state buffers
4. latches
5. address decoders
6. digital or binary
7. optical

## CHAPTER 27

### SELF-TEST

1. programmable and flexible
2. a keyboard or a video terminal
3. reduce
4. input and output modules
5. software

## CHAPTER 28

### SELF-TEST

1. no
2. arm
3. end effector
4. power supply
5. peripheral tooling
6. degrees of freedom
7. pitch
8. yaw
9. twist
10. right
11. angular
12. rectilinear
13. angular or rotary
14. rectilinear
15. full servo
16. Painting: hydraulic, full servo, fully articulated. Pick-and-place palletizing: non-servo, fully articulated

## CHAPTER 29

### SELF-TEST

1. training
2. brushless, permanent magnet, excited dc motor
3. hydraulic
4. compressibility
5. cylinders

## CHAPTER 30

### SELF-TEST

1. hazardous or repetitious
2. fire
3. fully articulated
4. accuracy

# Appendix B

# AC Motor Selection Guide

# MOTOR SELECTION

**The proper selection and application of an electric motor involves a large number of variables, which affect the installation, operation and servicing of the motor. The following items should be specified to properly select an A-C motor; power supply, horsepower rating, speed, enclosure, mounting, environment and load as reflected to the motor.**

## AMBIENT TEMPERATURE AND ALTITUDE

The motors in this catalog are designed for use in ambient temperatures of 0°C. to 40°C. and an altitude not exceeding 3300 feet (100 meters). Environments exceeding these conditions may require special motor selection or rerating of standard motors.

## ENCLOSURE

The motor enclosure with its associated features is generally determined by the environment in which it will operate if a normal life is expected. Reliance offers a variety of open and totally enclosed motors to meet special environments.

**Frames:** The frame determines the critical mounting dimensions of the motor. All motors of the same frame size and type have identical mounting dimensions regardless of their electrical characteristics and are therefore interchangeable. These dimensions are based on NEMA standards.

## HORSEPOWER RATING

Reliance motors are available from ⅛ to 20,000 HP. The horsepower and motor size are determined by the load running and starting requirements. Some horsepower ratings not shown in this catalog are available on a production basis; contact your local Reliance distributor or Reliance sales office for details on availability.

## MOUNTING

The standard motors shown in the catalog are either foot mounted or have a C-face. Special industry flanges not shown in this catalog are available. The conduit boxes are located for standard F-1 mounting with some smaller frames having F-1/F-2 configurations. (NEMA F-1 mounting is defined as the conduit box being on the left side when facing the shaft.)

## POWER SUPPLY

**Phase:** Reliance Electric offers motors for single phase (type CS) and polyphase, or three phase, (type P) power supplies. Normally the larger motors, above 1 HP, are type P.

**Frequency:** The standard motors shown in this catalog are specified for operation with a 60 Hz (cycles per second) power supply. Many motors can be rerated for operation at 50 Hz, using Reliance sticker # RE491A4. Your Reliance distributor or sales office can supply the sticker and details on rerating. In addition, Reliance can design and build motors on a production basis specifically designed for operation with 50 Hz power.

**Voltage:** Reliance polyphase A-C motors are wound for operation on 230/460 volts or 460 volts only or 200 or 575 volts (as marked in this catalog). 200 and 575 volt motors not cataloged are available in many cases at no additional cost. 2300 or 4160 volt motors are also available on some 447T and 449T frames.

In accordance with NEMA, Reliance motors will operate ±10% of the nameplate voltage at maintained frequency. When nominal starting torques are required, a 230 volt motor can be used on a 208 volt 4-wire power supply and stickers are available through Reliance sales offices and Reliance distributors for rating the motor to the 208 power supply. Starting and running torques of a 230 volt motor will be reduced by approximately 20% when used with 208 volt systems.

## TORQUE-LOAD

The full load torque of a motor expressed in foot pounds is equal to $\frac{\text{HP} \times 5250}{\text{Full Load Speed}}$ = torque in ft. lbs.

Because different loads present different requirements; in terms of starting (breakaway), minimum (pull-up), breakdown (pull-out) and full load torque, Reliance offers motors with 4 different speed-torque characteristics. In this catalog most motors are design "B" with some design "C" ratings shown. Design "D" motors are available on a production basis for various ratings. The characteristics for each design are shown in the following chart, and on page 49.

| NEMA Design | Starting Torque | Starting Current | Breakdown Torque | Full Load Slip Torque | Typical Applications |
|---|---|---|---|---|---|
| A | Normal | High | High | Low | Mach. Tools, Fans |
| B | Normal | Normal | Normal | Normal | Same as Design "A" |
| C | High | Normal | Normal | Normal | Loaded compressor<br>Loaded conveyor |
| D | Very High | Low | — | High | Punch press or Hoists |

## SPEED

The operating speed of an A-C motor is a function of its design, load and frequency of the power supply. Standard 60 Hz motors are available with synchronous speeds of 3600, 1800, 1200 or 900 RPM. For slower speed and multi-speed motors consult your Reliance sales office or distributor.

# D-C V★S DRIVES

## SELECTION & APPLICATION GUIDE

## MEASURING REQUIRED TORQUE

If the amount of torque required to drive a machine cannot be determined from the builder, it can be easily measured by one of two methods: (1) Use a torque wrench to turn the driven shaft and observe the indicated torque. (2) Fasten a pulley to the driven shaft; secure one end of a cord to the pulley; and wrap the cord around the pulley. Figure 3. Attach a simple spring scale to the exposed end. Pull on the scale enough to turn the shaft, and observe the highest reading on the scale. Multiply this value by the radius of the pulley to determine the torque.

**Figure 3**

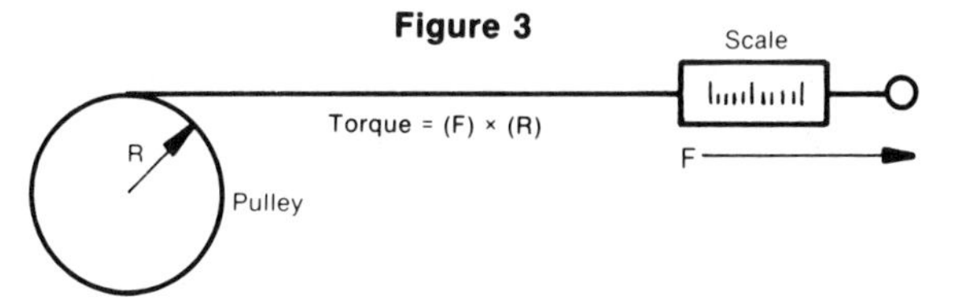


## CALCULATING ACCELERATING TORQUE FOR ROTARY MOTION

High inertia loads frequently require higher torques for acceleration than is required to maintain a desired running speed.

The formula to calculate acceleration torque of a rotating member:

$$T = \frac{(WK^2)N}{308t}$$

Where:

T = Accelerating torque (lb-ft)

$WK^2$ = Total inertia ($lb\text{-}ft^2$) that the motor must accelerate. This value includes motor armature, reducer and load.

N = Change in speed required (RPM).

t = Time to accelerate load (seconds).

The same formula can also be used to determine the minimum accelerating time of a given drive:

$$t = \frac{(WK^2)\ (N)}{308T}$$

NOTE: Many drives have 150% load capability for 1 minute, which may allow the required additional accelerating torque to be obtained without increasing the drive horsepower rating.

## CALCULATING ACCELERATING FORCE FOR LINEAR MOTION

The following formula may be useful to calculate the **approximate** accelerating force required for linear motion. However, before sizing the drive, add the torque required to accelerate the motor armature, gears, pulleys, etc., to the linear-motion accelerating force converted to torque.

$$\text{Acceleration Force (F)} = \frac{WV}{1933t}$$

Where:

W = Weight (lb)
V = Change in Velocity (FPM)
t = Time (seconds) to accelerate weight

## CALCULATING HORSEPOWER

For rotating objects:

$$Hp = \frac{TN}{63{,}000}$$ where: T = Torque (lb-in), N = Speed (RPM)

or:

$$Hp = \frac{TN}{5250}$$ where: T = Torque (lb-ft), N = Speed (RPM)

For objects in linear motion:

$$Hp = \frac{FV}{396{,}000}$$ where: F = Force (lb), V = Velocity (IPM)

or:

$$Hp = \frac{FV}{33{,}000}$$ where: F = Force (lb), V = Velocity (FPM)

For pumps:

$$Hp = \frac{(GPM) \times (\text{Head in feet}) \times (\text{Specific Gravity})}{3960 \times (\text{Efficiency of Pump})}$$

For fans and blowers:

$$Hp = \frac{CFM \times (\text{Pressure in Pounds/Sq. Ft})}{33{,}000 \times \text{Efficiency}}$$

## OTHER USEFUL FORMULAE

Torque = Force x Radius

Reflected $WK^2$ through a reducer (gear or belt)

$$= \frac{WK^2 \text{ of Load}}{(\text{Reduction Ratio})^2}$$

$$RPM = \frac{FPM}{.262 \times \text{Diameter (Inches)}}$$

# HORSEPOWER OUTPUT AT VARIOUS SPEEDS FOR CONSTANT TORQUE DRIVES

| Drive HP Rating at 1750 Base Speed | HP Rating at Various Motor Output rpm | | | | | | | Output Torque at all Speeds (lb-in. at Motor Shaft) |
|---|---|---|---|---|---|---|---|---|
| | 1750 | 875 | 350 | 175 | 87.5 | 43.75 | 24.00 | |
| 1/4 | .250 | .125 | .0500 | .0250 | .0125 | .0063 | .0034 | 9.0 |
| 1/3 | .333 | .167 | .0667 | .0333 | .0167 | .0083 | .0046 | 12.0 |
| 1/2 | .500 | .250 | .100 | .0500 | .0250 | .0125 | .0069 | 18.0 |
| 3/4 | .750 | .375 | .150 | .0750 | .0375 | .0188 | .0103 | 27.0 |
| 1 | 1.00 | .500 | .200 | .100 | .0500 | .0250 | .0137 | 36.0 |
| 1 1/2 | 1.50 | .750 | .300 | .150 | .0750 | .0375 | .0206 | 54.0 |
| 2 | 2.00 | 1.00 | .400 | .200 | .1000 | .0500 | .0274 | 72.0 |
| 3 | 3.00 | 1.50 | .600 | .300 | .1500 | .0750 | .0411 | 108.0 |

EFFECTIVE January 2, 1986

# GENERAL ENGINEERING DATA

## TORQUE: NEMA DESIGN CLASSES

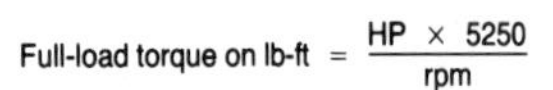

Full-load torque on lb-ft $= \dfrac{HP \times 5250}{rpm}$

Motor torque is defined at four points as shown by Figure 1.

1. Breakaway or starting
2. Minimum or "pull-up"
3. Breakdown or "pull-out"
4. Full load

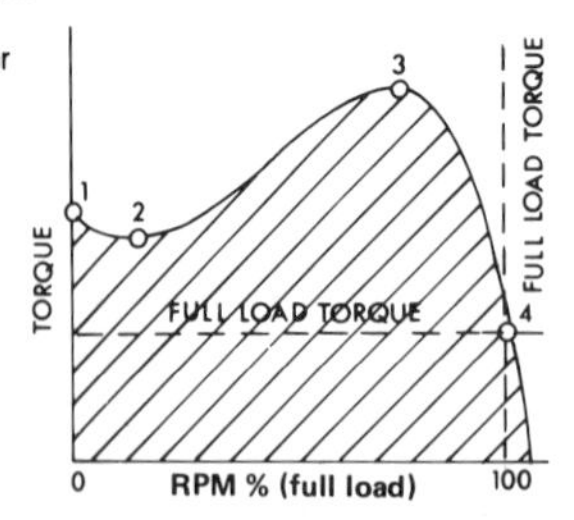


The sectioned area under the curve represents a motor's accelerating torque from zero to full speed. Torque, horsepower, and speed requirements demanded in drives for most machines can be met with one of four design classes of squirrel-cage polyphase induction motors described by National Electrical Manufacturers Association standards.

All four are suitable for across-the-line starting.

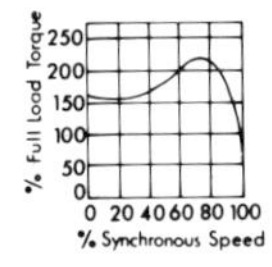


**Design B Motors** are the standard general purpose design. They have low starting current-normal torque, and normal slip. Their field of application is very broad and includes fans, blowers, pumps, and machine tools. Torque values are listed in this Modification Section.

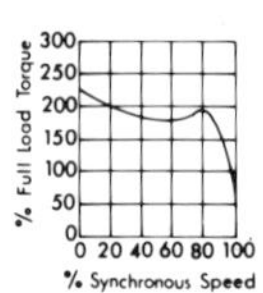


**Design C Motors** have high breakaway torque, low starting current, and normal slip. The higher breakaway torque makes this motor advantageous for "hard-to-start" applications, such as plunger pumps, conveyors, and compressors.

**Design D Motors** have a high breakaway torque combined with high slip. Breakaway torque for 4, 6 and 8 pole motors is 275% or more of full load torque. Two slip groups are described.

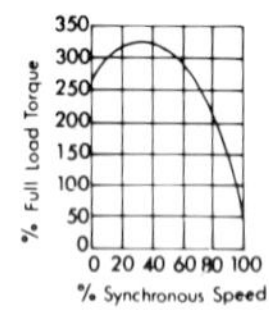


**5-8% and 8-13% Slip Design D Motors** are recommended for punch presses, shears, and other high inertia machinery, where it is desired to make use of the energy stored in a flywheel under heavy fluctuating load conditions. They are also used for multimotor conveyor drives where motors operate in mechanical parallel.

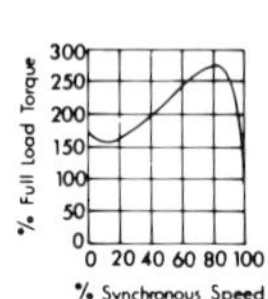


**Design A** covers a wide varitey of motors similar to design B except that their breakdown torque and starting current are higher.

## HIGH INERTIA LOADS

$$t = \frac{WK^2 \times rpm}{308 \times Ts} \qquad WK^2 = \text{inertia in lb-ft}^2$$

t = accelerating time in sec.
T = Av. accelerating torque lb-ft

$$T = \frac{WK^2 \times rpm}{308 \times t}$$

$$\text{Inertia reflected to motor} = \text{Load inertia} \left(\frac{\text{Load rpm}}{\text{Motor rpm}}\right)^2$$

## RULES OF THUMB (Approximation)

**At 1800 rpm,** a motor develops a 3 lb-ft per hp
**At 1200 rpm,** a motor develops a 4.5 lb-ft per hp
**At 575 volts,** a 3-phase motor draws 1 amp per hp
**At 460 volts,** a 3-phase motor draws 1.25 amp per hp
**At 230 volts,** a 3-phase motor draws 2.5 amp per hp
**At 230 volts,** a single-phase motor draws 5 amp per hp
**At 115 volts,** a single-phase motor draws 10 amp per hp

## MECHANICAL FORMULAS

$$\text{Torque in Lb-ft} = \frac{HP \times 5250}{rpm} \qquad HP = \frac{Torque \times rpm}{5250}$$

$$rpm = \frac{120 \times \text{Frquency}}{\text{No. of Poles}}$$

## ELECTRICAL FORMULAS

| To Find | Alternating Current | |
|---|---|---|
| | Single-Phase | Three-Phase |
| Amperes when horsepower is known | $\frac{HP \times 746}{E \times Eff \times pf}$ | $\frac{HP \times 746}{1.73 \times E \times Eff \times pf}$ |
| Amperes when kilowatts are known | $\frac{Kw \times 1000}{E \times pf}$ | $\frac{Kw \times 1000}{1.73 \times E \times pf}$ |
| Amperes when kva are known | $\frac{Kva \times 1000}{E}$ | $\frac{Kva \times 1000}{1.73 \times E}$ |
| Kilowatts | $\frac{I \times E \times pf}{1000}$ | $\frac{1.73 \times I \times E \times pf}{1000}$ |
| Kva | $\frac{I \times E}{1000}$ | $\frac{1.73 \times I \times E}{1000}$ |
| Horsepower = (Output) | $\frac{I \times E \times Eff \times pf}{746}$ | $\frac{1.73 \times I \times E \times Eff \times pf}{746}$ |

I = Amperes; E = Volts; Eff = Efficiency; pf = Power factor; Kva = Kilovolt-amperes; Kw = Kilowatts.

## TEMPERATURE CONVERSION

Deg C = (Deg F − 32) × 5/9
Deg F = (Deg C × 9/5) + 32

## MOUNTING POSITIONS

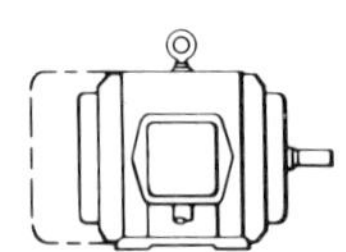

**ASSEMBLY F-1 STANDARD MOUNTING**

**ASSEMBLY F-2**

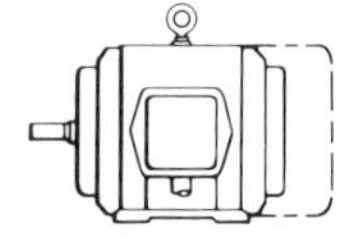

# ENERGY EFFICIENT

# EVALUATING MOTOR EFFICIENCY

## Designed For Superior Energy Performance

Efficiency is a measure of the effectiveness with which a motor converts electrical energy into mechanical energy. In other words, watts in vs. watts out. In the conversion process, watts are lost — transformed into heat which is dissipated through the frame, or lost through friction and windage. To improve efficiency, watts loss must be reduced. This is true even in higher horsepower motors, which are typically more efficient than low horsepower models.

In the Duty Master® XE, Reliance Electric technology has succeeded in reducing watts loss, as a result of energy conversion, as much as 50 percent over standard industrial motor designs. This excellent reduction is the achievement of optimized design, improved material selection and meticulous quality control in the five areas of watts loss shown in the illustration at right and explained below.

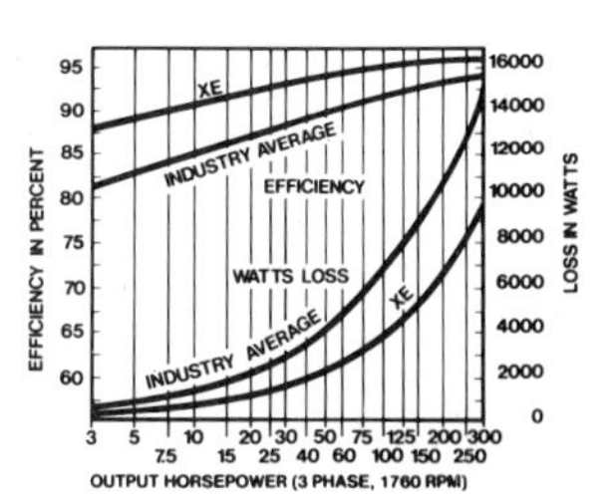


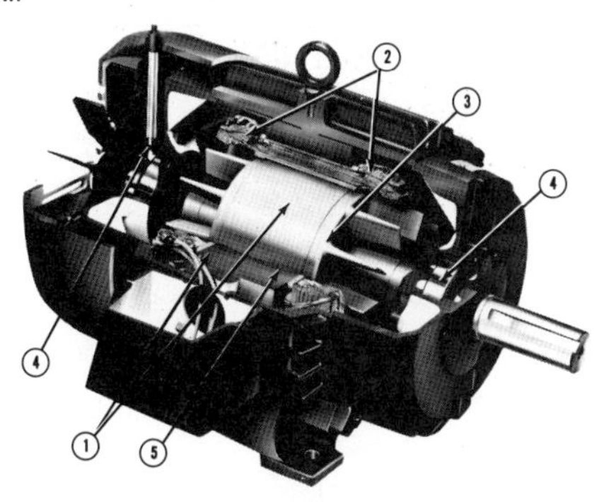


| WATTS LOSS AREA | EFFICIENCY IMPROVEMENT |
|---|---|
| 1. Iron | Use of thinner gauge, lower loss core steel reduces eddy current losses. Longer core adds more steel to the design which reduces losses due to lower operation flux densities. |
| 2. Stator $I^2$ R | Use of more copper and larger conductors increases cross sectional area of stator windings. This lowers resistance (R) of the windings and reduces losses due to current flow (I). |
| 3. Rotor $I^2$ R | Use of larger rotor conductor bars increases size of cross section, lowering conductor resistance (R) and losses due to current flow (I). |
| 4. Friction & Windage | Use of low loss fan design (on fan cooled models) reduces losses due to air movement. Exclusive PLS, Positive Lubrication System, with its open bearing construction, reduces loss contributed by bearing friction. |
| 5. Stray Load Loss | Use of optimized design and strict quality control procedures minimizes stray load losses. |

## Energy Savings Justify Initial Investment

When it comes to investments, you look for the most efficient use of your money. When you invest in an energy efficient Duty Master® XE Motor you get it. In many applications, an initial purchase price premium can be justified based on energy cost savings. The basis for this justification depends on the user's situation. Factors such as running hours, cost of electricity, payback period, tax rate, cost of capital and service life affect the premium price justification, and vary with the individual user.

**There are three steps in determining the justified price premium for a particular XE motor.**

**Step No. 1** — *Determination of "P," the justified price premium per kilowatt saved.*

This is your constant, particular to your plant's energy cost, motor operation time and payback requirements.

For a single payback analysis, you need to know the number of hours per year that the motor will operate, your cost of energy in dollars per kilowatt hour, the number of years that you are willing to operate the motor until it reaches the break-even point, and your corporate tax rate. With these numbers, you can calculate a "justified price premium" per kilowatt saved using the formula below:

**Where:** Hrs/Yr. = **Hours of motor operation per year**
$/KWH = **Cost of electricity (dollars per kilowatt hour)**
Yrs. = **Maximum acceptable years to payback or break-even**
T = **Tax rate**

**Then:** P = **Justified Price Premium/KW saved**
= **Hrs/Yr. × $/KWH × Yrs. × (1-T)**

As an example, let's assume that the motor is going to operate for three 8-hour shifts per day, on a 6-day work week, for 50 weeks per year. Thus total hours per year are 7200. Also, assume power costs are $0.04/KWH. If your tax rate is 40 percent, and you are willing to wait three years for the motor to save enough in power costs to justify it price premium, you obtain a justified price premium of $518 for each kilowatt saved.

P = **Hrs/Yr. × $/KWH × Years × (1-T)**
= **7200 × $.04 × 3 × (1.40)**
= **$518**

To determine your own value of "P," simply insert your company's electrical rate, required payback period and tax rate in the formula. Or you can use alternate methods for determing "P," such as a Present Value Analysis. This method takes into account the service life of the motor, required rate of return and depreciation benefits.

No matter how you determine the value of "P," remember that it represents only the justified price premium per kilowatt saved. In order to determine the justified price premium you can pay for the motor, you have to multiply kilowatts saved by "P."

**Step No. 2** — *Determination of kilowatts saved by a Duty Master® XE Motor.*

You can determine kilowatts saved by a Duty Master® XE Motor over standard industrial designs by using the simple formula below:

$$\text{Kilowatts Saved} = \text{HP} \times .746 \left[\frac{1}{\text{Std. Ind. Motor Eff.}} - \frac{1}{\text{XE Motor Eff.}}\right]$$

**FOR EXAMPLE:** To determine kilowatts saved by a 25 hp Duty Master® XE with a 93% efficiency vs. industry average motor efficiency of 88.0%, calculate as follows:

$$\text{Kilowatts Saved} = \text{HP} \times .746 \left[\frac{1}{.880} - \frac{1}{.930}\right]$$

$$= 1.139$$

**Step No. 3** — *Determination of the justified price premium for a particular XE motor.*

Multiplying the kilowatts saved in this formula by the value of "P" established in the previous formula yields a justified price premium of $590 for the Duty Master® XE motor vs. standard industrial designs.

**Justified Price Premium for Motor** = **Kilowatts Saved** × **P**
= **1.139** × **$518**
= **$590**

# MOTOR FRAME DIMENSIONS

Note — Refer to drawing at right when using this table. **(in inches)**

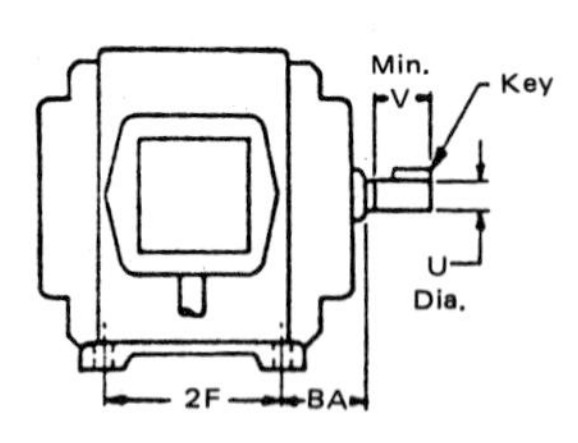


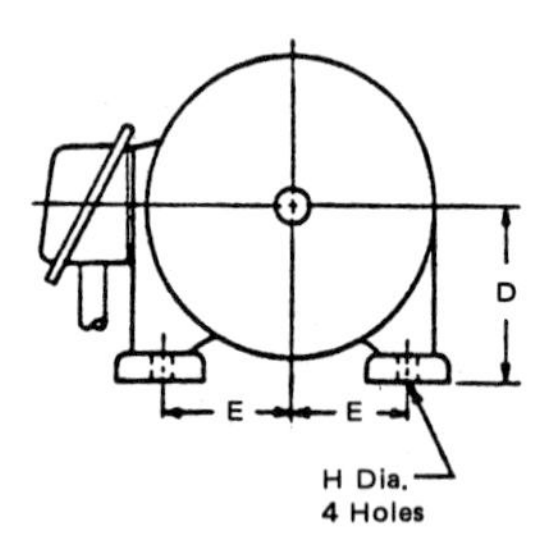


**These dimensions are for estimating purposes. Refer to listed dimension sheet number for additional dimensions.**

| Frame Size | D | E | 2F | H Dia. (4) Holes | U Dia. | BA | V Min. | Key |
|---|---|---|---|---|---|---|---|---|
| 48 | 3 | 2 | 2¾ | 11/32 | ½ | 2½ | — | 3/64 Flat |
| 56 | 3½ | 2 7/16 | 3 | 11/32 | ⅝ | 2¾ | — | 3/16 × 3/16 × 1⅜ |
| 143T | 3½ | 2¾ | 4 | 11/32 | ⅞ | 2¼ | 2 | 3/16 × 3/16 × 1⅜ |
| 145T | 3½ | 2¾ | 5 | 11/32 | ⅞ | 2¼ | 2 | 3/16 × 3/16 × 1⅜ |
| 182 | 4½ | 3¾ | 4½ | 13/32 | ⅞ | 2¾ | 2 | 3/16 × 3/16 × 1⅜ |
| 182T | 4½ | 3¾ | 4½ | 13/32 | 1⅛ | 2¾ | 2½ | ¼ × ¼ × 1¾ |
| 184 | 4½ | 3¾ | 5½ | 13/32 | ⅞ | 2¾ | 2 | 3/16 × 3/16 × 1⅜ |
| 184T | 4½ | 3¾ | 5½ | 13/32 | 1⅛ | 2¾ | 2½ | ¼ × ¼ × 1¾ |
| 213 | 5¼ | 4¼ | 5½ | 13/32 | 1⅛ | 3½ | 2¾ | ¼ × ¼ × 2 |
| 213T | 5¼ | 4¼ | 5½ | 13/32 | 1⅜ | 3½ | 3⅛ | 5/16 × 5/16 × 2⅜ |
| 215 | 5¼ | 4¼ | 7 | 13/32 | 1⅛ | 3½ | 2¾ | ¼ × ¼ × 2 |
| 215T | 5¼ | 4¼ | 7 | 13/32 | 1⅜ | 3½ | 3⅛ | 5/16 × 5/16 × 2⅜ |
| 254U | 6¼ | 5 | 8¼ | 17/32 | 1⅜ | 4¼ | 3½ | 5/16 × 5/16 × 2¾ |
| 254T | 6¼ | 5 | 8¼ | 17/32 | 1⅝ | 4¼ | 3¾ | ⅜ × ⅜ × 2⅞ |
| 256U | 6¼ | 5 | 10 | 17/32 | 1⅜ | 4¼ | 3½ | 5/16 × 5/16 × 3¾ |
| 256T | 6¼ | 5 | 10 | 17/32 | 1⅝ | 4¼ | 3¾ | ⅜ × ⅜ × 2⅞ |
| 284U | 7 | 5½ | 9½ | 17/32 | 1⅝ | 4¾ | 4⅝ | ⅜ × ⅜ × 3¾ |
| 284T | 7 | 5½ | 9½ | 17/32 | 1⅞ | 4¾ | 4⅜ | ½ × ½ × 3¼ |
| 284TS | 7 | 5½ | 9½ | 17/32 | 1⅝ | 4¾ | 3 | ⅜ × ⅜ × 1⅞ |
| 286U | 7 | 5½ | 11 | 17/32 | 1⅝ | 4¾ | 4⅝ | ⅜ × ⅜ × 3¾ |
| 286T | 7 | 5½ | 11 | 17/32 | 1⅞ | 4¾ | 4⅜ | ½ × ½ × 3¼ |
| 286TS | 7 | 5½ | 11 | 17/32 | 1⅝ | 4¾ | 3 | ⅜ × ⅜ × 1⅞ |
| 324U | 8 | 6¼ | 10½ | 21/32 | 1⅞ | 5¼ | 5⅜ | ½ × ½ × 4¼ |
| 324T | 8 | 6¼ | 10½ | 21/32 | 2⅛ | 5¼ | 5 | ½ × ½ × 3⅞ |
| 324TS | 8 | 6¼ | 10½ | 21/32 | 1⅞ | 5¼ | 3½ | ½ × ½ × 2 |
| 326U | 8 | 6¼ | 12 | 21/32 | 1⅞ | 5¼ | 5⅜ | ½ × ½ × 4¼ |
| 326T | 8 | 6¼ | 12 | 21/32 | 2⅛ | 5¼ | 5 | ½ × ½ × 3⅞ |
| 326TS | 8 | 6¼ | 12 | 21/32 | 1⅞ | 5¼ | 3½ | ½ × ½ × 2 |
| 364U | 9 | 7 | 11¼ | 21/32 | 2⅛ | 5⅞ | 6⅛ | ½ × ½ × 5 |
| 364US | 9 | 7 | 11¼ | 21/32 | 1⅞ | 5⅞ | 3½ | ½ × ½ × 2 |
| 364T | 9 | 7 | 11¼ | 21/32 | 2⅜ | 5⅞ | 5⅝ | ⅝ × ⅝ × 4¼ |
| 364TS | 9 | 7 | 11¼ | 21/32 | 1⅞ | 5⅞ | 3½ | ½ × ½ × 2 |
| 365U | 9 | 7 | 12¼ | 21/32 | 2⅛ | 5⅞ | 6⅛ | ½ × ½ × 5 |
| 365US | 9 | 7 | 12¼ | 21/32 | 1⅞ | 5⅞ | 3½ | ½ × ½ × 2 |
| 365T | 9 | 7 | 12¼ | 21/32 | 2⅜ | 5⅞ | 5⅝ | ⅝ × ⅝ × 4¼ |
| 365TS | 9 | 7 | 12¼ | 21/32 | 1⅞ | 5⅞ | 3½ | ½ × ½ × 2 |
| 404U | 10 | 8 | 12¼ | 13/16 | 2⅜ | 6⅝ | 6⅞ | ⅝ × ⅝ × 5½ |
| 404US | 10 | 8 | 12¼ | 13/16 | 2⅛ | 6⅝ | 4 | ½ × 4 × 2¾ |
| 404T | 10 | 8 | 12¼ | 13/16 | 2⅞ | 6⅝ | 7 | ¾ × ¾ × 5⅝ |
| 404TS | 10 | 8 | 12¼ | 13/16 | 2⅛ | 6⅝ | 4 | ½ × ½ × 2¾ |
| 405U | 10 | 8 | 13¾ | 13/16 | 2⅜ | 6⅝ | 6⅞ | ⅝ × ⅝ × 5½ |
| 405US | 10 | 8 | 13¾ | 13/16 | 2⅛ | 6⅝ | 4 | ½ × ½ × 2¾ |
| 405T | 10 | 8 | 13¾ | 13/16 | 2⅞ | 6⅝ | 7 | ¾ × ¾ × 5⅝ |
| 405TS | 10 | 8 | 13¾ | 13/16 | 2⅛ | 6⅝ | 4 | ½ × ½ × 2¾ |
| 444U | 11 | 9 | 14½ | 13/16 | 2⅞ | 7½ | 8⅜ | ¾ × ¾ × 7 |
| 444US | 11 | 9 | 14½ | 13/16 | 2⅛ | 7½ | 4 | ½ × ½ × 2¾ |
| 444T | 11 | 9 | 14½ | 13/16 | 3⅜ | 7½ | 8¼ | ⅞ × ⅞ × 6⅞ |
| 444TS | 11 | 9 | 14½ | 13/16 | 2⅜ | 7½ | 4½ | ⅝ × ⅝ × 3 |
| 445U | 11 | 9 | 16½ | 13/16 | 2⅞ | 7½ | 8⅜ | ¾ × ¾ × 7 |
| 445US | 11 | 9 | 16½ | 13/16 | 2⅛ | 7½ | 4 | ½ × ½ × 2¾ |
| 445T | 11 | 9 | 16½ | 13/16 | 3⅜ | 7½ | 8¼ | ⅞ × ⅞ × 6⅞ |
| 445TS | 11 | 9 | 16½ | 13/16 | 2⅜ | 7½ | 4½ | ⅝ × ⅝ × 3 |
| 447T | 11 | 9 | 20 | 13/16 | 3⅜ | 7½ | 8¼ | ⅞ × ⅞ × 6⅞ |
| 447TS | 11 | 9 | 20 | 13/16 | 2⅜ | 7½ | 4½ | ⅝ × ⅝ × 3 |
| 449T | 11 | 9 | 25 | 13/16 | 3⅜ | 7½ | 8¼ | ⅞ × ⅞ × 6⅞ |
| 449TS | 11 | 9 | 25 | 13/16 | 2⅜ | 7⅓ | 4½ | ⅝ × ⅝ × 3 |

## NEMA C-Face Mounting Dimensions (in inches)

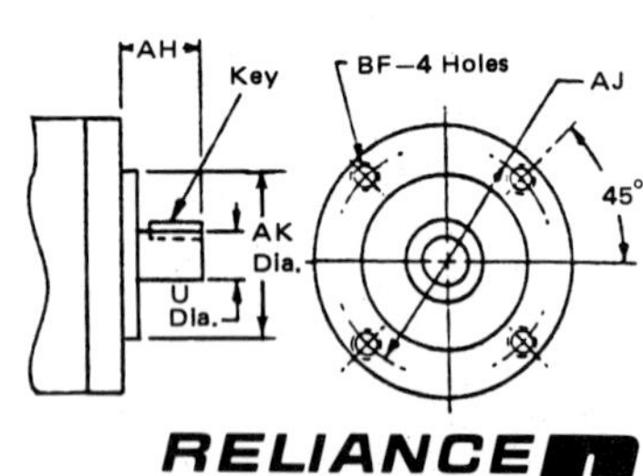


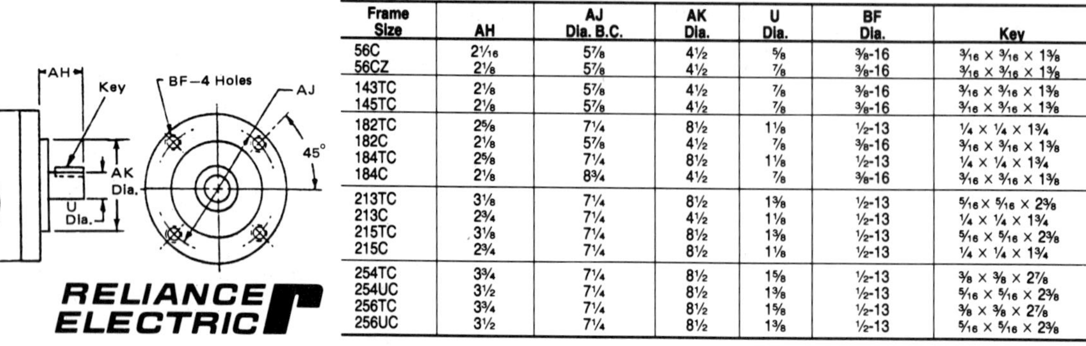

| Frame Size | AH | AJ Dia. B.C. | AK Dia. | U Dia. | BF Dia. | Key |
|---|---|---|---|---|---|---|
| 56C | 2 1/16 | 5⅞ | 4½ | ⅝ | ⅜-16 | 3/16 × 3/16 × 1⅜ |
| 56CZ | 2⅛ | 5⅞ | 4½ | ⅞ | ⅜-16 | 3/16 × 3/16 × 1⅜ |
| 143TC | 2⅛ | 5⅞ | 4½ | ⅞ | ⅜-16 | 3/16 × 3/16 × 1⅜ |
| 145TC | 2⅛ | 5⅞ | 4½ | ⅞ | ⅜-16 | 3/16 × 3/16 × 1⅜ |
| 182TC | 2⅝ | 7¼ | 8½ | 1⅛ | ½-13 | ¼ × ¼ × 1¾ |
| 182C | 2⅛ | 5⅞ | 4½ | ⅞ | ⅜-16 | 3/16 × 3/16 × 1⅜ |
| 184TC | 2⅝ | 7¼ | 8½ | 1⅛ | ½-13 | ¼ × ¼ × 1¾ |
| 184C | 2⅛ | 8¾ | 4½ | ⅞ | ⅜-16 | 3/16 × 3/16 × 1⅜ |
| 213TC | 3⅛ | 7¼ | 8½ | 1⅜ | ½-13 | 5/16 × 5/16 × 2⅜ |
| 213C | 2¾ | 7¼ | 4½ | 1⅛ | ½-13 | ¼ × ¼ × 1¾ |
| 215TC | 3⅛ | 7¼ | 8½ | 1⅜ | ½-13 | 5/16 × 5/16 × 2⅜ |
| 215C | 2¾ | 7¼ | 8½ | 1⅛ | ½-13 | ¼ × ¼ × 1¾ |
| 254TC | 3¾ | 7¼ | 8½ | 1⅝ | ½-13 | ⅜ × ⅜ × 2⅞ |
| 254UC | 3½ | 7¼ | 8½ | 1⅜ | ½-13 | 5/16 × 5/16 × 2⅜ |
| 256TC | 3¾ | 7¼ | 8½ | 1⅝ | ½-13 | ⅜ × ⅜ × 2⅞ |
| 256UC | 3½ | 7¼ | 8½ | 1⅜ | ½-13 | 5/16 × 5/16 × 2⅜ |

RELIANCE ELECTRIC

# Appendix C

# DC VS Drives

# D-C V★S DRIVES

SELECTION & APPLICATION GUIDE

## GENERAL INFORMATION ON D-C V★S DRIVES

### APPLICATION ENGINEERING INFORMATION

#### HOW A D-C DRIVE WORKS

A basic d-c drive consists of three elements: (1) Operator's control station determines when the drive starts and stops and the speed control potentiometer gives a reference signal to the drive controller. (2) Drive controller converts in-plant a-c power to controlled d-c power. The d-c output voltage is determined by the voltage of the reference signal. (3) Drive motor, the speed of which is determined by the armature voltage, Fig. 1. Thus, the higher the reference voltage from the speed potentiometer, the higher the d-c armature voltage and the higher the motor speed. This basic description assumes the voltage to the motor field is constant or the motor has a permanent magnet field.

**Figure 1**

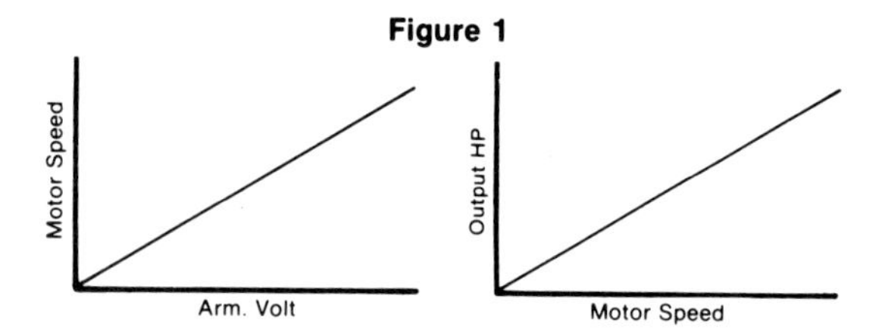


Drives are also available which, in addition to controlling armature voltage, also control field voltage by using a field regulator. Reducing the motor field voltage causes the motor to increase speed. However, "weakening the field" by reducing the field voltage also reduces the motor torque. Thus, the drive delivers constant horsepower when operating in the field weakened range, Figure 2. Drives with this capability are frequently called "Voltage and Field Drives."

**Figure 2**

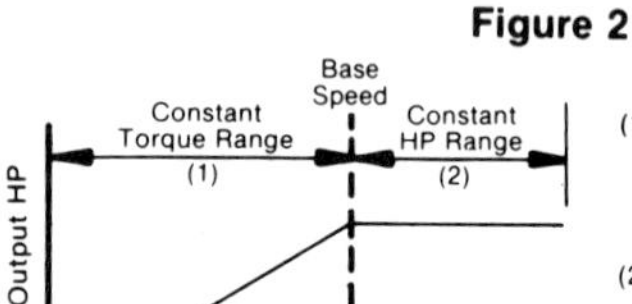


(1) In the constant torque range the field voltage is held constant and the armature voltage is varied.

(2) In the constant horsepower range the armature voltage is held constant and the field voltage is varied.

### SELECTION EXAMPLE

#### APPLICATION

A d-c drive is needed to power a screw conveyor from 13 to 150 rpm. The load is essentially constant torque, at 175 lb.-in. Operator's control will be remote from drive controller, and two emergency stop pushbuttons are required. Delivery is needed from stock. Available plant power is 460V a-c 60 Hz. The speed should be held constant within 1% of base speed regardless of load, temperature, or any other variable.

#### PROCEDURE

1. Select TIGear Reducer from A-400.
2. TIGear table shows ¾ hp is needed for the reducer to produce the required torque at this needed speed.
3. A MinPak Plus V★S drive rated ¾ hp will satisfy the requirements; can have the operator's control station at a remote location from the drive controller; can be used with emergency stop pushbutton since it uses "three wire" start-stop circuit and has dynamic braking available. When used with a transformer, it can be used with emergency stop pushbutton since it uses "three wire" start-stop circuit and has dynamic braking available. When used with a transformer, it can be oeprated from 460V; and can be used with tachometer feedback to achieve the maximum speed deviation of 1% due to load change.
4. The MinPak Plus Model No. 14C10 and motor, model T56H1022 is selected from D3-12. The diamond beside the model number denotes that it is a stock rating.
5. To enable the operator to control the MinPak Plus from a remote location, a remote operator's adapter kit Model 14C220 is selected. Refer to page D3-20. Since the remote operator's controls are used, select a Blank Panel Model 14C200 from page D3-20.
6. A tachometer-generator R45G1000 and mounting kit 413327-W are selected from page D3-47 and D3-48. A tachometer-Feedback Kit Model No. 14C221 to interface the tachometer-generator and the MinPak Plus regulator is selected for D3-22.
7. Since the ¾ hp MinPak Plus operates from 115 v a-c, a transformer will be needed to reduce the plant 460 v to 115 v. This transformer is selected from page D4-16, number 77530-10V, since delivery from stock is needed.
8. The two emergency stop pushbuttons are selected from page D4-7. "Pushbutton 1½ Inch Mushroom" for "Mounted Only" and two pushbutton enclosures are selected from page D3-79 each 1-element.

# Appendix D

# Heater Selection Guide

# Selector Guide By Order Of Application

TEMPCO® ELECTRIC HEATER CORPORATION

| HEATING APPLICATIONS | CARTRIDGE HEATERS | BAND HEATERS | CAST-IN-HEATERS | STRIP HEATERS | TUBULAR HEATERS | IMMERSION HEATERS | DUCT HEATERS | RADIANT QUARTZ HEATERS | SILICONE RUBBER HEATERS | DRUM HEATERS | HEATING CABLE | SOLID-STATE TEMPERATURE CONTROLS | SCR SOLID-STATE POWER CONTROLS | THERMOCOUPLES | THERMOSTATIC CONTROLS | ELECTRIC THERMOSTATS | SURFACE MOUNT THERMOSTATS | VARIABLE POWER TRANSFORMERS | POWER MODULE | MAGNETIC CONTACTORS | MERCURY RELAYS | HAND HELD PYROMETERS | HEATER ACCESSORIES |
|---|---|---|---|---|---|---|---|---|---|---|---|---|---|---|---|---|---|---|---|---|---|---|---|
| AIR & GAS | • | | | | • | • | • | • | | | | | | • | • | • | | • | • | • | • | | • |
| APPLIANCES | • | • | • | • | • | • | | • | • | | | | | • | • | • | • | | • | | | | • |
| CASTINGS | • | • | • | • | • | | | | | | | • | • | • | • | | • | | | | | • | • |
| COMFORT & SPACE | | | | • | • | • | • | • | | | | | | • | • | • | | | | | | | • |
| DRYING & CURING | • | | | • | • | | • | • | • | | • | • | • | • | • | • | | • | • | • | • | | • |
| FOOD EQUIP. | • | • | • | • | • | • | | • | • | | | • | • | • | • | • | • | | • | • | • | | • |
| MEDICAL EQUIP. | • | • | • | • | • | • | | | • | | | • | • | • | • | • | • | • | • | • | • | | • |
| MELTING PROCESSES | • | | | • | • | • | | | | • | • | • | • | • | • | • | • | • | • | • | • | | • |
| MOLDS & DIES | • | | • | • | • | | | | • | | | • | • | • | • | • | • | • | • | • | • | • | • |
| OVENS & TUNNELS | | | | • | • | | • | • | | | | • | • | • | • | • | | • | • | • | • | • | • |
| PACKAGING MCHY. | • | | • | • | • | | | • | • | | • | • | • | • | • | • | • | • | • | • | • | • | • |
| PIPELINES | | • | • | • | • | | | | • | | • | | | • | • | • | • | • | • | • | • | • | • |
| PLASTIC MCHY. | • | • | • | • | • | • | | • | • | | • | • | • | • | • | • | • | • | • | • | • | • | • |
| PLATEN | • | | • | • | • | | | | • | | | • | • | • | • | • | • | • | • | • | • | • | • |
| PREHEATING | • | | • | • | • | • | | • | | • | | • | • | • | • | • | • | • | • | • | • | • | • |
| ROLLS & CYLINDERS | • | • | • | • | • | | | | • | • | • | • | • | • | • | • | • | | • | • | • | • | • |
| SUPERHEATING | | | | | • | • | | | | | | • | • | • | | | | | | • | • | | • |
| TANKS-PROCESS | • | • | | | • | • | | | | • | | • | • | • | • | • | | | | • | • | | • |
| TANKS-STORAGE | | • | | • | • | • | | | | • | | • | • | | | | | | | • | • | | • |
| VACUUM FORMING | | | • | • | • | | | • | • | | | • | • | • | • | • | • | • | • | • | • | | • |

Courtesy of Tempco Electric Heater Corporation.

# Engineering Data

## HEAT REQUIREMENT CALCULATIONS

There are two basic heat energy requirements to be considered in the sizing of heaters for a particular application.

1. **Start-Up-Heat** is the heat energy required to bring a process up to operating temperature.
2. **Operating Heat** is the heat energy required to maintain the desired operating temperature thru normal work load cycles. The larger of these two heat energy values will be the wattage required for the application.

A safety factor is usually added to allow for unknown or unexpected operating conditions. The safety factor is dependent on the accuracy of the wattage calculation. A figure of 10% is adequate for small systems closely calculated, while 20% additional wattage is more common, and figures of 25% to 35% should be considered for larger systems with many unknown conditions existing.

**Start-Up-Heat** requirements will include one or more of the following calculations depending on the application.

1. **Wattage required to heat material:**

$$\frac{\text{Weight of material (lbs.)} \times \text{Specific Heat (Btu/lb °F)} \times \text{Temperature rise (°F)}}{\text{3.412 btu/watt hr.} \times \text{Heatup time (hr.)}} = \text{watts}$$

2. **Wattage required to heat container or tank:**

$$\frac{\text{Weight of container (lbs.)} \times \text{Specific Heat (Btu/lb °F)} \times \text{Temperature rise (°F)}}{\text{3.412 btu/watt hr.} \times \text{Heatup time (hr.)}} = \text{watts}$$

3. **Wattage required to heat hardware in container:**

$$\frac{\text{Weight of hardware (lbs.)} \times \text{Specific Heat (Btu/lb °F)} \times \text{Temperature rise (°F)}}{\text{3.412 btu/watt hr.} \times \text{Heatup time (hr.)}} = \text{watts}$$

4. **Wattage required to melt a solid to a liquid at constant temperature:**

$$\frac{\text{Heat of fusion (Btu/lb)} \times \text{Weight of material melted (lb/hr)}}{\text{3.412 Btu/watt hr.}} = \text{watts}$$

**Heat of Fusion:** The amount of heat required to change one pound of a given substance from solid to liquid state without change in temperature is termed the heat of fusion.
It requires 144 Btu to change one pound of ice at 32°F to one pound of water at 32°F, the heat of fusion of ice being 144 Btu per pound.
A change of state is usually accompanied by a change of specific heat. The specific heat of ice is 0.5, while that of water is 1.0.

5. **Wattage required to change a liquid to a vapor state at constant temperature:**

$$\frac{\text{Heat of vaporization (Btu/lb)} \times \text{Weight of material vaporized (lb/hr)}}{\text{3.412 Btu/watt hr.}} = \text{watts}$$

**Heat of Vaporization:** The amount of heat required to change one pound of a given substance from liquid to vapor state without a change in temperature is termed the heat of vaporization. It requires 965 Btu to change one pound of water at 212°F to one pound of steam at 212°F.

6. **Wattage to counteract liquid surface losses:** See Graph 3 for loss rates of water and oils.

$$\frac{\text{Total liquid surface area (sq. ft.)} \times \text{Loss rate at final temperature (watts/sq. ft.)}}{2} = \text{watts}$$

7. **Wattage to counteract losses from container walls,** platen surfaces, etc. See Graph 2 for losses from metal surfaces. See Graph 1 for losses from insulated surfaces.

$$\frac{\text{Total surface area (sq. ft.)} \times \text{Loss rate at final temperature (watts/sq. ft.)}}{2} = \text{watts}$$

*Courtesy of Tempco Electric Heater Corporation.*

# Engineering Data

**Operating heat** requirements will include one or more of the following calculations. Any additional losses particular to the application should also be estimated and included.

1. **Wattage to counteract losses from open liquid surfaces:** See Graph 3 for loss rates of water and oils.

   Total liquid surface area (sq. ft.) × Loss rate at operating temperature (watts/sq. ft.) = watts

2. **Wattage to counteract container or platen surface losses,** either insulated (See Graph 1) or uninsulated (See Graph 2)

   Total surface area (sq. ft.) × Loss rate at operating temperature (watts/sq. ft.) = watts

3. **Wattage required to heat material transferred in and out of the system.** (Metal dipped in heated tanks, air flows, make-up liquids, etc.)

$$\frac{\text{Weight of material heated (lb/hr)} \times \text{Specific Heat (btu/lb °F)} \times \text{Temperature change of material (°F)}}{\text{3.412 btu/watt hr.}} = \text{watts}$$

4. **Heatup of racks of containers,** etc. transferred in and out of the system:

$$\frac{\text{Weight of items heated (lb/hr)} \times \text{Specific Heat (btu/lb °F)} \times \text{Temperature change (°F)}}{\text{3.412 btu/watt hr.}} = \text{watts}$$

**Start-up or operating heat** calculations which include a material change of state should be calculated in two parts, from original temperature to change of state temperature and from change temperature to final temperature, since the specific heat may vary between the two states. The same applies for calculations covering large temperature changes in materials whose specific heat varies with temperature.

**Specific Heat:** The heat necessary to increase the temperature of all other substances has been referred to water as a standard. The ratio of the amount of heat required to increase the temperature of one pound of any substance by one degree to the amount necessary to increase one pound of water is known as the specific heat of that substance.

## HEAT LOSS INFORMATION

**GRAPH 1** HEAT LOSSES THROUGH INSULATED WALLS. (BASED ON STANDARD THERMAL INSULATIONS)

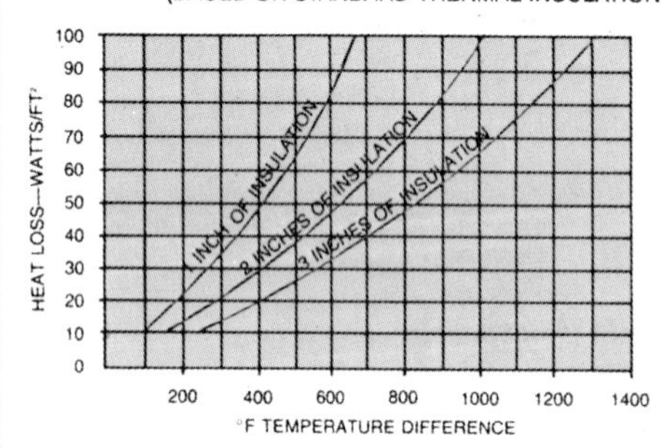


**GRAPH 2** HEAT LOSSES FROM UNINSULATED METAL SURFACES

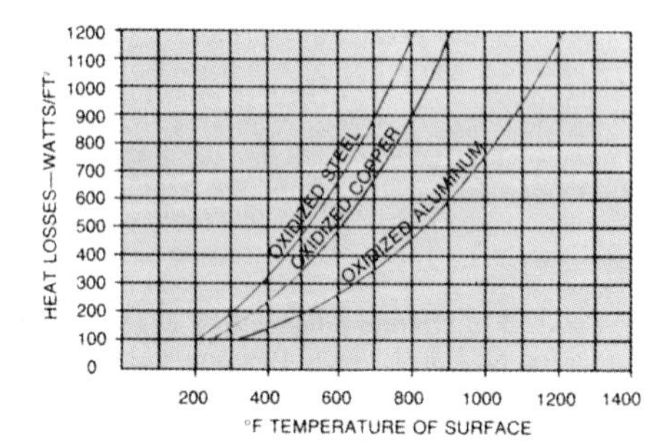


Figures are for vertical walls. Multiply by 110% for horizontal top surface, by 55% for horizontal bottom surface, and by 83% for average of both.

**GRAPH 3** HEAT LOSSES FROM SURFACE OF WATER AND OIL

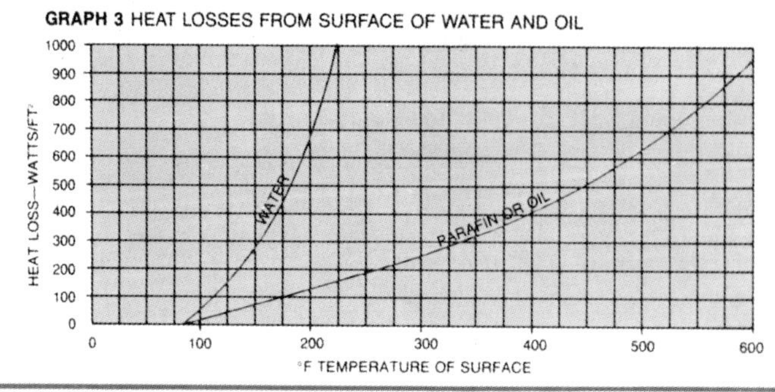


*Courtesy of Tempco Electric Heater Corporation.*

# Engineering Data

## Amperage Conversion Table

| Watts | Volts Single Phase | | | Volts 3 Phase Balanced Load | | Watts |
|---|---|---|---|---|---|---|
| | 120 | 240 | 480 | 240 | 480 | |
| 100 | .83 | .42 | .21 | .24 | .13 | 100 |
| 150 | 1.25 | .63 | .31 | .36 | .18 | 150 |
| 200 | 1.67 | .83 | .42 | .49 | .25 | 200 |
| 250 | 2.08 | 1.04 | .52 | .61 | .30 | 250 |
| 300 | 2.50 | 1.25 | .63 | .73 | .37 | 300 |
| 350 | 2.92 | 1.46 | .73 | .85 | .43 | 350 |
| 400 | 3.33 | 1.67 | .84 | .97 | .49 | 400 |
| 450 | 3.75 | 1.88 | .93 | 1.10 | .55 | 450 |
| 500 | 4.17 | 2.08 | 1.04 | 1.20 | .60 | 500 |
| 600 | 5.00 | 2.50 | 1.25 | 1.45 | .73 | 600 |
| 700 | 5.83 | 2.92 | 1.46 | 1.70 | .85 | 700 |
| 750 | 6.25 | 3.13 | 1.56 | 1.81 | .91 | 750 |
| 800 | 6.67 | 3.33 | 1.67 | 1.92 | .97 | 800 |
| 900 | 7.50 | 3.75 | 1.87 | 2.17 | 1.09 | 900 |
| 1000 | 8.33 | 4.17 | 2.10 | 2.41 | 1.21 | 1000 |
| 1100 | 9.17 | 4.58 | 2.30 | 2.65 | 1.33 | 1100 |
| 1200 | 10.0 | 5.00 | 2.51 | 2.90 | 1.45 | 1200 |
| 1250 | 10.4 | 5.21 | 2.61 | 3.00 | 1.50 | 1250 |
| 1300 | 10.8 | 5.42 | 2.71 | 3.13 | 1.57 | 1300 |
| 1400 | 11.7 | 5.83 | 2.91 | 3.38 | 1.69 | 1400 |
| 1500 | 12.5 | 6.25 | 3.12 | 3.62 | 1.82 | 1500 |
| 1600 | 13.3 | 6.67 | 3.34 | 3.86 | 1.93 | 1600 |
| 1700 | 14.2 | 7.08 | 3.54 | 4.10 | 2.05 | 1700 |
| 1750 | 14.6 | 7.29 | 3.65 | 4.22 | 2.10 | 1750 |
| 1800 | 15.0 | 7.50 | 3.75 | 4.34 | 2.17 | 1800 |
| 1900 | 15.8 | 7.92 | 3.96 | 4.58 | 2.29 | 1900 |
| 2000 | 16.7 | 8.33 | 4.17 | 4.82 | 2.41 | 2000 |
| 2200 | 18.3 | 9.17 | 4.59 | 5.30 | 2.65 | 2200 |
| 2500 | 20.8 | 10.4 | 5.21 | 6.01 | 3.00 | 2500 |
| 2750 | 23.0 | 11.5 | 5.73 | 6.63 | 3.32 | 2750 |
| 3000 | 25.0 | 12.5 | 6.25 | 7.23 | 3.62 | 3000 |
| 3500 | 29.2 | 14.6 | 7.30 | 8.45 | 4.23 | 3500 |
| 4000 | 33.3 | 16.7 | 8.33 | 9.64 | 4.82 | 4000 |
| 4500 | 37.5 | 18.8 | 9.38 | 10.84 | 5.42 | 4500 |
| 5000 | 41.7 | 20.8 | 10.42 | 12.1 | 6.01 | 5000 |
| 6000 | 50.0 | 25.0 | 12.50 | 14.50 | 7.25 | 6000 |
| 7000 | 58.3 | 29.2 | 14.59 | 16.9 | 8.5 | 7000 |
| 8000 | 66.7 | 33.3 | 16.67 | 19.3 | 9.65 | 8000 |
| 9000 | 75.0 | 37.5 | 18.75 | 21.7 | 10.85 | 9000 |
| 10000 | 83.3 | 41.7 | 20.85 | 24.1 | 12.1 | 10000 |

## SELECTION OF HOOK-UP LEADWIRE GAUGE

APPROXIMATE AMPERAGE CARRYING CAPACITIES FOR HIGH TEMPERATURE INSULATED COPPER, NICKEL CLAD COPPER AND NICKEL (GRADE A) BASED ON AMBIENT TEMPERATURE OF 86°F (30°C). MAXIMUM CONDUCTOR TEMPERATURE OF 482°F (250°C).

| Conductor Size AWG | Copper | Ni-Clad Copper | Nickel |
|---|---|---|---|
| 24 | 7.5 | 9.5 | 4.5 |
| 22 | 10 | 12 | 6 |
| 20 | 13 | 16 | 8 |
| 18 | 17 | 23 | 11 |
| 16 | 22 | 29 | 14 |
| 14 | 30 | 40 | 20 |
| 12 | 40 | 55 | 26 |
| 10 | 50 | 75 | 35 |
| 8 | 65 | 95 | 45 |

Determine the ambient temperature the wire will be exposed to. The conductor amperage carrying capacity of any given wire gauge will decrease as the temperature increases. To assist you in determining the correct wire gauge size, use this formula.

FORMULA: Ampacity of wire being used at room temperature X correction factor for ambient temperature (From chart) = Amperage carrying capacity at higher ambient.

SAMPLE: 18 Ga. wire with 27% N.C.C. conductors in 257°F ambient temperature would be $.75 \times 23 = 17.25$ Max. Amps.

| Ambient Temperature | | Correction Factors |
|---|---|---|
| 140°F | (60°C) | .93 |
| 176°F | (80°C) | .88 |
| 212°F | (100°C) | .82 |
| 257°F | (125°C) | .75 |
| 302°F | (150°C) | .67 |
| 392°F | (200°C) | .48 |
| 437°F | (225°C) | .33 |

## WATT DENSITY CALCULATIONS

**BAND HEATERS**

$$\text{WATTS/IN}^2 = \frac{\text{WATTAGE}}{\text{DIA.} \times 3.1416 \times \text{WIDTH}}$$

**CARTRIDGE AND TUBULAR HEATERS**

$$\text{WATTS/IN}^2 = \frac{\text{WATTAGE}}{\text{DIA.} \times 3.1416 \times \text{HEATED LENGTH}}$$

**MICA STRIP HEATERS**

$$\text{WATTS/IN}^2 = \frac{\text{WATTAGE}}{\text{HEATED LENGTH} \times \text{WIDTH}}$$

**CHANNEL STRIP HEATERS**

$$\text{WATTS/IN}^2 = \frac{\text{WATTAGE}}{\text{HEATED LENGTH} \times 3.625}$$

*Courtesy of Tempco Electric Heater Corporation.*

# Engineering Data
# Physical Properties of Materials

## PROPERTIES OF METALS

| Material | *Density lbs./in³ | Average specific heat, c, Btu/(lb)(°F) | *Thermal Conductivity k, (Btu)(in.)/(hr.)(sq. ft.)(°F) | Melting point, °F (lowest) | Latent heat of fusion, r, Btu/lb. | *Thermal expansion in./in./°F × 10⁻⁶ |
|---|---|---|---|---|---|---|
| Aluminum | .098 | .24 | 1540 | 1190 | 169 | 13.1 |
| Brass, yellow | .306 | .096 | 830 | 1710 | | 11.2 |
| Copper | .322 | .095 | 2680 | 1981 | 91.1 | 9.8 |
| Gold | .698 | .032 | 2030 | 1945 | 29.0 | 7.9 |
| Incoloy 800 | .290 | .13 | 80 | 2500 | | 7.9* |
| Inconel 600 | .304 | .126 | 103 | 2500 | | 5.8* |
| Invar 36% Ni | .293 | .126 | 73 | 2600 | | 0.6 |
| Iron, cast | .260 | .12 | 346 | 2150 | | 6.0 |
| Lead, solid | .410 | .032 | 240 | 620 | 11.3 | 16.4 |
| Lead, liquid | .387 | .037 | 108 | | | |
| Magnesium | .063 | .27 | 1106 | 1202 | 160 | 14 |
| Molybdenum | .369 | .071 | 980 | 4750 | 126 | 2.94 |
| Monel 400 | .319 | .11 | 151 | 2400 | | 6.4* |
| Nickel 200 | .321 | .12 | 520 | 2615 | 133 | 5.8* |
| Nichrome (80% Ni-20% Cr) | .302 | .11 | 104 | 2550 | | 7.3 |
| Palladium 99.5% | .432 | .062 | 490 | 2830 | | 6.5 |
| Platinum | .775 | .035 | 480 | 3225 | 49 | 4.9 |
| Silver | .379 | .057 | 2900 | 1760 | 38 | 10.8 |
| Solder (50% Pb-50% Sn) | .323 | .051 | 310 | 361 | 17 | 13.1 |
| Steel, mild | .284 | .122 | 460 | 2760 | | 6.7 |
| Steel, stainless 304 | .286 | .12 | 105 | 2550 | | 9.6 |
| Steel, stainless 430 | .275 | .11 | 155 | 2650 | | 6.0 |
| Tantalum | .60 | .035 | 375 | 5425 | | 3.57 |
| Tin, solid | .263 | .065 | 455 | 450 | 26.1 | 13 |
| Tin, liquid | .253 | .052 | 218 | | | |
| Titanium 99.0% | .164 | .13 | 112 | 3035 | | 4.7 |
| Tungsten | .697 | .040 | 1130 | 6170 | 79 | 2.45 |
| Type metal (85% Pb-15% Sb) | .387 | .040 | | 500 | 14± | |
| Zinc | .258 | .096 | 740 | 787 | 43.3 | 22.1 |
| Zirconium | .234 | .067 | 145 | 3350 | 108 | 3.22 |

## PROPERTIES OF NON-METALLIC SOLIDS

| Material | *Density lbs./in³ | Average specific heat, c, Btu/(lb)(°F) | *Thermal Conductivity k, (Btu)(in.)/(hr.)(sq. ft.)(°F) | Melting point, °F (lowest) | Latent heat of fusion, r, Btu/lb. | *Thermal expansion in./in./°F × 10⁻⁶ |
|---|---|---|---|---|---|---|
| Asbestos cement board | .070 | .25± | 5.2 | | | |
| Asphalt | .076 | .40 | 5.3 | | | |
| Boron Nitride | .082 | .33 | 125 | 5430 | | 1-4 |
| Brickwork | .076 | .22 | 3-7 | | | 3-6 |
| Carbon | .080 | .28 | 165 | 6700 | | 0.3-2.4 |
| Cellulose Acetate | .047 | .3-.5 | 1.2-2.3 | | | 61-83 |
| Cellulose acetate butyrate | .043 | .3-.4 | 1.2-2.3 | | | 61-94 |
| Delrin | .051 | .35 | 1.6 | | | 45 |
| Glass, crown | .101 | .161 | 7.5 | | | 5 |
| Ice | .0324 | .53 | 11 | 32 | 144 | 28.3 |
| Mica | .102 | .21 | 3.0 | | | 18 |
| Mg0 (Compacted) | .112 | .209 | 20 | | | 7.7 |
| Nylon | .040 | .4 | 1.5 | | | 61-63 |
| Paper | .034 | .45 | .82 | | | |
| Paraffin | .032 | .69 | 1.6 | 133 | 63 | |
| Phenolic, general purpose | .046 | .40 | .6-1.2 | | | 44-61 |
| Polyethylene | .035 | .55 | 2.3 | | | 94 |
| Polystyrene | .038 | .32 | .7-1.0 | | | 33-44 |
| Rubber | .044 | .44 | 1.1 | | | 340 |
| Steatite | .094 | .20 | 17.5-23 | | | 4.5-5.5 |
| Sulfur | .075 | .175 | 1.9 | 246 | 17 | 36 |
| Teflon | .078 | .25 | 1.7 | | | 55 |
| Vinyl | .046 | .3-.5 | .8-2.0 | | | 28-100 |
| Wood, oak | .029 | .57 | 1.1 | | | |

*at or near room temperature ± Approximate

## PROPERTIES OF LIQUIDS

| Liquid | *Density lbs./Gal. | Average specific heat, c, Btu/(lb)(°F) | *Thermal Conductivity k, (Btu)(in.)/(hr.)(sq. ft.)(°F) | Boiling point °F | Heat of vaporization, r, Btu/lb. |
|---|---|---|---|---|---|
| Acetic acid, 20%** | 8.6 | .91 | 3.7 | 214± | 810± |
| Alcohol (ethyl) | 6.6 | .60 | 1.3 | 173 | 367 |
| Brine (25% NiCi)** | 9.9 | .81 | 4.0 | 221± | 728± |
| Caustic Soda (18% NaOH)** | 10.0 | .84 | 3.9 | 221± | 795± |
| Dowtherm A | 8.8 | .44 | .96 | 496 | 42.2 |
| Freon 12 | 11.0 @ 70 psig | .23 | .67 | −21.6 | 62 |
| Glycerine | 10.5 | .61 | 2.0 | 556 | |
| Hydrochloric acid 10% | 8.9 | .93 | 3.9 | 221 | |
| NaK (78% K) | 6.2*** | .21*** | 167.0 | 1446 | |
| Nitric acid, 7%** | 8.6 | .92 | 3.8 | 220± | 918± |
| Oils (petroleum) | 7.8 | .43 | 1.0 | | |
| Paraffin (melted) | 6.3 | .71± | 1.0 | | |
| Potassium (K) | 6.0*** | .18*** | 320.0 | 1400 | 893 |
| Sodium (Na) | 6.8*** | .30*** | 580.0 | 1621 | 1810 |
| Sulfuric Acid 10% | 9.9 | .92 | 4.0 | 216 | |
| Therminol FR-2 | 12.1 | .30 | .70 | 648± | |
| Turpentine | 7.3 | .41 | .90 | 318 | 123 |
| Vegetable oil | 7.7 | .43± | 1.1 | | |
| Water | 8.3 | 1.0 | 4.2 | 212 | 970 |

*At or near room temperature
**Percent concentration by weight in $H_2O$ solution
***At 1000°F
± Approximate

## PROPERTIES OF GASES

| Gas | *Density lbs/ft.³ | *Specific heat, c, Btu/lb./°F | *Thermal conductivity k, (Btu)(in.)/(hr.)(sq. ft.)(°F) |
|---|---|---|---|
| Air at 80°F | .073 | .240 | .18 |
| at 400°F | .046 | .245 | .27 |
| Ammonia | .044 | .523 | .16 |
| Argon | .102 | .125 | .12 |
| Carbon dioxide | .113 | .199 | .12 |
| Carbon monoxide | .072 | .248 | .18 |
| Chlorine | .184 | .115 | .06 |
| Helium | .011 | 1.25 | 1.10 |
| Hydrogen | .0052 | 3.39 | .13 |
| Methane | .041 | .528 | .25 |
| Nitrogen | .072 | .248 | .19 |
| Oxygen | .082 | .218 | .18 |
| Sulphur dioxide | .172 | .152 | .07 |

*At or near room temperature

## AIR DENSITY TABLE (LB/CU. FT.)

| Temp °F | Specific Heat | GAUGE PRESSURE | | | | | | |
|---|---|---|---|---|---|---|---|---|
| | | 0 | 10 | 50 | 100 | 200 | 250 | 300 |
| 0 | .240 | .086 | .145 | .380 | .674 | 1.261 | 1.555 | 1.848 |
| 10 | .240 | .085 | .142 | .372 | .659 | 1.234 | 1.522 | 1.808 |
| 20 | .240 | .083 | .139 | .364 | .646 | 1.208 | 1.490 | 1.771 |
| 30 | .240 | .081 | .136 | .357 | .632 | 1.184 | 1.459 | 1.735 |
| 40 | .240 | .079 | .133 | .350 | .620 | 1.160 | 1.430 | 1.700 |
| 50 | .240 | .078 | .131 | .343 | .608 | 1.137 | 1.402 | 1.667 |
| 60 | .240 | .076 | .128 | .336 | .596 | 1.115 | 1.375 | 1.635 |
| 70 | .240 | .075 | .126 | .330 | .585 | 1.094 | 1.349 | 1.600 |
| 80 | .240 | .074 | .124 | .324 | .574 | 1.074 | 1.324 | 1.574 |
| 90 | .240 | .072 | .121 | .318 | .563 | 1.055 | 1.300 | 1.546 |
| 100 | .240 | .071 | .119 | .312 | .553 | 1.036 | 1.277 | 1.518 |
| 120 | .240 | .068 | .115 | .301 | .534 | 1.000 | 1.233 | 1.466 |
| 140 | .240 | .066 | .111 | .291 | .516 | .967 | 1.192 | 1.417 |
| 160 | .241 | .064 | .108 | .282 | .500 | .936 | 1.153 | 1.371 |
| 180 | .241 | .062 | .104 | .273 | .484 | .906 | 1.117 | 1.328 |
| 200 | .242 | .060 | .101 | .265 | .470 | .879 | 1.084 | 1.288 |
| 220 | .242 | .058 | .098 | .257 | .456 | .853 | 1.052 | 1.250 |
| 240 | .242 | .057 | .095 | .250 | .443 | .829 | 1.022 | 1.215 |
| 260 | .243 | .055 | .093 | .243 | .430 | .806 | .993 | 1.181 |
| 280 | .243 | .054 | .090 | .236 | .419 | .784 | .966 | 1.149 |
| 300 | .244 | .052 | .088 | .230 | .408 | .763 | .941 | 1.119 |
| 320 | .244 | .051 | .086 | .224 | .397 | .744 | .917 | 1.090 |
| 340 | .244 | .050 | .083 | .219 | .387 | .725 | .894 | 1.063 |
| 360 | .246 | .048 | .081 | .213 | .378 | .707 | .872 | 1.037 |
| 380 | .246 | .047 | .079 | .208 | .369 | .691 | .851 | 1.012 |
| 400 | .247 | .046 | .078 | .203 | .360 | .674 | .832 | .989 |
| 420 | .247 | .045 | .076 | .199 | .352 | .659 | .813 | .966 |
| 440 | .247 | .044 | .074 | .194 | .344 | .644 | .795 | .945 |
| 460 | .248 | .043 | .073 | .190 | .337 | .630 | .777 | .924 |
| 480 | .248 | .042 | .071 | .186 | .330 | .617 | .761 | .905 |
| 500 | .249 | .041 | .070 | .182 | .323 | .604 | .745 | .886 |
| 520 | .249 | .041 | .068 | .178 | .316 | .592 | .730 | .868 |
| 540 | .249 | .040 | .067 | .175 | .310 | .580 | .715 | .850 |
| 560 | .250 | .039 | .065 | .171 | .304 | .569 | .701 | .834 |
| 580 | .251 | .038 | .064 | .168 | .298 | .558 | .688 | .818 |
| 600 | .252 | .037 | .063 | .165 | .292 | .547 | .675 | .802 |
| 620 | .252 | .037 | .062 | .162 | .287 | .537 | .662 | .787 |
| 640 | .252 | .036 | .061 | .159 | .281 | .527 | .650 | .773 |
| 660 | .253 | .035 | .060 | .156 | .277 | .518 | .639 | .759 |
| 680 | .253 | .035 | .059 | .153 | .272 | .509 | .627 | .746 |
| 700 | .254 | .034 | .058 | .151 | .267 | .500 | .616 | .733 |
| 720 | .254 | .034 | .057 | .148 | .263 | .492 | .606 | .721 |
| 740 | .255 | .033 | .056 | .146 | .258 | .483 | .596 | .709 |
| 760 | .256 | .033 | .055 | .143 | .254 | .475 | .586 | .697 |
| 780 | .256 | .032 | .054 | .141 | .250 | .468 | .577 | .686 |
| 800 | .257 | .032 | .053 | .139 | .246 | .460 | .568 | .675 |
| 820 | .257 | .031 | .052 | .137 | .242 | .453 | .559 | .664 |
| 840 | .257 | .031 | .051 | .134 | .238 | .446 | .550 | .654 |
| 860 | .258 | .030 | .051 | .132 | .235 | .439 | .542 | .644 |
| 880 | .259 | .030 | .050 | .130 | .231 | .433 | .534 | .634 |
| 900 | .260 | .029 | .049 | .129 | .228 | .427 | .526 | .625 |
| 920 | .260 | .029 | .048 | .127 | .225 | .420 | .518 | .616 |
| 940 | .260 | .028 | .048 | .125 | .221 | .414 | .511 | .607 |
| 960 | .261 | .028 | .047 | .123 | .218 | .408 | .504 | .599 |
| 980 | .261 | .028 | .046 | .121 | .215 | .403 | .497 | .590 |
| 1000 | .262 | .027 | .046 | .120 | .212 | .397 | .490 | .582 |
| 1020 | .262 | .027 | .045 | .118 | .209 | .392 | .483 | .574 |
| 1040 | .263 | .026 | .044 | .117 | .207 | .387 | .477 | .567 |
| 1060 | .264 | .026 | .044 | .115 | .204 | .382 | .470 | .559 |
| 1080 | .264 | .026 | .043 | .114 | .201 | .377 | .464 | .552 |
| 1100 | .265 | .025 | .043 | .112 | .199 | .372 | .458 | .545 |
| 1120 | .265 | .025 | .042 | .111 | .196 | .367 | .453 | .538 |
| 1140 | .265 | .025 | .042 | .109 | .194 | .363 | .447 | .531 |
| 1160 | .266 | .025 | .041 | .108 | .191 | .358 | .441 | .525 |
| 1180 | .266 | .024 | .041 | .107 | .189 | .354 | .436 | .518 |
| 1200 | .267 | .024 | .040 | .105 | .187 | .349 | .431 | .512 |

*Courtesy of Tempco Electric Heater Corporation.*

# Engineering Data

## Percent of Rated Wattage for Various Applied Voltages

| Applied Voltage | Rated Voltage | | | | | | | | | | | | | |
|---|---|---|---|---|---|---|---|---|---|---|---|---|---|---|
| | 110 | 115 | 120 | 208 | 220 | 230 | 240 | 277 | 380 | 415 | 440 | 460 | 480 | 550 |
| 110 | 100% | 91% | 84% | 28% | 25% | 23% | 21% | 16% | 8.4% | 7% | 6.2% | 5.7% | 5.2% | 4% |
| 115 | 109% | 100% | 92% | 31% | 27% | 25% | 23% | 17% | 9.0% | 7.6% | 6.7% | 6.2% | 5.7% | 4.3% |
| 120 | 119% | 109% | 100% | 33% | 30% | 27% | 25% | 19% | 10% | 8.4% | 7.4% | 6.8% | 6.3% | 4.8% |
| 208 | | | 300% | 100% | 89% | 82% | 75% | 56% | 30% | 25% | 22% | 20% | 19% | 14% |
| 220 | | | | 112% | 100% | 91% | 84% | 63% | 34% | 28% | 25% | 23% | 21% | 16% |
| 230 | | | | 122% | 109% | 100% | 92% | 69% | 37% | 31% | 27% | 25% | 23% | 17% |
| 240 | | | | 133% | 119% | 109% | 100% | 75% | 40% | 33% | 30% | 27% | 25% | 19% |
| 277 | | | | | | | 133% | 100% | 53% | 45% | 40% | 36% | 33% | 25% |
| 380 | | | | | | | | 188% | 100% | 84% | 74% | 68% | 63% | 47% |
| 415 | | | | | | | | | 119% | 100% | 89% | 81% | 75% | 57% |
| 440 | | | | | | | | | | 112% | 100% | 91% | 84% | 64% |
| 460 | | | | | | | | | | 123% | 109% | 100% | 92% | 70% |
| 480 | | | | | | | | | | | 119% | 109% | 100% | 76% |
| 550 | | | | | | | | | | | 156% | 143% | 131% | 100% |

To determine the resultant wattage on a voltage not shown in the chart above, use the following formula:

$$\text{Actual Wattage} = \text{Rated Wattage} \times \frac{(\text{Applied Voltage})^2}{(\text{Rated Voltage})^2}$$

**CAUTION**

Applying higher voltage than the actual rated voltage on heating elements will increase the watt density ($W/INCH^2$), which can result in premature heater failure and/or damage the material being heated.

## TEMPERATURE CONVERSION TABLE

| °C | Temp. | °F | °C | Temp. | °F | °C | Temp. | °F |
|---|---|---|---|---|---|---|---|---|
| -17.8 | 0 | 32.0 | 371.1 | 700 | 1292.0 | 760.0 | 1400 | 2552.0 |
| 10.0 | 50 | 122.0 | 398.9 | 750 | 1382.0 | 787.8 | 1450 | 2642.0 |
| 37.8 | 100 | 212.0 | 426.7 | 800 | 1472.0 | 815.6 | 1500 | 2732.0 |
| 65.6 | 150 | 302.0 | 454.4 | 850 | 1562.0 | 843.3 | 1550 | 2822.0 |
| 93.3 | 200 | 392.0 | 482.2 | 900 | 1652.0 | 871.1 | 1600 | 2912.0 |
| 121.1 | 250 | 482.0 | 510.0 | 950 | 1742.0 | 898.9 | 1650 | 3002.0 |
| 148.9 | 300 | 572.0 | 537.8 | 1000 | 1832.0 | 926.7 | 1700 | 3092.0 |
| 176.7 | 350 | 662.0 | 565.6 | 1050 | 1922.2 | 954.4 | 1750 | 3162.0 |
| 204.4 | 400 | 752.0 | 593.3 | 1100 | 2012.0 | 982.2 | 1800 | 3272.0 |
| 232.2 | 450 | 842.0 | 621.1 | 1150 | 2102.0 | 1010.0 | 1850 | 3362.0 |
| 260.0 | 500 | 932.0 | 648.9 | 1200 | 2192.0 | 1037.8 | 1900 | 3452.0 |
| 287.7 | 550 | 1022.0 | 676.7 | 1250 | 2282.0 | 1065.6 | 1950 | 3542.0 |
| 315.6 | 600 | 1112.0 | 704.4 | 1300 | 2372.0 | 1093.3 | 2000 | 3632.0 |
| 343.3 | 650 | 1202.0 | 732.2 | 1350 | 2462.0 | 1148.9 | 2100 | 3812 |

**RECOMMENDED MAXIMUM OPERATING TEMPERATURES FOR VARIOUS SHEATH MATERIALS:**

| Sheath Material | Max. Operating Temperature | |
|---|---|---|
| Copper | 350°F | 176.7°C |
| Aluminum | 500° | 260.0° |
| Brass | 750° | 398.9° |
| Steel | 750° | 398.9° |
| Stainless Steels | 1500° | 815.6° |
| Incoloy | 1600° | 871.1° |
| Inconel | 1700° | 926.7° |

## INTERPOLATION FACTORS

| °C | | °F | °C | | °F |
|---|---|---|---|---|---|
| 0.55 | 1 | 1.8 | 3.33 | 6 | 10.8 |
| 1.11 | 2 | 3.6 | 3.88 | 7 | 12.6 |
| 1.66 | 3 | 5.4 | 4.44 | 8 | 14.4 |
| 2.22 | 4 | 7.2 | 5.00 | 9 | 16.2 |
| 2.77 | 5 | 9.0 | 5.55 | 10 | 18.0 |

## TEMPERATURE CONVERSION

°F = 9/5 °C + 32
°C = 5/9 (°F − 32)

°R = °F + 460
K = °C + 273

Sheath Temperature of Tubular Elements at Various Watt Densities in Free or Forced Air at 80°F.

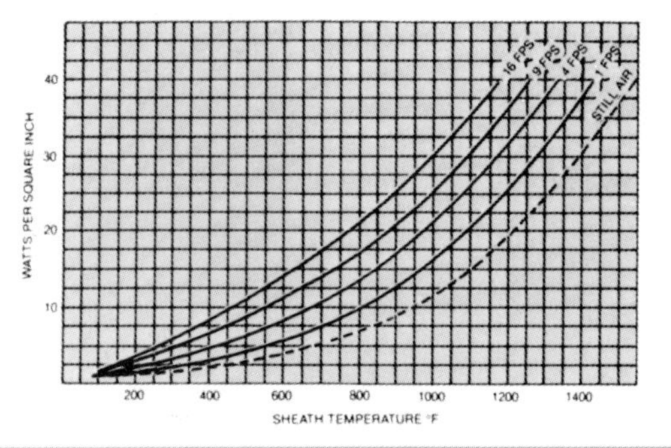


Allowable Watt Density of Tubular Elements Operating at 200° to 1400°F Sheath Temperature for Various Temperatures in Distributed Air Velocity of 16 Fps.

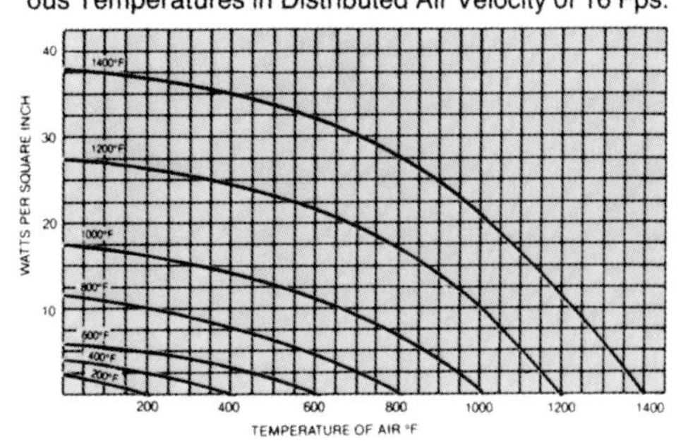


*Courtesy of Tempco Electric Heater Corporation.*

# Engineering Data
# Recommended Sheath Materials For

The following sheath material selection guide is based on available data and recommendations from chemical suppliers as well as industrial user experience. Due to the many variables in chemical processes, this guide should only be used as a starting point in your selection of a suitable sheath material. Final selection should be based on your experience and knowledge of your process.

## EXAMPLES OF PROCESS VARIABLES THAT CAN AFFECT HEATER SHEATH SELECTION

1. Solution chemistry
2. Solution contamination
3. Temperature
4. Flow rate (velocity) past heater
5. Heater watt density
6. Heating cycle (time-on, time-off)
7. Galvanic behavior
8. Degree of aeration

## NOTES TO SHEATH MATERIAL SELECTION GUIDE

1. This solution contains a mixture of various chemical compounds whose identity and proportions are unknown or subject to change. Check with chemical supplier to confirm suitability of sheath material chosen.
2. Caution—Flammable material
3. Chemical composition varies widely. Check supplier for specific recommendations.
4. Direct immersion heaters not practical. Use clamp-on heaters on outside surface of cast iron pot.
5. Element surface loading should not exceed 20 watts per square inch.
6. For concentrations greater than 15%, element surface loading should not exceed 20 watts per square inch.
7. See suggested watt density chart.
8. Remove crusts at liquid level.
9. Clean often.
10. Passivate stainless steel, Inconel and Incoloy.

## MAXIMUM RECOMMENDED WATT DENSITIES FOR VARIOUS MATERIALS

| Material Being Heated | Maximum Operating Temperature °F | Maximum Watt Density W/IN² |
|---|---|---|
| Acid Solutions | 180 | 40 |
| Alkaline Solutions, Oakite | 212 | 40 |
| Ammonia Pltg. Solution | 50 | 25 |
| Asphalt, Tar or Heavy Compounds | 200-500 | 2-5 |
| Caustic Soda 2% | 210 | 45 |
| 10% | 210 | 25 |
| 75% | 180 | 25 |
| Degreasing Solution Vapor | 275 | 20 |
| Electroplating Solution | 180 | 40 |
| Ethylene Glycol | 300 | 30 |
| Fatty Acids | 150 | 20 |
| Fuel Oils | | |
| Light Grade | 180 | 25–30 circu. |
| Heavy (Bunker C) | 160 | 8 |
| Gasoline | 300 | 2–5 |
| Glycerine | 50 | 40 |
| Machine Oil SAE 30 | 250 | 15–20 non-circ. |
| Metal Melting Pot | 500–900 | 20–27 |
| Mineral Oil | 400 | 16 |
| Molasses | 100 | 4–5 |
| Molten Tin | 600 | 20 |
| Oil Draw Bath | 600 | 20 |
| Paraffin or Wax | 150 | 16 |
| Potassium Hydroxide | 160 | 25 |
| Propylene Glycol | 150 | 20 |
| Steel Tubing Cast Into Aluminum | 500–750 | 50 |
| Steel Tubing Cast Into Iron | 750–1000 | 55 |
| Trichlorethylene | 150 | 20 |
| Water (Process) | 35–150 | 100–125 circ.<br>75–100 non-circ. |
| | 212 | 75 circ.<br>50 non-circ. |

## SHEATH MATERIAL ANALYSIS

| Sheath Material | Chemical Composition | | | | | | | | | | | | | | | | |
|---|---|---|---|---|---|---|---|---|---|---|---|---|---|---|---|---|---|
| | Al | C | Co | Cr | Cu | Fe | Mn | Mo | Ni | P | S | Si | Ta | Ti | V | W | Notes |
| Steel—1010 Carbon | | .08/.13 | | | | Bal | .3/.6 | | | .04 * | .05 * | | | | | | |
| Cast Iron | | | | | | | | | | | | | | | | | |
| Grey | | 2/4 | | | | Bal | .4/.75 | | | .04/1 | .04/.1 | .2/3 | | | | | |
| Nickel Resist | | 3 | | 3.2 | 6.48 | Bal | 1.0 | | 18/22 | | | 1.6 | | | | | |
| **Stainless Steels** | | | | | | | | | | | | | | | | | |
| 304 | | .08 * | | 18/20 | | Bal | 2 * | | 8/12 | | | 1 * | | | | | |
| 316 | | .08 * | | 16/18 | | Bal | 2 * | 2/3 | 10/14 | | | 1 * | | | | | |
| 321 | | .08 * | | 17/19 | | Bal | 2 * | | 9/12 | | | 1 * | | | | | Ti * = 5 × C |
| 347 | | .08 * | | 17/19 | | Bal | 2 * | | 9/13 | | | 1 * | | | | | Cb */Ta * = 10 × C |
| Type 20 | | .07 * | | 20 | 4 | Bal | .7 | 2.5 | 29 | | | 1.5 * | | | | | |
| Carpenter 20Cb-3 | | .06 * | | 19/20 | 3/4 | Bal | 2 * | 2/3 | 32.5/35 | .05 * | .035 * | 1 * | | | | | Cb/Ta = 8 × C or 1% max. |
| **Nickel Alloys** | | | | | | | | | | | | | | | | | |
| Incoloy 800 | .15/.6 | .1 | | 19/23 | .75 | Bal | 1.5 | | 30/35 | | .015 | 1.0 | | .15/.16 | | | |
| Hastelloy C/C276 | | .02 * | 2.5 * | 14.5/16.5 | | 4/7 | 1.0 * | 15/17 | Bal | .04 * | .03 * | .08 * | | | .35 * | 3/4.5 | |
| Hastelloy B/B2 | | .02 * | 1.0 * | 1.0 * | | 2 * | 1.0 * | 26/30 | Bal | .04 * | .03 * | .10 * | | | | | |
| Monel 400 | | .3 * | | | Bal | 2.5 * | 2 * | | 63/70 | | .024 * | .5 * | | | | | |
| Inconel 600 | | .15 * | | 14/17 | .5 * | 6/10 | 1 * | | | | .015 * | .5 * | | | | | Nickel + Cobalt = 72% min. |

* MAXIMUM PERCENTAGE

*Courtesy of Tempco Electric Heater Corporation.*

# Engineering Data
# Immersion Heating Applications

| MEDIA BEING HEATED | ELEMENT SHEATH MATERIAL | | | | | | | | | | | | | | | | | | |
|---|---|---|---|---|---|---|---|---|---|---|---|---|---|---|---|---|---|---|---|
| | IRON & STEEL | CAST IRON GRAY | CAST IRON NI. RESIST | ALUMINUM | COPPER | LEAD | MONEL 400 | NICKEL 200 | 304, 321, 347 STN. STL. | 316 STN STL. | TYPE 20 STN STL. | INCOLOY 800 | INCONEL 600 | TITANIUM | HASTELLOY B | QUARTZ | GRAPHITE | TEFLON | *NOTES |
| Acetaldehyde | | | | | A | | | | | A | A | | | | | | | | Note 2 |
| Acetic Acid, Crude | X | | C | F | F | X | F | F | F | F | | C | C | | | | | | |
| Pure | | | X | A | F | F | A | F | | | | C | C | | | | | | |
| Vapors | | | X | C | F | X | F | F | | | | C | C | F | | | | | |
| 150 PSI; 400°F | | | | C | F | X | F | F | | | | C | C | | | | | | |
| Aerated | X | X | X | C | X | X | X | X | X | F | F | | X | A | | | | | |
| No Air | | X | X | C | F | X | A | F | C | F | F | | X | A | | | | | |
| Acetone | C | X | F | F | A | A | A | A | A | A | A | A | A | A | A | A | A | | Note 2 |
| Actane™ 70 | | | | | | | | | | | | | | | | | A | A | Note 1 |
| Actane™ 80 | | | | | | | | | | | | | | | | | A | A | Note 1 |
| Actane™ Salt | | | | | | | | | | | | | | | | | A | | Note 1 |
| Alboloy Process | A | | | | | | | | | | | | | | | | | | |
| Alcoa™ R5 Bright Dip | | | | | | | | | | | | | | | | A | | A | Note 1 |
| Allyl Alcohol | | A | A | F | A | F | A | A | A | A | A | A | A | A | | | | | |
| Alcohol | F | F | | F | A | A | A | A | F | A | A | A | A | A | A | A | A | | Note 2 |
| Alcorite™ | | | | | | | | | | | | | | | A | | | | Note 1 |
| Alkaline Cleaners | | | | | | | | | A | | | | | | | | | | Note 1 |
| Alkaline Soaking Cleaners | A | | | | | | | | | | | | | | | | | | Note 1 |
| Alodine™ | | | | | | | | | | A | | | | | | | | | Note 1 |
| Aluminum (Molten) | | | | | | | | | CONSULT TEMPCO | | | | | | | | | | |
| Aluminum Acetate | X | X | | | F | A | F | F | F | A | A | | F | A | A | | | | |
| Aluminum Bright Dip | | | | | | | | | | | | | | | | A | | A | Note 1 |
| Aluminum Chloride | X | X | | X | X | X | X | X | X | X | X | X | X | X | A | A | A | A | Note 1 |
| Aluminum Cleaners | C | C | | X | X | X | A | A | A | A | F | A | A | F | | X | X | | Notes 1, 9 |
| Aluminum Potassium Sulfate (Alum) | | X | X | X | A | F | F | F | X | C | F | | F | F | | | | | |
| Aluminum Sulfate | X | X | X | X | X | F | X | X | F | F | F | X | X | A | | A | A | | Note 1 |
| Ammonia | X | X | | C | X | C | X | X | X | X | X | C | F | A | A | A | A | | |
| Ammonia (Anhydrous) (Gas) | F | | | | X | | | | A | A | | | | | | | | | |
| Cold | C | | A | A | A | F | A | A | A | A | A | | A | A | | | | | |
| Hot | C | | C | | A | X | A | A | C | C | A | | A | | | | | | |
| Ammonia and Oil | A | | | | | | | | | | | | | | | | | | |
| Ammonium Acetate | A | F | F | A | X | X | A | A | A | A | A | A | A | | | | | | |
| Ammonium Chloride | X | X | F | X | X | X | F | F | X | C | C | C | C | A | | A | A | A | |
| Ammonium Hydroxide | F | F | F | C | X | F | X | A | A | A | A | A | A | A | | X | A | | |
| Ammonium Nitrate | F | X | C | F | X | X | X | X | A | A | A | X | X | X | | A | A | | |
| Ammonium Persulfate | X | X | | X | X | C | X | X | F | F | F | | X | | | A | A | A | |
| Ammonium Sulfate | X | X | F | X | X | F | F | F | C | F | F | F | F | A | | A | A | | |
| Amyl Acetate | F | | | | A | | A | A | A | A | A | A | | A | | | | | |
| Amyl Alcohol | A | F | F | C | A | | A | F | A304 | A | A | A | A | A | | A | | | Note 2 |

Corrosion Resistance Ratings: A = Good F = Fair C = Depends on Conditions X = Unsuitable
Blank = Data Not Available

* SEE PAGE 418 FOR NOTES TO SHEATH MATERIAL SELECTION GUIDE

*Courtesy of Tempco Electric Heater Corporation.*

# Engineering Data
# Recommended Sheath Materials for

| MEDIA BEING HEATED | ELEMENT SHEATH MATERIAL | | | | | | | | | | | | | | | | | |
|---|---|---|---|---|---|---|---|---|---|---|---|---|---|---|---|---|---|---|
| | IRON & STEEL | CAST IRON GRAY | CAST IRON NI. RESIST | ALUMINUM | COPPER | LEAD | MONEL 400 | NICKEL 200 | 304, 321, 347 STN. STL. | 316 STN STL. | TYPE 20 STN STL. | INCOLOY 800 | INCONEL 600 | TITANIUM | HASTELLOY B | QUARTZ | GRAPHITE | TEFLON | *NOTES |
| Aniline | F | A | | F | X | F | F | F | A304 | A | A | F | F | A | | A | A | | |
| Aniline, Oil | A | | | X | X | | | | A | A | | | | | | | | | |
| Aniline, Dyes | | | | | | | A | | A | A | | | | | | | | | |
| Anodizing Solutions (10%) | | | | | | | | | | | | | | | | | | | |
| Chromic Acid 96°F | C | | | | | | | | A | A | | | | A | | | | | |
| Sulfuric Acid 70°F | | | | | | A | | | | | A | | | | | | | | |
| Sodium Hydroxide Alkaline | A | | | | A | | | A | | A | A | A | | A | | | | | |
| Nigrosine Black Dye | | | | | | | F | F | | | | | | | | | | | |
| Nickel Acetate | | | | | | C | A | F | | | | | | | | | | | |
| ARP™ 28 | | | | | | | | | | | | | | | | | A | A | Note 1 |
| ARP™ 80 Blackening Salt | | | | | | | | | | | | | | | | | A | | Note 1 |
| Arsenic Acid | X | X | | X | X | X | X | X | C | F | F | X | X | X | | A | A | A | |
| Asphalt | A | A | | X | X | X | X | A | A | A | A | A | A | A | | A | A | | |
| Barium Chloride | | | | X | | | | A | F | F | | | A | | | | | | |
| Barium Hydroxide | F | F | | X | X | X | F | A | F | A | A | F | F | X | | A | A | | |
| Barium Sulfate | F | F | F | | F | F | F | F | F | F | F | F | F | A | | A | A | | |
| Barium Sulfite | | | | | | | | | F | | | | | | | | | | |
| Black Nickel | | | | | | | | | | | | | | | | A | | A | Note 5 |
| Black Oxide | | | | | | | | | A | | | | | | | | | | Note 5 |
| Bleaching Solution 1½ lb. Oxalic Acid per Gallon of $H_2O$ at 212°F | | | | | | | A | | F | | | | | | | | | | |
| Bonderizing™ (Zinc Phosphate) | C | | F | | | | | | A | A | | | | | | | | | |
| Boric Acid | X | X | | X | C | C | C | C | C | C | C | C | C | A | A | A | A | A | |
| Brass Cyanide | | | | | | | | | A | | | | | | | | | | Note 1 |
| Bright Nickel | | | | | | | | | | | | | | A | | A | | | Notes 1,5 |
| Brine (Salt Water) | | | | | | | A | | | | | | F | | | | | | |
| Bronze Plating | A | | | | | | | | A | | | | | | | | | | Note 1 |
| Butanol | A | A | | F | A | A | A | A | A | A | A | A | A | A | | A | A | A | Note 2 |
| Cadmium Black | | | | | | | | | | | | | | | | A | | | Note 1 |
| Cadmium Fluoborate | | | | | | | | | | | | | | | | | A | A | Note 1 |
| Cadmium Plating | | | | | | | | | A | | | A | A | | | | | | Note 1 |
| Calcium Chlorate | F | F | | F | C | C | F | F | F | F | F | F | F | | | A | | | |
| Calcium Chloride | F | F | | C | F | X | F | F | F | F | F | F | F | A | A | A | A | A | |
| Carbon Dioxide—Dry Gas | X | X | A | A | A | F | A | A | A | A | A | A | A | X | | A | X | X | |
| Carbon Dioxide—Wet Gas | X | X | C | A | X | F | A | A | A | A | A | A | A | X | | A | X | X | |
| Carbon Tetrachloride | X | X | C | X | C | A | A | A | C | F | F | A | A | A | | A | | | |
| Carbonic Acid | C | C | | C | C | X | C | C | A | F | A | F | A | A | | A | A | A | |
| Castor Oil | A | A | | A | A | A | A | A | A | A | A | A | A | A | | A | A | A | |
| Caustic Etch | A | A | | X | X | | A | A | A | A | X | X | X | A | | X | A | X | |

Corrosion Resistance Ratings: A = Good F = Fair C = Depends on Conditions X = Unsuitable
Blank = Data Not Available

* SEE PAGE 418 FOR NOTES TO SHEATH MATERIAL SELECTION GUIDE

*Courtesy of Tempco Electric Heater Corporation.*

# Engineering Data
# Immersion Heating Applications

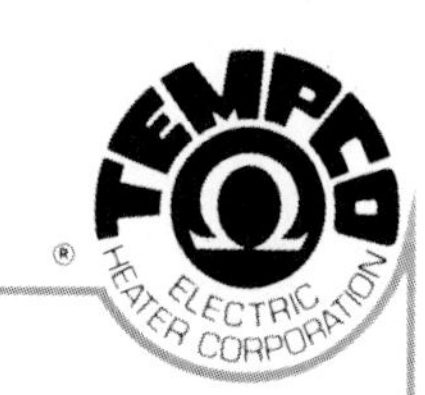


| MEDIA BEING HEATED | IRON & STEEL | CAST IRON GRAY | CAST IRON NI. RESIST | ALUMINUM | COPPER | LEAD | MONEL 400 | NICKEL 200 | 304, 321, 347 STN. STL. | 316 STN STL. | TYPE 20 STN STL. | INCOLOY 800 | INCONEL 600 | TITANIUM | HASTELLOY B | QUARTZ | GRAPHITE | TEFLON | *NOTES |
|---|---|---|---|---|---|---|---|---|---|---|---|---|---|---|---|---|---|---|---|
| Caustic Soda (Lye) (Sodium Hydroxide) 2% | F | F | F | X | F | X | A | A | X | F | A | A | A | A | | | | | |
| 10-30%, 210°F | F | F | A | X | F | X | A | A | A | A | A | A | A | A | | | | | |
| 76%, 180°F | X | X | X | X | X | X | F | A | F | F | F | A | A | F | | | | | |
| Chlorine Gas: Dry | X | X | F | X | X | X | F | C | C | C | F | C | F | F | | A | F | F | Note 2 |
| Wet | X | X | X | X | X | X | X | X | X | X | X | X | X | X | | A | X | X | Note 2 |
| Chloroacetic Acid | X | X | | X | X | X | F | F | X | X | | C | C | A | | A | A | A | |
| Chromic Acetate | | | | | | | | | | | | | | | | A | | | Note 1 |
| Chromic Acid | X | C | X | X | X | F | X | X | X | X | X | X | X | A | | A | A | X | |
| Chrome Plating | X | X | | X | X | F | X | X | X | X | X | X | X | A | | A | A | X | |
| Chromylite | | | | | | | | | | | | | | | | A | | | Note 1 |
| Citric Acid | X | X | C | C | C | X | F | F | C | C | F | F | F | A | A | A | A | A | |
| Clear Chromate | | | | | | | | | | A | | | | | | | | | Note 1 |
| Cobalt Acetate at 130°F | | | | | | | F | F | A | A | | F | F | | | | | | |
| Cobalt Nickel | | | | | | | | | | | | | | | | A | | | Notes 1, 6 |
| Cobalt Plating | | | | | | | | | A | | | | | | | A | | | Note 1 |
| Coconut Oil | | | | | | | F | A | | | | | | | | | | | |
| Cod Liver Oil | | | | A | | | | A | A | A | A | A | A | | | | | | |
| Copper Acid | | | | | | | | | | | | | | A | | A | | | Note 1 |
| Copper Bright | | | | | | | | | A | A | | | | | | | | | Note 1 |
| Copper Bright Acid | | | | | | | | | | | | | | | | A | | | |
| Copper Chloride | X | X | | C | X | C | X | X | X | X | X | X | X | A | | A | A | A | |
| Copper Cyanide | A | A | | X | X | | C | X | F | F | F | X | X | | | A | A | A | |
| Copper Fluoborate | | | | | | | F | F | F | F | F | F | F | | | | A | A | |
| Copper Nitrate | X | X | X | X | X | | X | X | F | F | F | X | X | | | A | A | A | |
| Copper Plating | A | | | | | | | | | | | | | | | | | | |
| Copper Pyrophosphate | | | | | | | | | A | | | | | | | | | | Note 1 |
| Copper Strike | A | A | | | | | | | A | | | | | | | | | | Note 1 |
| Copper Sulfate | X | X | F | X | C | A | X | X | F | F | A | C | X | A | | A | A | A | |
| Creosote | A | F | F | C | F | X | F | F | F | F | F | F | F | | | A | | | Note 2 |
| Cresylic Acid | C | C | | C | C | X | F | F | F | A | A | C | F | F | | A | A | A | Note 2 |
| Deoxidine™ | | | | | | | | | A | | | | | | | | | | |
| Deoxlyte™ | | | | | | | | | A | | | | | | | | | | |
| Deoxidizer (Etching) | | | | | | | | | | | | | | | | A | | | Note 1 |
| Deoxidizer (3AL-13) | | | | | | | | | A | A | | | | | | | | | Note 1, Non-Chromate |
| Dichromic Seal | X | X | | | | | | | | | | | | | | | | | |
| Diethylene Glycol | F | A | | F | F | A | F | F | A | A | A | F | F | A | | A | A | A | |
| Diphenyl 300°-350°F | A | A | A | A | A | A | A | A | A | | A | | A | | | | | | |
| Disodium Phosphate | A | | | | | | | | | | | | | | | | | | |
| Diversey™ DS9333 | | | | | | | | | | | | | | | | A | | | Note 1 |

Corrosion Resistance Ratings: A = Good F = Fair C = Depends on Conditions X = Unsuitable
Blank = Data Not Available

* SEE PAGE 418 FOR NOTES TO SHEATH MATERIAL SELECTION GUIDE

*Courtesy of Tempco Electric Heater Corporation.*

# Engineering Data
# Recommended Sheath Materials for

| MEDIA BEING HEATED | ELEMENT SHEATH MATERIAL | | | | | | | | | | | | | | | | | | *NOTES |
|---|---|---|---|---|---|---|---|---|---|---|---|---|---|---|---|---|---|---|---|
| | IRON & STEEL | CAST IRON GRAY | CAST IRON NI. RESIST | ALUMINUM | COPPER | LEAD | MONEL 400 | NICKEL 200 | 304, 321, 347 STN. STL. | 316 STN STL. | TYPE 20 STN STL. | INCOLOY 800 | INCONEL 600 | TITANIUM | HASTELLOY B | QUARTZ | GRAPHITE | TEFLON | |
| Diversey™ 99 | A | | | | | | | | | | | | | | | | | | |
| Diversey™ 511 | | | | | | | | | | | | | | | | A | | | Notes 1, 5 |
| Diversey™ 514 | | | | | | | | | | | | | | | | | A | A | Note 1 |
| Dowtherm™ A | A | | | | | | | | | | | | | | | | | | |
| Electro Polishing | | | | | | | | | | | | | | | | A | | | Note 1 |
| Electroless Nickel | | | | | | | | | | | | | | A | | A | | | Note 1 |
| Electroless Tin (Acid) | | | | | | | | | | | | | | | | A | | | Note 1 |
| (Alkaline) | | | | | | | | | | A | | | | A | | | | | Note 1 |
| Enthone Acid-80 | | | | | | | | | | | | | | | | | A | A | Note 1 |
| Ether | F | F | | F | F | F | F | F | F | F | A | F | F | A | | A | | | Note 2 |
| Ethyl Chloride | F | F | | F | A | F | F | A | F | F | A | F | A | A | | A | A | A | Note 2 |
| Ethylene Glycol | A | F | | A | F | X | F | F | F | F | F | F | F | A | | A | A | A | Note 5 |
| Fatty Acids | X | X | | A | X | X | F | F | F | A | A | F | F | A | | A | A | | |
| Ferric Chloride | X | X | X | X | X | X | X | X | X | X | X | X | X | A | | A | A | A | |
| Ferric Nitrate | X | X | | X | X | | X | X | F | F | A | X | X | | | A | A | | |
| Ferric Sulfate | X | X | X | X | X | A | X | C | F | F | F | C | C | A | | A | A | | |
| Fluorine Gas, Dry | C | X | | X | X | X | A | A | C | C | C | C | A | A | | C | X | | |
| Formaldehyde | X | X | F | F | F | X | F | F | A | A | A | F | F | A | | A | A | | |
| Formic Acid | X | X | | X | F | X | C | C | X | X | A | F | C | X | | A | A | | |
| Freon | A | A | A | A | A | A | A | A | A | A | A | A | A | | | | | | |
| Fuel Oil | A | A | | A | A | A | F | F | A | A | A | F | F | A | | | | | Notes 2, 3, 7 |
| Fuel Oil-Acid | X | X | | X | X | A | C | C | C | F | A | C | C | A | | | | | Notes 2, 3, 7 |
| Gasoline-Refined | A | A | A | A | A | A | F | F | A | A | A | F | F | | | A | A | | Notes 2, 5 |
| Gasoline-Sour | C | C | | C | C | A | X | X | F | F | A | X | X | | | A | A | | Notes 2, 3, 5 |
| Glycerine, Glycerol | F | C | F | A | F | F | A | A | A | A | A | A | A | | | A | A | | |
| Gold Acid | A | | | | | | | | | | | | | A | | A | | | Note 1 |
| Gold Cyanide | | | | | | | | | A | A | | | | | | | | | Note 1 |
| Grey Nickel | | | | | | | | | | | | | | A | | A | | A | Notes 1, 5 |
| Holdens 310A Tempering Bath | | | | | | | | A | | | | | | | | | | | |
| Hot Seal Sodium Dichromate | | | | | | | | | | A | | | | | | | | | Note 1 |
| Houghtone Mar Tempering Salt | C | | | | | | | C | | | | | | | | | | | |
| Hydrocarbons-Aliphatic | A | A | | A | A | | A | A | A | A | A | A | A | | | A | A | | Note 2 |
| Hydrocarbons-Aromatic | A | A | | A | A | | A | A | A | A | A | A | A | | | A | A | | Note 2 |
| Hydrochloric Acid ⟨150°F | X | X | X | X | X | X | X | X | X | X | X | X | X | X | | A | A | | |
| ⟩150°F | X | X | | X | X | X | X | X | X | X | X | X | X | X | | A | A | A | |
| Hydrocyanic Acid | X | X | | F | X | X | F | F | F | F | F | F | F | | | A | A | | |
| Hydrofluoric Acid, Cold ⟨65% | X | X | X | X | X | X | C | X | X | X | X | X | X | X | | X | A | A | Note 5 |
| ⟩65% | F | X | X | X | X | X | C | X | X | X | | X | X | X | | | | | |
| Hydrofluoric Acid, Hot ⟨65% | X | | | X | X | X | C | X | X | | | | | | | | | | |
| ⟩65% | X | | | X | X | X | C | X | X | X | | X | X | X | | | | | |

Corrosion Resistance Ratings: A = Good F = Fair C = Depends on Conditions X = Unsuitable
Blank = Data Not Available

* SEE PAGE 418 FOR NOTES TO SHEATH MATERIAL SELECTION GUIDE

*Courtesy of Tempco Electric Heater Corporation.*

# Engineering Data
# Immersion Heating Applications

| MEDIA BEING HEATED | ELEMENT SHEATH MATERIAL | | | | | | | | | | | | | | | | | |
|---|---|---|---|---|---|---|---|---|---|---|---|---|---|---|---|---|---|---|
| | IRON & STEEL | CAST IRON GRAY | CAST IRON NI. RESIST | ALUMINUM | COPPER | LEAD | MONEL 400 | NICKEL 200 | 304, 321, 347 STN. STL. | 316 STN STL. | TYPE 20 STN STL. | INCOLOY 800 | INCONEL 600 | TITANIUM | HASTELLOY B | QUARTZ | GRAPHITE | TEFLON | *NOTES |
| Hydrogen Peroxide | X | X | X | A | X | X | C | F | F | F | F | F | F | A | | A | X | | |
| Indium | | | | | | | | | | | | | | | | A | | A | Note 1 |
| Iridite™ #4-75, #4-73, #14, #14-2, #14-9, #18-P | | | | | | | | | | A | | | | | | | | | Note 1 |
| Iridite™ #1, #2, #3, #4-C, #4PC&S, #4P-4, #4-80, #4L-1, #4-2, #4-2A, #4-2P, #5P-1, #7-P, #8, #8-P, #8-2, #12-P, #15, #17P, #18P | | | | | | | | | | | | | | | | A | | | Note 1 |
| Iridite™ Dyes-#12L-2, #40, #80 | | | | | | | | | | | | | | | | A | | A | Note 1 |
| Irilac™ | | | | | | | | | | | | | | | | A | | A | Note 1 |
| Iron Fluoborate | | | | | | | | | | | | | | | | | A | A | Note 1 |
| Iron Phosphate (Parkerizing) | C | | F | | | | | | A | A | | | | | | | | | |
| Isoprep™ Deoxidizer #187, #188 | | | | | | | | | | A | | | | | | | | | Note 1 |
| Isoprep™ #191 Acid Salts | | | | | | | | | | | | | | | | | A | A | Note 1 |
| Isoprep™ Acid Aluminum Cleaner #186 | | | | | | | | | | A | | | | | | | | | Note 1 |
| Isopropanol | C | | | | A | | A | A | A | A | A | | A | | | | | | |
| Jetal™ | | | | | | | | | A | | | | | | | | | | Note 1 |
| Kerosene | A | | | A | A | | A | A | A | A | A | A | A | | | | A | | Note 2 |
| Kolene | | | | | | | | A | | | | | | | | | | | |
| Lacquer Solvent | F | A | A | A | F | A | F | F | A | A | A | F | F | A | | A | | | Note 2 |
| Lead Acetate | X | X | | X | X | X | A | A | A | A | A | A | A | A | | A | A | | |
| Lead Acid Salts | | | | | | | | | A | | | | | | | | | | Note 1 |
| Lime Saturated Water | F | F | | X | F | X | F | F | F | A | F | F | F | | | X | A | | |
| Linseed Oil | X | A | | F | F | X | F | F | A | A | A | F | F | | | A | X | | Note 2 |
| Magnesium Chloride | X | C | F | X | F | X | F | A | F | F | A | F | A | A | | A | A | | |
| Magnesium Hydroxide | A | A | A | F | A | A | F | A | A | A | A | A | A | | | A | A | | |
| Magnesium Nitrate | F | F | | F | F | C | F | F | F | F | F | F | X | F | | A | A | | |
| Magnesium Sulfate | F | F | F | F | F | A | A | A | F | F | A | F | A | A | | A | A | | |
| MacDermid™ M629 | | | | | | | | | | | | | | | | | A | A | Note 1 |
| Mercuric Chloride | X | X | X | X | X | X | X | X | X | X | X | X | X | F | | A | A | | |
| Mercury | A | A | A | X | X | X | F | F | F | A | A | A | F | X | | A | | | |
| Methyl Alcohol (Methanol) | F | F | | C | F | F | A | A | F | A | A | F | A | A | | A | A | | Note 2 |
| Methyl Bromide | C | C | | X | F | F | F | F | A | A | A | F | F | A | | A | | | |
| Methyl Chloride | C | C | | X | A | C | C | C | C | C | C | C | C | A | | A | A | | |
| Methylene Chloride | X | C | | C | C | F | C | F | C | F | A | C | F | A | | A | A | | |
| Mineral Oil | A | A | | A | A | A | A | A | A | A | A | A | A | A | | A | A | | |

Corrosion Resistance Ratings: A = Good F = Fair C = Depends on Conditions X = Unsuitable
Blank = Data Not Available

* SEE PAGE 418 FOR NOTES TO SHEATH MATERIAL SELECTION GUIDE

*Courtesy of Tempco Electric Heater Corporation.*

# Engineering Data
# Recommended Sheath Materials for

| MEDIA BEING HEATED | IRON & STEEL | CAST IRON GRAY | CAST IRON NI. RESIST | ALUMINUM | COPPER | LEAD | MONEL 400 | NICKEL 200 | 304, 321, 347 STN. STL. | 316 STN STL. | TYPE 20 STN STL. | INCOLOY 800 | INCONEL 600 | TITANIUM | HASTELLOY B | QUARTZ | GRAPHITE | TEFLON | *NOTES |
|---|---|---|---|---|---|---|---|---|---|---|---|---|---|---|---|---|---|---|---|
| Muriato | | | | | | | | | | | | | | | | A | | A | Note 1 |
| Naptha | A | F | F | A | A | A | A | A | A | A | A | A | A | A | | A | A | A | Note 2 |
| Napthalene | A | A | A | F | F | A | F | F | A | A | A | F | F | A | | | | | Note 2 |
| Nickel Acetate Seal | | | | | | | | | | A | | | | | | | | | Note 1 |
| Nickel Chloride | X | X | X | X | X | C | C | X | X | C | C | C | F | F | | A | A | A | Notes 1, 5 |
| Nickel Plate-Bright | | | | | | A | | | | | | | | A | | A | | A | Notes 1, 5 |
| Nickel Plate-Dull | | | | | | A | | | | | | | | | | A | | A | Notes 1, 5 |
| Nickel Plate-Watts Solution | | | | | | | | | | | | | | A | | A | | A | Notes 1, 5 |
| Nickel Sulfate | X | X | X | X | F | F | C | F | F | F | F | C | F | | | A | A | A | |
| Nickel Copper Strike (Cyanide Free) | | | | | | | | | A | A | | | | | | | | | Note 1 |
| Nitric Acid, Crude | X | | | | X | X | X | X | C | C | | X | X | | | A | | A | |
| Concentrated | X | | | | X | X | X | X | F | F | | X | X | | | A | | A | |
| Diluted | X | | | | X | X | X | X | A | A | | X | X | | | A | | A | |
| Nitric Hydrochloric Acid | X | X | | X | X | X | X | X | X | X | X | X | X | X | | A | A | A | |
| Nitric 6% Phosphoric Acid | | | | | | | | | | C | | | | | | A | | A | Note 1 |
| Nitric Sodium Chromate | | | | | | | | | | A | | | | | | A | | A | Note 1 |
| Nitrobenzene | A | A | A | A | F | X | A | A | A | A | A | A | A | A | | A | | | Note 2 |
| Oakite™ #67 | | | | | | | | | A | | | | | | | | | | Note 1 |
| Oakite™ #20, 23, 24, 30, 51, 90 | A | | | | | | | | | | | | | | | | | | |
| Oleic Acid | C | C | C | C | C | X | F | F | C | F | A | F | A | F | | A | A | A | |
| Oxalic Acid | X | X | X | F | F | X | C | F | X | X | F | X | F | X | | A | A | A | |
| Paint Stripper (High Alkaline Type) | A | | | | | | | | | | | | | | | | | | Note 1 |
| Paint Stripper (Solvent Type) | | | | | | | | | | A | | | | | | | | | Notes 1, 2 |
| Paraffin | A | A | | A | A | | F | | A | A | A | | | | | | | | Notes 2, 7 |
| Parkerizing™ (See Iron Phosphate) | | | | | | | | | | | | | | | | | | | |
| Perchloroethylene | F | F | | C | F | F | A | A | F | F | F | F | A | A | | A | | | |
| Perm-A-Clor™ | | | | | | | | | A | | | | | | | | | | |
| Petroleum-Crude <500°F | F | F | A | A | C | C | A | C | A | A | A | | | | | A | A | | Notes 2, 3, 7 |
| >500°F | A | | A | A | X | X | X | X | A | | | | | | | | | | |
| >1000°F | X | | | X | X | X | X | X | A347 | | | | | | | | | | |
| Phenol | F | F | | F | | X | F | | C | F | F | F | F | A | A | | | | |
| Phosphate | | | | | | | | | | A | | | | | | | | X | Notes 1, 5, 9 |
| Phosphate Cleaner | | | | | | | | | A | | | | | | | | | X | Notes 1, 5, 9 |
| Phosphatizing | | | | | | | | | | A | | | | | | | | X | Notes 1, 5, 9 |
| Phosphoric Acid, Crude | C | | | X | X | C | X | X | C | | | | | | | | | | |
| Pure <45% | X | X | X | C | C | C | F | C | C | C | F | A | A | X | | | | | |

Corrosion Resistance Ratings: A = Good F = Fair C = Depends on Conditions X = Unsuitable
Blank = Data Not Available

* SEE PAGE 418 FOR NOTES TO SHEATH MATERIAL SELECTION GUIDE

*Courtesy of Tempco Electric Heater Corporation.*

# Engineering Data
# Immersion Heating Applications

| MEDIA BEING HEATED | ELEMENT SHEATH MATERIAL | | | | | | | | | | | | | | | | | | |
|---|---|---|---|---|---|---|---|---|---|---|---|---|---|---|---|---|---|---|---|
| | IRON & STEEL | CAST IRON GRAY | CAST IRON NI. RESIST | ALUMINUM | COPPER | LEAD | MONEL 400 | NICKEL 200 | 304, 321, 347 STN. STL. | 316 STN STL. | TYPE 20 STN STL. | INCOLOY 800 | INCONEL 600 | TITANIUM | HASTELLOY B | QUARTZ | GRAPHITE | TEFLON | *NOTES |
| ›45% Cold | X | X | X | X | F | C | F | C | A | F | F | A | | X | | | | | |
| ›45% Hot | X | X | X | X | C | X | C | X | X | X | F | A | F | X | | | | | |
| Photo Fixing Bath | | | | | | | C | | A | | | | | | | | | | |
| Picric Acid | X | X | | X | X | X | X | X | F | F | F | C | C | | | A | A | A | |
| Potassium Sulfate | | | | | | | | | | | | | | | | A | | A | Note 1 |
| Potassium Bichromate | C | F | F | F | | F | F | F | A347 | A | A | F | | F | A | A | | A | |
| Potassium Chloride | C | X | F | X | C | C | F | F | C | F | A | C | F | A | | A | A | | |
| Potassium Cyanide | C | X | F | X | X | X | C | F | F | F | F | F | F | X | | A | C | A | |
| Potassium Dichromate | | | | | | | | | A347 | | | | | | | | | | |
| Potassium Hydrochloric | | | | | | | | | | | | | | | | A | | A | Note 1 |
| Potassium Hydroxide | X | X | | X | C | X | F | A | C | C | C | C | F | X | | X | A | A | |
| Potassium Nitrate | F | F | F | A | F | F | F | F | F | F | F | F | F | A | | A | A | | |
| Potassium Sulfate | C | C | C | A | F | A | A | F | A | A | A | F | F | A | | A | A | A | |
| Prestone™ 350°F | A | | | | | | A | | | | | | | | | | | | |
| R5 Bright Dip For Copper Polish at 180°F | | | | | | | | | | A | | | | | | | | | |
| Reynolds Brightener | | | | | | | | | | | | | | | | A | | A | Note 1 |
| Rhodium Hydroxide | | | | | | | | | | | | | | | | A | | A | |
| Rochelle Salt Cyanide | A | | | | | | | | A | | | | | | | | | | Note 1 |
| Ruthenium Plating | | | | | | | | | | | | | | | | A | | A | Note 1 |
| Silver Bromide | X | X | | X | X | | C | C | X | X | C | | | A | | A | A | A | |
| Silver Cyanide | C | C | | X | X | | F | | A | A | A | A | | | | A | | | |
| Silver Lume | | | | | | | | | A | | | | | | | | | | Note 1 |
| Silver Nitrate | X | X | | X | X | X | X | X | C | C | F | C | C | A | | A | A | | |
| Soap Solutions | A | A | A | X | C | | A | | A | A | A | | | | | | | | Note 3 |
| Sodium-Liquid Metal | C | X | | X | X | X | F | A | A | | | A | A | | | X | X | | |
| Sodium Bisulfate | X | X | X | C | F | C | C | F | X | X | A | | F | | | | | | |
| Sodium Bromide | F | C | | X | F | F | F | F | C | F | F | F | F | | | A | A | A | |
| Sodium Carbonate | C | C | | X | A | X | F | F | F | F | A | F | F | A | | C | A | A | |
| Sodium Chlorate | X | X | | F | A | F | A | A | F | F | F | F | A | A | | A | A | A | |
| Sodium Chloride | C | X | F | X | F | F | A | F | X | X | C | F | A | C | | A | A | | |
| Sodium Citrate | X | X | | X | X | X | | | F | F | F | | | | | A | A | A | |
| Sodium Cyanide | C | F | C | X | X | X | C | C | A | A | A | A | A | C | | A | C | | |
| Sodium Dichromate (Sodium Bichromate) | F | F | F | C | X | | | | F | F | F | | | C | | A | | | |
| Sodium Hydroxide (See Caustic Soda) | | | | | | | | | | | | | | | | | | | |
| Sodium Hypochlorite | X | X | X | X | X | X | X | X | X | X | F | X | X | A | A | A | A | A | |
| Sodium Nitrate | F | F | A | C | C | C | F | F | A | A | A | A | A | A | | A | A | | |
| Sodium Peroxide | F | A | F | C | X | X | F | F | F | F | F | | F | | | | | | |

Corrosion Resistance Ratings: A = Good F = Fair C = Depends on Conditions X = Unsuitable
Blank = Data Not Available

* SEE PAGE 418 FOR NOTES TO SHEATH MATERIAL SELECTION GUIDE

# Engineering Data
# Recommended Sheath Materials for

| MEDIA BEING HEATED | ELEMENT SHEATH MATERIAL | | | | | | | | | | | | | | | | | | |
|---|---|---|---|---|---|---|---|---|---|---|---|---|---|---|---|---|---|---|---|
| | IRON & STEEL | CAST IRON GRAY | CAST IRON NI. RESIST | ALUMINUM | COPPER | LEAD | MONEL 400 | NICKEL 200 | 304, 321, 347 STN. STL. | 316 STN STL. | TYPE 20 STN STL. | INCOLOY 800 | INCONEL 600 | TITANIUM | HASTELLOY B | QUARTZ | GRAPHITE | TEFLON | *NOTES |
| Sodium Phosphate | C | C | F | X | F | F | A | C | F | A | F | F | A | A | | A | A | A | |
| Sodium Salicylate | F | C | F | | F | | F | F | F | F | F | F | F | | | A | A | A | |
| Sodium Silicate | A | F | A | X | F | X | A | A | A | A | A | A | A | A | | A | A | A | Note 4 |
| Sodium Stannate | C | C | C | | | | F | F | F | F | F | F | F | | | A | | A | |
| Sodium Sulfate | F | C | | F | F | F | F | F | X | F | F | F | F | C | | A | A | A | |
| Sodium Sulfide | C | X | C | C | X | A | F | F | X | C | C | C | C | C | | C | A | A | |
| Solder Bath | X | X | X | X | X | X | X | X | X | X | X | X | X | X | | X | X | X | Note 4 |
| Soybean Oil | | | | | | | | | A | | | | | | | | | | |
| Stannostar™ | | | | | | | | | | | | | | | | A | | A | Note 1 |
| Steam ‹500°F | A | | | A | A | C | A | A | A | | | A | A | | | | | | |
| 500-1000°F | C | | | C | C | X | C | C | A | | | A | A | | | | | | |
| ›1000°F | X | | | X | X | | X | X | A | | | A | A | | | | | | |
| Stearic Acid | C | C | C | C | X | X | F | F | C | A | A | F | F | F | | A | A | | |
| Sugar Solution | A | A | | A | A | A | A | A | A | A | A | A | A | A | | A | A | A | Note 7 |
| Sulfamate Nickel | | | | | | | | | | | | | | A | | A | | A | Note 1 |
| Sulfamic Acid | X | X | | X | | | | | X | X | | | | | | A | | A | |
| Sulfur | C | X | C | A | X | X | F | C | C | F | F | A | A | A | | A | A | | |
| Sulfur Chloride | X | X | C | X | X | F | X | C | C | X | C | C | F | | | A | X | A | |
| Sulfur Dioxide | C | C | | C | C | F | X | X | C | F | F | C | C | A | | A | A | | |
| Sulfuric Acid ‹10% Cold | X | | X | C | A | F | F | C | X | C | F | | X | | | | | | |
| Hot | X | X | X | C | X | X | X | X | X | X | X | | F | | | | | | |
| 10-75% Cold | X | | | X | F | F | C | C | X | X | F | | X | X | | | | | |
| Hot | X | | | X | X | F | C | X | X | X | C | | X | X | | | | | |
| 75-95% Cold | F | F | F | X | F | F | X | X | F | F | F | | | X | | | | | |
| Hot | X | X | X | X | X | C | X | X | X | X | X | | | X | | | | | |
| Fuming | C | X | C | X | X | X | X | X | F | C | C | C | C | | | | | | |
| Sulfurous Acid | X | X | | C | X | A | X | X | X | C | F | | C | A | | | | | |
| Tannic Acid | C | C | | C | C | X | C | C | C | A | A | | A | A | | A | | | |
| Tar | A | | | A | | | | | A | | | A | A | | | | | | |
| Tartaric Acid | | X | F | C | | C | F | C | C | A | F | | F | F | | | | | |
| Tetrachlorethylene | F | F | | C | F | F | A | A | F | F | F | F | A | A | | A | | | |
| Thermoil Granodine™ | F | | | | | | | | | | | | | | | | | | |
| Therminol™ FR1 8-12 W/Sq. In. 640°F | A | | | | | | | | | | | | | | | | | | |
| Tin (Molten) | F | F | | X | X | X | X | X | F | F | X | | X | A | | | X | X | Note 4 |
| Tin-Nickel Plating | | | | | | | | | | | | | | | | A | | A | Note 1 |
| Tin Plating-Acid | | | | | | | | | | | | | | | | | A | A | Note 1 |
| Tin Plating-Alkaline | A | | | | | | | | A | | | | | | | | | | Note 1 |
| Toluene | A | A | A | A | C | A | A | A | A | A | A | A | A | A | | | | | |
| Triad Solvent | C | | | | | | | | | | | | | | | | | | |

Corrosion Resistance Ratings: A = Good F = Fair C = Depends on Conditions X = Unsuitable
Blank = Data Not Available

* SEE PAGE 418 FOR NOTES TO SHEATH MATERIAL SELECTION GUIDE

# Engineering Data
# Immersion Heating Applications

| MEDIA BEING HEATED | ELEMENT SHEATH MATERIAL | | | | | | | | | | | | | | | | | | |
|---|---|---|---|---|---|---|---|---|---|---|---|---|---|---|---|---|---|---|---|
| | IRON & STEEL | CAST IRON GRAY | CAST IRON NI. RESIST | ALUMINUM | COPPER | LEAD | MONEL 400 | NICKEL 200 | 304, 321, 347 STN. STL. | 316 STN STL. | TYPE 20 STN STL. | INCOLOY 800 | INCONEL 600 | TITANIUM | HASTELLOY B | QUARTZ | GRAPHITE | TEFLON | *NOTES |
| Trichloroethane | A | C | C | F | F | F | F | F | A | F | F | F | F | A | | A | A | | |
| Trichloroethyene | F | C | C | F | C | X | C | C | F | F | F | F | A | A | | A | A | | |
| Triethylene Glycol | A | A | A | A | A | A | A | A | A | A | A | A | A | A | | A | | | |
| Trioxide (Pickle) | | | | | | | | | | | | | | | | A | | A | Note 1 |
| Trisodium Phosphate | A | A | | X | C | X | C | C | C | C | C | | | | | X | F | X | |
| Turco™ 2623 | A | | | | | | | | | | | | | | | | | | |
| Turco™ 4008, 4181, 4338 | | | | | | | | | | A | | | | | | | | | Note 1 |
| Turco™ Ultrasonic Solution | | | | | | | | | | A | | | | | | | | | Note 1 |
| Turpentine | C | C | C | A | F | A | A | A | A | A | A | | A | | | | | | |
| Ubac™ | | | | | | | | | | | | | | | | A | | | Note 1 |
| Udylite #66 | | | | | | | | | | | | | | A | | A | | A | Notes 1, 5 |
| Unichrome™ CR-110 | | | | | | | | | | | | | | | | A | | A | Note 1 |
| Unichrome™ 5RHS | | | | | | | | | | | | | | | | A | | A | Note 1 |
| Urea Ammonia Liquor 48°F | A | | | | | | | | | | | | | | | | | | |
| Vegetable Oil | C | | C | F | X | X | A | A | A | A | A | A | | | | | | | |
| Vinegar | C | | | C | | | A | | F | A | | | | | | | | | |
| Water, Acid Mine Containing Oxidizing Salts | X | | C | C | C | C | X | C | A | | | | | | | | | | |
| No Oxidizing Salts | C | | A | A | | | A | | X | | | | | | | | | | |
| Water, Deionized | X | X | | X | X | | A | A | A | A | A | A | A | | | | | | Note 10 |
| Demineralized | X | X | | X | X | | A | A | A | A | A | A | A | | | | | | Note 10 |
| Distilled | X | X | | | X | X | C | A | | | | A | A | | | | | | Note 10 |
| Potable | X | C | A | A | A | X | A | A | C | F | A | A | A | A | | A | | | |
| Return Condensate | A | | A | A | A | A | | | A | A | | A | | | | | | | |
| Sea | X | X | A | X | X | A | A | | C | C | A | F | F | A | | A | A | | |
| Watt's Nickel Strike | | | | | | | | | | | | | | | | A | | | Note 1 |
| Whiskey and Wines | X | | C | | A | | A | A | A | A | A | A | A | | | | | | Note 2 |
| Wood's Nickel Strike | | | | | | | | | | | | | | | | A | | | Note 1 |
| Yellow Dichromate | | | | | | | | | | A | | | | | | A | | | Note 1 |
| X-Ray Solution | | | | | | | | | A | | | | | | | | | | |
| Zinc (Molten) | | | | X | X | X | X | X | X | X | X | X | X | X | | | | X | |
| Zinc Chloride | C | C | C | X | X | | F | F | X | X | F | X | F | C | | A | A | A | |
| Zinc Phosphate | | | | | | | | | | A | | | | | | | | X | Notes 1, 5 |
| Zinc Plating Acid | | | | | | | | | | | | | | | | A | | | Note 1 |
| Zinc Plating Cyanide | A | | | | | | | | A | | | | | | | | | | Note 1 |
| Zinc Sulphate | C | X | A | C | F | A | F | C | C | C | C | | F | A | | | | | |
| Zincate™ | A | | | | | | | | A | | | | | | | | | | Note 1 |

Corrosion Resistance Ratings: A = Good F = Fair C = Depends on Conditions X = Unsuitable
Blank = Data Not Available

* SEE PAGE 418 FOR NOTES TO SHEATH MATERIAL SELECTION GUIDE

*Courtesy of Tempco Electric Heater Corporation.*

# Engineering Data
# Metric Conversion Factors

## DECIMAL AND MILLIMETER EQUIVALENTS

| Fraction | DECIMALS | MILLIMETERS |
|---|---|---|
| 1/64 | 0.015625 | 0.397 |
| 1/32 | .03125 | 0.794 |
| 3/64 | .046875 | 1.191 |
| 1/16 | .0625 | 1.588 |
| 5/64 | .078125 | 1.984 |
| 3/32 | .09375 | 2.381 |
| 7/64 | .109375 | 2.778 |
| 1/8 | .1250 | 3.175 |
| 9/64 | .140625 | 3.572 |
| 5/32 | .15625 | 3.969 |
| 11/64 | .171875 | 4.366 |
| 3/16 | .1875 | 4.763 |
| 13/64 | .203125 | 5.159 |
| 7/32 | .21875 | 5.556 |
| 15/64 | .234375 | 5.953 |
| 1/4 | .2500 | 6.350 |
| 17/64 | .265625 | 6.747 |
| 9/32 | .28125 | 7.144 |
| 19/64 | .296875 | 7.541 |
| 5/16 | .3125 | 7.938 |
| 21/64 | .328125 | 8.334 |
| 11/32 | .34375 | 8.731 |
| 23/64 | .359375 | 9.128 |
| 3/8 | .3750 | 9.525 |
| 25/64 | .390625 | 9.922 |
| 13/32 | .40625 | 10.319 |
| 27/64 | .421875 | 10.716 |
| 7/16 | .4375 | 11.113 |
| 29/64 | .453125 | 11.509 |
| 15/32 | .46875 | 11.906 |
| 31/64 | .484375 | 12.303 |
| 1/2 | .5000 | 12.700 |

1 mm = .03937″

| Fraction | DECIMALS | MILLIMETERS |
|---|---|---|
| 33/64 | 0.515625 | 13.097 |
| 17/32 | .53125 | 13.494 |
| 35/64 | .546875 | 13.891 |
| 9/16 | .5625 | 14.288 |
| 37/64 | .578125 | 14.684 |
| 19/32 | .59375 | 15.081 |
| 39/64 | .609375 | 15.478 |
| 5/8 | .6250 | 15.875 |
| 41/64 | .640625 | 16.272 |
| 21/32 | .65625 | 16.669 |
| 43/64 | .671875 | 17.066 |
| 11/16 | .6875 | 17.463 |
| 45/64 | .703125 | 17.859 |
| 23/32 | .71875 | 18.256 |
| 47/64 | .734375 | 18.653 |
| 3/4 | .7500 | 19.050 |
| 49/64 | .765625 | 19.447 |
| 25/32 | .78125 | 19.844 |
| 51/64 | .796875 | 20.241 |
| 13/16 | .8125 | 20.638 |
| 53/64 | .828125 | 21.034 |
| 27/32 | .84375 | 21.431 |
| 55/64 | .859375 | 21.828 |
| 7/8 | .8750 | 22.225 |
| 57/64 | .890625 | 22.622 |
| 29/32 | .90625 | 23.019 |
| 59/64 | .921875 | 23.416 |
| 15/16 | .9375 | 23.813 |
| 61/64 | .953125 | 24.209 |
| 31/32 | .96875 | 24.606 |
| 63/64 | .984375 | 25.003 |
| 1 | 1.000 | 25.400 |

.001″ = .0254 mm

| MM | INCHES | MM | INCHES |
|---|---|---|---|
| .1 | .0039 | 46 | 1.8110 |
| .2 | .0079 | 47 | 1.8504 |
| .3 | .0118 | 48 | 1.8898 |
| .4 | .0158 | 49 | 1.9291 |
| .5 | .0197 | 50 | 1.9685 |
| .6 | .0236 | 51 | 2.0079 |
| .7 | .0276 | 52 | 2.0472 |
| .8 | .0315 | 53 | 2.0866 |
| .9 | .0354 | 54 | 2.1260 |
| 1 | .0394 | 55 | 2.1654 |
| 2 | .0787 | 56 | 2.2047 |
| 3 | .1181 | 57 | 2.2441 |
| 4 | .1575 | 58 | 2.2835 |
| 5 | .1969 | 59 | 2.3228 |
| 6 | .2362 | 60 | 2.3622 |
| 7 | .2756 | 61 | 2.4016 |
| 8 | .3150 | 62 | 2.4409 |
| 9 | .3543 | 63 | 2.4803 |
| 10 | .3937 | 64 | 2.5197 |
| 11 | .4331 | 65 | 2.5591 |
| 12 | .4724 | 66 | 2.5984 |
| 13 | .5118 | 67 | 2.6378 |
| 14 | .5512 | 68 | 2.6772 |
| 15 | .5906 | 69 | 2.7165 |
| 16 | .6299 | 70 | 2.7559 |
| 17 | .6693 | 71 | 2.7953 |
| 18 | .7087 | 72 | 2.8346 |
| 19 | .7480 | 73 | 2.8740 |
| 20 | .7874 | 74 | 2.9134 |
| 21 | .8268 | 75 | 2.9528 |
| 22 | .8661 | 76 | 2.9921 |
| 23 | .9055 | 77 | 3.0315 |
| 24 | .9449 | 78 | 3.0709 |
| 25 | .9843 | 79 | 3.1102 |
| 26 | 1.0236 | 80 | 3.1496 |
| 27 | 1.0630 | 81 | 3.1890 |
| 28 | 1.1024 | 82 | 3.2283 |
| 29 | 1.1417 | 83 | 3.2677 |
| 30 | 1.1811 | 84 | 3.3071 |
| 31 | 1.2205 | 85 | 3.3465 |
| 32 | 1.2598 | 86 | 3.3858 |
| 33 | 1.2992 | 87 | 3.4252 |
| 34 | 1.3386 | 88 | 3.4646 |
| 35 | 1.3780 | 89 | 3.5039 |
| 36 | 1.4173 | 90 | 3.5433 |
| 37 | 1.4567 | 91 | 3.5827 |
| 38 | 1.4961 | 92 | 3.6220 |
| 39 | 1.5354 | 93 | 3.6614 |
| 40 | 1.5748 | 94 | 3.7008 |
| 41 | 1.6142 | 95 | 3.7402 |
| 42 | 1.6535 | 96 | 3.7795 |
| 43 | 1.6929 | 97 | 3.8189 |
| 44 | 1.7323 | 98 | 3.8583 |
| 45 | 1.7717 | 99 | 3.8976 |
| | | 100 | 3.9370 |

### LENGTH

| Symbol | When You Know | Multiply by | To Find | Symbol |
|---|---|---|---|---|
| in | INCHES | 2.54 | CENTIMETERS | cm |
| ft | FEET | 30.48 | CENTIMETERS | cm |
| yd | YARDS | 0.9 | METERS | m |
| mi | MILES | 1.6 | KILOMETERS | km |

*Courtesy of Tempco Electric Heater Corporation.*

# Appendix E

# Temperature Conversion Chart

## Instructions For Use:

1. Start in the "Temp" column and find the temperature you wish to convert.
2. If the temperature to be converted is in degrees C, scan to the right column for the degrees F equivalent.
3. If the temperature to be converted is in degrees F, scan to the left column for the degrees C equivalent.

| °C | TEMP. | °F | °C | TEMP. | °F | °C | TEMP. | °F |
|---|---|---|---|---|---|---|---|---|
| −101 | −150 | −238 | −25 | −13 | 8.6 | −3.9 | 25 | 77 |
| −95.6 | −140 | −220 | −24.4 | −12 | 10.4 | −3.3 | 26 | 78.8 |
| −90 | −130 | −202 | −23.9 | −11 | 12.2 | −2.8 | 27 | 80.6 |
| −84.4 | −120 | −184 | −23.3 | −10 | 14 | −2.2 | 28 | 82.4 |
| −78.9 | −110 | −166 | −22.8 | − 9 | 15.8 | −1.7 | 29 | 84.2 |
| −73.3 | −100 | −148 | −22.2 | − 8 | 17.6 | −1.1 | 30 | 86 |
| −67.8 | − 90 | −130 | −21.7 | − 7 | 19.4 | −0.6 | 31 | 87.8 |
| −62.2 | − 80 | −112 | −21.1 | − 6 | 21.2 | 0 | 32 | 89.6 |
| −56.7 | − 70 | −94 | −20.6 | − 5 | 23 | 0.6 | 33 | 91.4 |
| −51.1 | − 60 | −76 | −20 | − 4 | 24.8 | 1.1 | 34 | 93.2 |
| −45.6 | − 50 | −58 | −19.4 | − 3 | 26.6 | 1.7 | 35 | 95 |
| −40 | − 40 | −40 | −18.9 | − 2 | 28.4 | 2.2 | 36 | 96.8 |
| −39.4 | − 39 | −38.2 | −18.3 | − 1 | 30.2 | 2.8 | 37 | 98.6 |
| −38.9 | − 38 | −36.4 | −17.8 | 0 | 32 | 3.3 | 38 | 100.4 |
| −38.3 | − 37 | −34.6 | −17.2 | 1 | 33.8 | 3.9 | 39 | 102.2 |
| −37.8 | − 36 | −32.8 | −16.7 | 2 | 35.6 | 4.4 | 40 | 104 |
| −37.2 | − 35 | −31 | −16.1 | 3 | 37.4 | 5 | 41 | 105.8 |
| −36.7 | − 34 | −29.2 | −15.6 | 4 | 39.2 | 5.6 | 42 | 107.6 |
| −36.1 | − 33 | −27.4 | −15 | 5 | 41 | 6.1 | 43 | 109.4 |
| −35.6 | − 32 | −25.6 | −14.4 | 6 | 42.8 | 6.7 | 44 | 111.2 |
| −35 | − 31 | −23.8 | −13.9 | 7 | 44.6 | 7.2 | 45 | 113 |
| −34.4 | − 30 | −22 | −13.3 | 8 | 46.4 | 7.8 | 46 | 114.8 |
| −33.9 | − 29 | −20.2 | −12.8 | 9 | 48.2 | 8.3 | 47 | 116.6 |
| −33.3 | − 28 | −18.4 | −12.2 | 10 | 50 | 8.9 | 48 | 118.4 |
| −32.8 | − 27 | −16.6 | −11.7 | 11 | 51.8 | 9.4 | 49 | 120.2 |
| −32.2 | − 26 | −14.8 | −11.1 | 12 | 53.6 | 10 | 50 | 122 |
| −31.7 | − 25 | −13 | −10.6 | 13 | 55.4 | 10.6 | 51 | 123.8 |
| −31.1 | − 24 | −11.2 | −10 | 14 | 57.2 | 11.1 | 52 | 125.6 |
| −30.6 | − 23 | − 9.4 | − 9.4 | 15 | 59 | 11.7 | 53 | 127.4 |
| −30 | − 22 | − 7.6 | − 8.9 | 16 | 60.8 | 12.2 | 54 | 129.2 |
| −29.4 | − 21 | − 5.8 | − 8.3 | 17 | 62.6 | 12.8 | 55 | 131 |
| −28.9 | − 20 | − 4 | − 7.8 | 18 | 64.4 | 13.3 | 56 | 132.8 |
| −28.3 | − 19 | − 2.2 | − 7.2 | 19 | 66.2 | 13.9 | 57 | 134.6 |
| −27.8 | − 18 | 0.4 | − 6.7 | 20 | 68 | 14.4 | 58 | 136.4 |
| −27.2 | − 17 | 1.4 | − 6.1 | 21 | 69.8 | 15 | 59 | 138.2 |
| −26.7 | − 16 | 3.2 | − 5.6 | 22 | 71.6 | 15.6 | 60 | 140 |
| −26.1 | − 15 | 5 | − 5.0 | 23 | 73.4 | 16.1 | 61 | 141.8 |
| −25.6 | − 14 | 6.8 | − 4.4 | 24 | 75.2 | 16.7 | 62 | 143.6 |

### CONVERSION FACTORS

°C = (°F − 32) × 5/9     0 Kelvin = −273.16°C

°F = (°C × 9/5) + 32     0° Rankine = −459.69°F

| °C | TEMP. | °F | °C | TEMP. | °F | °C | TEMP. | °F |
|---|---|---|---|---|---|---|---|---|
| 17.2 | 63 | 145.4 | 43.3 | 110 | 230 | 254 | 490 | 914 |
| 17.8 | 64 | 147.2 | 48.9 | 120 | 248 | 260 | 500 | 932 |
| 18.3 | 65 | 149 | 54.4 | 130 | 266 | 288 | 550 | 1022 |
| 18.9 | 66 | 150.8 | 60 | 140 | 284 | 316 | 600 | 1112 |
| 19.4 | 67 | 152.6 | 65.6 | 150 | 302 | 343 | 650 | 1202 |
| 20 | 68 | 154.4 | 71.1 | 160 | 320 | 371 | 700 | 1292 |
| 20.6 | 69 | 156.2 | 76.7 | 170 | 338 | 399 | 750 | 1382 |
| 21.1 | 70 | 158 | 82.2 | 180 | 356 | 427 | 800 | 1472 |
| 21.7 | 71 | 159.8 | 87.8 | 190 | 374 | 454 | 850 | 1562 |
| 22.2 | 72 | 161.6 | 93.3 | 200 | 392 | 482 | 900 | 1652 |
| 22.8 | 73 | 163.4 | 98.9 | 210 | 410 | 510 | 950 | 1742 |
| 23.3 | 74 | 165.2 | 104 | 220 | 428 | 538 | 1000 | 1832 |
| 23.9 | 75 | 167 | 110 | 230 | 446 | 566 | 1050 | 1922 |
| 24.4 | 76 | 168.8 | 116 | 240 | 464 | 593 | 1100 | 2012 |
| 25 | 77 | 170.6 | 121 | 250 | 482 | 621 | 1150 | 2102 |
| 25.6 | 78 | 172.4 | 127 | 260 | 500 | 649 | 1200 | 2192 |
| 26.1 | 79 | 174.2 | 132 | 270 | 518 | 677 | 1250 | 2282 |
| 26.7 | 80 | 176 | 138 | 280 | 536 | 704 | 1300 | 2372 |
| 27.2 | 81 | 177.8 | 143 | 290 | 554 | 732 | 1350 | 2462 |
| 27.8 | 82 | 179.6 | 149 | 300 | 572 | 760 | 1400 | 2552 |
| 28.3 | 83 | 181.4 | 154 | 310 | 590 | 788 | 1450 | 2642 |
| 28.9 | 84 | 183.2 | 160 | 320 | 608 | 816 | 1500 | 2732 |
| 29.4 | 85 | 185 | 166 | 330 | 626 | 843 | 1550 | 2822 |
| 30 | 86 | 186.8 | 171 | 340 | 644 | 871 | 1600 | 2912 |
| 30.6 | 87 | 188.6 | 177 | 350 | 662 | 899 | 1650 | 3002 |
| 31.1 | 88 | 190.4 | 182 | 360 | 680 | 927 | 1700 | 3092 |
| 31.7 | 89 | 192.2 | 188 | 370 | 698 | 954 | 1750 | 3182 |
| 32.2 | 90 | 194 | 193 | 380 | 716 | 982 | 1800 | 3272 |
| 32.8 | 91 | 195.8 | 199 | 390 | 734 | 1010 | 1850 | 3362 |
| 33.3 | 92 | 197.6 | 204 | 400 | 752 | 1038 | 1900 | 3452 |
| 33.9 | 93 | 199.4 | 210 | 410 | 770 | 1066 | 1950 | 3542 |
| 34.4 | 94 | 201.2 | 216 | 420 | 788 | 1093 | 2000 | 3632 |
| 35 | 95 | 203 | 221 | 430 | 806 | 1149 | 2100 | 3812 |
| 35.6 | 96 | 204.8 | 227 | 440 | 824 | 1204 | 2200 | 3992 |
| 36.1 | 97 | 206.6 | 232 | 450 | 842 | 1260 | 2300 | 4172 |
| 36.7 | 98 | 208.4 | 238 | 460 | 860 | 1316 | 2400 | 4352 |
| 37.2 | 99 | 210.2 | 243 | 470 | 878 | 1371 | 2500 | 4532 |
| 37.8 | 100 | 212 | 249 | 480 | 896 | | | |

# Appendix F

# Thermocouple Selection Guide

## Thermocouple Characteristics

Over the years specific pairs of thermocouple alloys have been developed to solve unique measurement problems. Idiosyncrasies of the more common thermocouples are discussed here.

We will use the term *standard wire error* to refer to the common commercial specification published in the *Annual Book of ASTM Standards*. It represents the allowable deviation between the actual thermocouple output voltage and the voltage predicted by the tables in NBS Monograph 125.

**Noble Metal Thermocouples** – The noble metal thermocouples, types B, R, and S, are all platinum or platinum-rhodium thermocouples and hence share many of the same characteristics.

**Diffusion** – Metallic vapor diffusion at high temperatures can readily change the platinum wire calibration, hence platinum wires should only be used inside a *non*-metallic sheath such as high-purity alumina. The one exception to this rule is a sheath made of platinum, and this option is prohibitively expensive.

**Stability** – The platinum-based couples are by far the most stable of all the common thermocouples. Type S is so stable that it is specified as the standard for temperature calibration between the antimony point (630.74°C) and the gold point (1064.43°C).

**Type B** – The B couple is the only common thermocouple that exhibits a double-valued ambiguity.

Due to the double-valued curve and the extremely low Seebeck coefficient at low temperatures, Type B is virtually useless below 50°C. Since the output is nearly zero from 0°C to 42°C, Type B has the unique advantage that the *reference* junction temperature is almost immaterial, as long as it is between 0° and 40°C. Of course, the *measuring* junction temperature is typically very high.

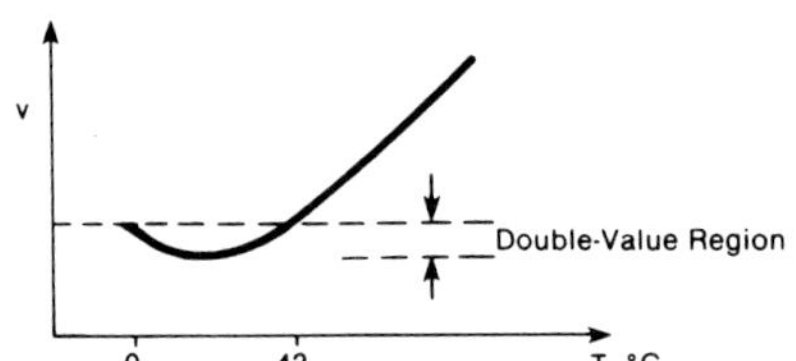


## Base Metal Thermocouples

Unlike the noble metal thermocouples, the base metal couples have no specified chemical composition. Any combination of metals may be used which results in a voltage vs. temperature curve fit that is within the standard wire errors. This leads to some rather interesting metal combinations. *Constantan*, for example, is not a specific metal alloy at all, but a generic name for a whole series of copper-nickel alloys. Incredibly, the *Constantan* used in a type T (copper-Constantan) thermocouple is not the same as the *Constantan* used in the type J (iron-Constantan) couple.[3]

**Type E** – Although Type E standard wire errors are not specified below 0°C, the type E thermocouple is ideally suited for low temperature measurements because of its high Seebeck coefficient (58$\mu$V/°C), low thermal conductivity and corrosion resistance.

The Seebeck coefficient for Type E is greater than all other standard couples, which makes it useful for detecting small temperature changes.

**Type J** - Iron, the positive element in a J couple, is an inexpensive metal rarely manufactured in pure form. J thermocouples are subject to poor conformance characteristics because of impurities in the iron. Even so, the J couple is popular because of its high Seebeck coefficient and low price.

The J couple should never be used above 760°C due to an abrupt magnetic transformation that can cause decalibration even when returned to lower temperatures.

**Type T** – This is the only couple with published standard wire errors for the temperature region below 0°C; however, type E is actually more suitable at very low temperatures because of its higher Seebeck coefficient and lower thermal conductivity.

Type T has the unique distinction of having one copper lead. This can be an advantage in a specialized monitoring situation where a temperature difference is all that is desired.

The advantage is that the copper thermocouple leads

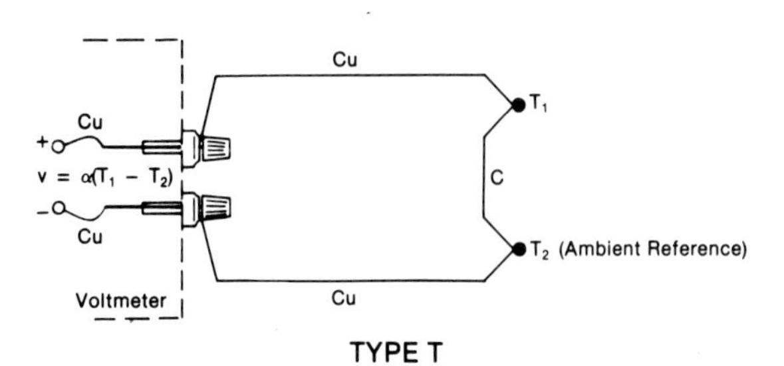


TYPE T

are the same metal as the dvm terminals, making lead compensation unnecessary.

**Types K & Nicrosil-Nisil** – The Nicrosil-Nisil thermocouple, type N, is similar to type K, but it has been designed to minimize some of the instabilities in the conventional Chromel-Alumel combination. Changes in the alloy content have improved the order/disorder transformations occurring at 500°C and a higher silicon content of the positive element improves the oxidation resistance at elevated temperatures. A full description with characteristic curves is published in NBS Monograph 161.[14]

**Tungsten** — The tungsten-rhenium thermocouples are normally used at high temperature in reducing or vacuum environments, but never in an oxidizing atmosphere because of the high reaction rates. Pure tungsten becomes very brittle when heated above its recrystallization temperature (about 1200°C). To make the wire easier to handle, rhenium alloys are used in both thermocouple legs. Types G (tungsten vs. tungsten-26% rhenium), C (tungsten 5% rhenium vs. tungsten 26% rhenium) and D (tungsten 3% rhenium vs. tungsten 25% rhenium) thermocouples are available in bare wire forms as well as complete probe assemblies. All materials conform to published Limits of Error.

# OUTPUT OF TYPE J THERMOCOUPLE IN mV

THERMOELECTRIC VOLTAGE IN ABSOLUTE MILLIVOLTS

| DEG C | 0 | 1 | 2 | 3 | 4 | 5 | 6 | 7 | 8 | 9 | 10 | DEG C |
|---|---|---|---|---|---|---|---|---|---|---|---|---|
| -210 | -8.096 | | | | | | | | | | | -210 |
| -200 | -7.890 | -7.912 | -7.934 | -7.955 | -7.976 | -7.996 | -8.017 | -8.037 | -8.057 | -8.076 | -8.096 | -200 |
| -190 | -7.659 | -7.683 | -7.707 | -7.731 | -7.755 | -7.778 | -7.801 | -7.824 | -7.846 | -7.868 | -7.890 | -190 |
| -180 | -7.402 | -7.429 | -7.455 | -7.482 | -7.508 | -7.533 | -7.559 | -7.584 | -7.609 | -7.634 | -7.659 | -180 |
| -170 | -7.122 | -7.151 | -7.180 | -7.209 | -7.237 | -7.265 | -7.293 | -7.321 | -7.348 | -7.375 | -7.402 | -170 |
| -160 | -6.821 | -6.852 | -6.883 | -6.914 | -6.944 | -6.974 | -7.004 | -7.034 | -7.064 | -7.093 | -7.122 | -160 |
| -150 | -6.499 | -6.532 | -6.565 | -6.598 | -6.630 | -6.663 | -6.695 | -6.727 | -6.758 | -6.790 | -6.821 | -150 |
| -140 | -6.159 | -6.194 | -6.228 | -6.263 | -6.297 | -6.331 | -6.365 | -6.399 | -6.433 | -6.466 | -6.499 | -140 |
| -130 | -5.801 | -5.837 | -5.874 | -5.910 | -5.946 | -5.982 | -6.018 | -6.053 | -6.089 | -6.124 | -6.159 | -130 |
| -120 | -5.426 | -5.464 | -5.502 | -5.540 | -5.578 | -5.615 | -5.653 | -5.690 | -5.727 | -5.764 | -5.801 | -120 |
| -110 | -5.036 | -5.076 | -5.115 | -5.155 | -5.194 | -5.233 | -5.272 | -5.311 | -5.349 | -5.388 | -5.426 | -110 |
| -100 | -4.632 | -4.673 | -4.714 | -4.755 | -4.795 | -4.836 | -4.876 | -4.916 | -4.956 | -4.996 | -5.036 | -100 |
| -90 | -4.215 | -4.257 | -4.299 | -4.341 | -4.383 | -4.425 | -4.467 | -4.508 | -4.550 | -4.591 | -4.632 | -90 |
| -80 | -3.785 | -3.829 | -3.872 | -3.915 | -3.958 | -4.001 | -4.044 | -4.087 | -4.130 | -4.172 | -4.215 | -80 |
| -70 | -3.344 | -3.389 | -3.433 | -3.478 | -3.522 | -3.566 | -3.610 | -3.654 | -3.698 | -3.742 | -3.785 | -70 |
| -60 | -2.892 | -2.938 | -2.984 | -3.029 | -3.074 | -3.120 | -3.165 | -3.210 | -3.255 | -3.299 | -3.344 | -60 |
| -50 | -2.431 | -2.478 | -2.524 | -2.570 | -2.617 | -2.663 | -2.709 | -2.755 | -2.801 | -2.847 | -2.892 | -50 |
| -40 | -1.960 | -2.008 | -2.055 | -2.102 | -2.150 | -2.197 | -2.244 | -2.291 | -2.338 | -2.384 | -2.431 | -40 |
| -30 | -1.481 | -1.530 | -1.578 | -1.626 | -1.674 | -1.722 | -1.770 | -1.818 | -1.865 | -1.913 | -1.960 | -30 |
| -20 | -0.995 | -1.044 | -1.093 | -1.141 | -1.190 | -1.239 | -1.288 | -1.336 | -1.385 | -1.433 | -1.481 | -20 |
| -10 | -0.501 | -0.550 | -0.600 | -0.650 | -0.699 | -0.748 | -0.798 | -0.847 | -0.896 | -0.945 | -0.995 | -10 |
| 0 | 0.000 | -0.050 | -0.101 | -0.151 | -0.201 | -0.251 | -0.301 | -0.351 | -0.401 | -0.451 | -0.501 | 0 |

| DEG C | 0 | 1 | 2 | 3 | 4 | 5 | 6 | 7 | 8 | 9 | 10 | DEG C |
|---|---|---|---|---|---|---|---|---|---|---|---|---|
| 0 | 0.000 | 0.050 | 0.101 | 0.151 | 0.202 | 0.253 | 0.303 | 0.354 | 0.405 | 0.456 | 0.507 | 0 |
| 10 | 0.507 | 0.558 | 0.609 | 0.660 | 0.711 | 0.762 | 0.813 | 0.865 | 0.916 | 0.967 | 1.019 | 10 |
| 20 | 1.019 | 1.070 | 1.122 | 1.174 | 1.225 | 1.277 | 1.329 | 1.381 | 1.432 | 1.484 | 1.536 | 20 |
| 30 | 1.536 | 1.588 | 1.640 | 1.693 | 1.745 | 1.797 | 1.849 | 1.901 | 1.954 | 2.006 | 2.058 | 30 |
| 40 | 2.058 | 2.111 | 2.163 | 2.216 | 2.268 | 2.321 | 2.374 | 2.426 | 2.479 | 2.532 | 2.585 | 40 |
| 50 | 2.585 | 2.638 | 2.691 | 2.743 | 2.796 | 2.849 | 2.902 | 2.956 | 3.009 | 3.062 | 3.115 | 50 |
| 60 | 3.115 | 3.168 | 3.221 | 3.275 | 3.328 | 3.381 | 3.435 | 3.488 | 3.542 | 3.595 | 3.649 | 60 |
| 70 | 3.649 | 3.702 | 3.756 | 3.809 | 3.863 | 3.917 | 3.971 | 4.024 | 4.078 | 4.132 | 4.186 | 70 |
| 80 | 4.186 | 4.239 | 4.293 | 4.347 | 4.401 | 4.455 | 4.509 | 4.563 | 4.617 | 4.671 | 4.725 | 80 |
| 90 | 4.725 | 4.780 | 4.834 | 4.888 | 4.942 | 4.996 | 5.050 | 5.105 | 5.159 | 5.213 | 5.268 | 90 |
| 100 | 5.268 | 5.322 | 5.376 | 5.431 | 5.485 | 5.540 | 5.594 | 5.649 | 5.703 | 5.758 | 5.812 | 100 |
| 110 | 5.812 | 5.867 | 5.921 | 5.976 | 6.031 | 6.085 | 6.140 | 6.195 | 6.249 | 6.304 | 6.359 | 110 |
| 120 | 6.359 | 6.414 | 6.468 | 6.523 | 6.578 | 6.633 | 6.688 | 6.742 | 6.797 | 6.852 | 6.907 | 120 |
| 130 | 6.907 | 6.962 | 7.017 | 7.072 | 7.127 | 7.182 | 7.237 | 7.292 | 7.347 | 7.402 | 7.457 | 130 |
| 140 | 7.457 | 7.512 | 7.567 | 7.622 | 7.677 | 7.732 | 7.787 | 7.843 | 7.898 | 7.953 | 8.008 | 140 |
| 150 | 8.008 | 8.063 | 8.118 | 8.174 | 8.229 | 8.284 | 8.339 | 8.394 | 8.450 | 8.505 | 8.560 | 150 |
| 160 | 8.560 | 8.616 | 8.671 | 8.726 | 8.781 | 8.837 | 8.892 | 8.947 | 9.003 | 9.058 | 9.113 | 160 |
| 170 | 9.113 | 9.169 | 9.224 | 9.279 | 9.335 | 9.390 | 9.446 | 9.501 | 9.556 | 9.612 | 9.667 | 170 |
| 180 | 9.667 | 9.723 | 9.778 | 9.834 | 9.889 | 9.944 | 10.000 | 10.055 | 10.111 | 10.166 | 10.222 | 180 |
| 190 | 10.222 | 10.277 | 10.333 | 10.388 | 10.444 | 10.499 | 10.555 | 10.610 | 10.666 | 10.721 | 10.777 | 190 |
| 200 | 10.777 | 10.832 | 10.888 | 10.943 | 10.999 | 11.054 | 11.110 | 11.165 | 11.221 | 11.276 | 11.332 | 200 |
| 210 | 11.332 | 11.387 | 11.443 | 11.498 | 11.554 | 11.609 | 11.665 | 11.720 | 11.776 | 11.831 | 11.887 | 210 |
| 220 | 11.887 | 11.943 | 11.998 | 12.054 | 12.109 | 12.165 | 12.220 | 12.276 | 12.331 | 12.387 | 12.442 | 220 |
| 230 | 12.442 | 12.498 | 12.553 | 12.609 | 12.664 | 12.720 | 12.776 | 12.831 | 12.887 | 12.942 | 12.998 | 230 |
| 240 | 12.998 | 13.053 | 13.109 | 13.164 | 13.220 | 13.275 | 13.331 | 13.386 | 13.442 | 13.497 | 13.553 | 240 |
| 250 | 13.553 | 13.608 | 13.664 | 13.719 | 13.775 | 13.830 | 13.886 | 13.941 | 13.997 | 14.052 | 14.108 | 250 |
| 260 | 14.108 | 14.163 | 14.219 | 14.274 | 14.330 | 14.385 | 14.441 | 14.496 | 14.552 | 14.607 | 14.663 | 260 |
| 270 | 14.663 | 14.718 | 14.774 | 14.829 | 14.885 | 14.940 | 14.995 | 15.051 | 15.106 | 15.162 | 15.217 | 270 |
| 280 | 15.217 | 15.273 | 15.328 | 15.383 | 15.439 | 15.494 | 15.550 | 15.605 | 15.661 | 15.716 | 15.771 | 280 |
| 290 | 15.771 | 15.827 | 15.882 | 15.938 | 15.993 | 16.048 | 16.104 | 16.159 | 16.214 | 16.270 | 16.325 | 290 |
| DEG C | 0 | 1 | 2 | 3 | 4 | 5 | 6 | 7 | 8 | 9 | 10 | DEG C |

THERMOELECTRIC VOLTAGE IN ABSOLUTE MILLIVOLTS

| DEG C | 0 | 1 | 2 | 3 | 4 | 5 | 6 | 7 | 8 | 9 | 10 | DEG C |
|---|---|---|---|---|---|---|---|---|---|---|---|---|
| 300 | 16.325 | 16.380 | 16.436 | 16.491 | 16.547 | 16.602 | 16.657 | 16.713 | 16.768 | 16.823 | 16.879 | 300 |
| 310 | 16.879 | 16.934 | 16.989 | 17.044 | 17.100 | 17.155 | 17.210 | 17.266 | 17.321 | 17.376 | 17.432 | 310 |
| 320 | 17.432 | 17.487 | 17.542 | 17.597 | 17.653 | 17.708 | 17.763 | 17.818 | 17.874 | 17.929 | 17.984 | 320 |
| 330 | 17.984 | 18.039 | 18.095 | 18.150 | 18.205 | 18.260 | 18.316 | 18.371 | 18.426 | 18.481 | 18.537 | 330 |
| 340 | 18.537 | 18.592 | 18.647 | 18.702 | 18.757 | 18.813 | 18.868 | 18.923 | 18.978 | 19.033 | 19.089 | 340 |
| 350 | 19.089 | 19.144 | 19.199 | 19.254 | 19.309 | 19.364 | 19.420 | 19.475 | 19.530 | 19.585 | 19.640 | 350 |
| 360 | 19.640 | 19.695 | 19.751 | 19.806 | 19.861 | 19.916 | 19.971 | 20.026 | 20.081 | 20.137 | 20.192 | 360 |
| 370 | 20.192 | 20.247 | 20.302 | 20.357 | 20.412 | 20.467 | 20.523 | 20.578 | 20.633 | 20.688 | 20.743 | 370 |
| 380 | 20.743 | 20.798 | 20.853 | 20.909 | 20.964 | 21.019 | 21.074 | 21.129 | 21.184 | 21.239 | 21.295 | 380 |
| 390 | 21.295 | 21.350 | 21.405 | 21.460 | 21.515 | 21.570 | 21.625 | 21.680 | 21.736 | 21.791 | 21.846 | 390 |
| 400 | 21.846 | 21.901 | 21.956 | 22.011 | 22.066 | 22.122 | 22.177 | 22.232 | 22.287 | 22.342 | 22.397 | 400 |
| 410 | 22.397 | 22.453 | 22.508 | 22.563 | 22.618 | 22.673 | 22.728 | 22.784 | 22.839 | 22.894 | 22.949 | 410 |
| 420 | 22.949 | 23.004 | 23.060 | 23.115 | 23.170 | 23.225 | 23.280 | 23.336 | 23.391 | 23.446 | 23.501 | 420 |
| 430 | 23.501 | 23.556 | 23.612 | 23.667 | 23.722 | 23.777 | 23.833 | 23.888 | 23.943 | 23.999 | 24.054 | 430 |
| 440 | 24.054 | 24.109 | 24.164 | 24.220 | 24.275 | 24.330 | 24.386 | 24.441 | 24.496 | 24.552 | 24.607 | 440 |
| 450 | 24.607 | 24.662 | 24.718 | 24.773 | 24.829 | 24.884 | 24.939 | 24.995 | 25.050 | 25.106 | 25.161 | 450 |
| 460 | 25.161 | 25.217 | 25.272 | 25.327 | 25.383 | 25.438 | 25.494 | 25.549 | 25.605 | 25.661 | 25.716 | 460 |
| 470 | 25.716 | 25.772 | 25.827 | 25.883 | 25.938 | 25.994 | 26.050 | 26.105 | 26.161 | 26.216 | 26.272 | 470 |
| 480 | 26.272 | 26.328 | 26.383 | 26.439 | 26.495 | 26.551 | 26.606 | 26.662 | 26.718 | 26.774 | 26.829 | 480 |
| 490 | 26.829 | 26.885 | 26.941 | 26.997 | 27.053 | 27.109 | 27.165 | 27.220 | 27.276 | 27.332 | 27.388 | 490 |
| 500 | 27.388 | 27.444 | 27.500 | 27.556 | 27.612 | 27.668 | 27.724 | 27.780 | 27.836 | 27.893 | 27.949 | 500 |
| 510 | 27.949 | 28.005 | 28.061 | 28.117 | 28.173 | 28.230 | 28.286 | 28.342 | 28.398 | 28.455 | 28.511 | 510 |
| 520 | 28.511 | 28.567 | 28.624 | 28.680 | 28.736 | 28.793 | 28.849 | 28.906 | 28.962 | 29.019 | 29.075 | 520 |
| 530 | 29.075 | 29.132 | 29.188 | 29.245 | 29.301 | 29.358 | 29.415 | 29.471 | 29.528 | 29.585 | 29.642 | 530 |
| 540 | 29.642 | 29.698 | 29.755 | 29.812 | 29.869 | 29.926 | 29.983 | 30.039 | 30.096 | 30.153 | 30.210 | 540 |
| 550 | 30.210 | 30.267 | 30.324 | 30.381 | 30.439 | 30.496 | 30.553 | 30.610 | 30.667 | 30.724 | 30.782 | 550 |
| 560 | 30.782 | 30.839 | 30.896 | 30.954 | 31.011 | 31.068 | 31.126 | 31.183 | 31.241 | 31.298 | 31.356 | 560 |
| 570 | 31.356 | 31.413 | 31.471 | 31.528 | 31.586 | 31.644 | 31.702 | 31.759 | 31.817 | 31.875 | 31.933 | 570 |
| 580 | 31.933 | 31.991 | 32.048 | 32.106 | 32.164 | 32.222 | 32.280 | 32.338 | 32.396 | 32.455 | 32.513 | 580 |
| 590 | 32.513 | 32.571 | 32.629 | 32.687 | 32.746 | 32.804 | 32.862 | 32.921 | 32.979 | 33.038 | 33.096 | 590 |
| 600 | 33.096 | 33.155 | 33.213 | 33.272 | 33.330 | 33.389 | 33.448 | 33.506 | 33.565 | 33.624 | 33.[illegible]3 | 600 |
| 610 | 33.683 | 33.742 | 33.800 | 33.859 | 33.918 | 33.977 | 34.036 | 34.095 | 34.155 | 34.214 | 34.273 | 610 |
| 620 | 34.273 | 34.332 | 34.391 | 34.451 | 34.510 | 34.569 | 34.629 | 34.688 | 34.748 | 34.807 | 34.867 | 620 |
| 630 | 34.867 | 34.926 | 34.986 | 35.046 | 35.105 | 35.165 | 35.225 | 35.285 | 35.344 | 35.404 | 35.464 | 630 |
| 640 | 35.464 | 35.524 | 35.584 | 35.644 | 35.704 | 35.764 | 35.825 | 35.885 | 35.945 | 36.005 | 36.066 | 640 |
| 650 | 36.066 | 36.126 | 36.186 | 36.247 | 36.307 | 36.368 | 36.428 | 36.489 | 36.549 | 36.610 | 36.671 | 650 |
| 660 | 36.671 | 36.732 | 36.792 | 36.853 | 36.914 | 36.975 | 37.036 | 37.097 | 37.158 | 37.219 | 37.280 | 660 |
| 670 | 37.280 | 37.341 | 37.402 | 37.463 | 37.525 | 37.586 | 37.647 | 37.709 | 37.770 | 37.831 | 37.893 | 670 |
| 680 | 37.893 | 37.954 | 38.016 | 38.078 | 38.139 | 38.201 | 38.262 | 38.324 | 38.386 | 38.448 | 38.510 | 680 |
| 690 | 38.510 | 38.572 | 38.633 | 38.695 | 38.757 | 38.819 | 38.882 | 38.944 | 39.006 | 39.068 | 39.130 | 690 |
| 700 | 39.130 | 39.192 | 39.255 | 39.317 | 39.379 | 39.442 | 39.504 | 39.567 | 39.629 | 39.692 | 39.754 | 700 |
| 710 | 39.754 | 39.817 | 39.880 | 39.942 | 40.005 | 40.068 | 40.131 | 40.193 | 40.256 | 40.319 | 40.382 | 710 |
| 720 | 40.382 | 40.445 | 40.508 | 40.571 | 40.634 | 40.697 | 40.760 | 40.823 | 40.886 | 40.950 | 41.013 | 720 |
| 730 | 41.013 | 41.076 | 41.139 | 41.203 | 41.266 | 41.329 | 41.393 | 41.456 | 41.520 | 41.583 | 41.647 | 730 |
| 740 | 41.647 | 41.710 | 41.774 | 41.837 | 41.901 | 41.965 | 42.028 | 42.092 | 42.156 | 42.219 | 42.283 | 740 |
| 750 | 42.283 | 42.347 | 42.411 | 42.475 | 42.538 | 42.602 | 42.666 | 42.730 | 42.794 | 42.858 | 42.922 | 750 |
| 760 | 42.922 | | | | | | | | | | | 760 |
| DEG C | 0 | 1 | 2 | 3 | 4 | 5 | 6 | 7 | 8 | 9 | 10 | DEG C |

# Melting Temperatures of Some Important Metals

Approximate melting points are given only as a guide for material selection since many factors including atmosphere, type of process, mounting, etc., all affect the operating maximum.

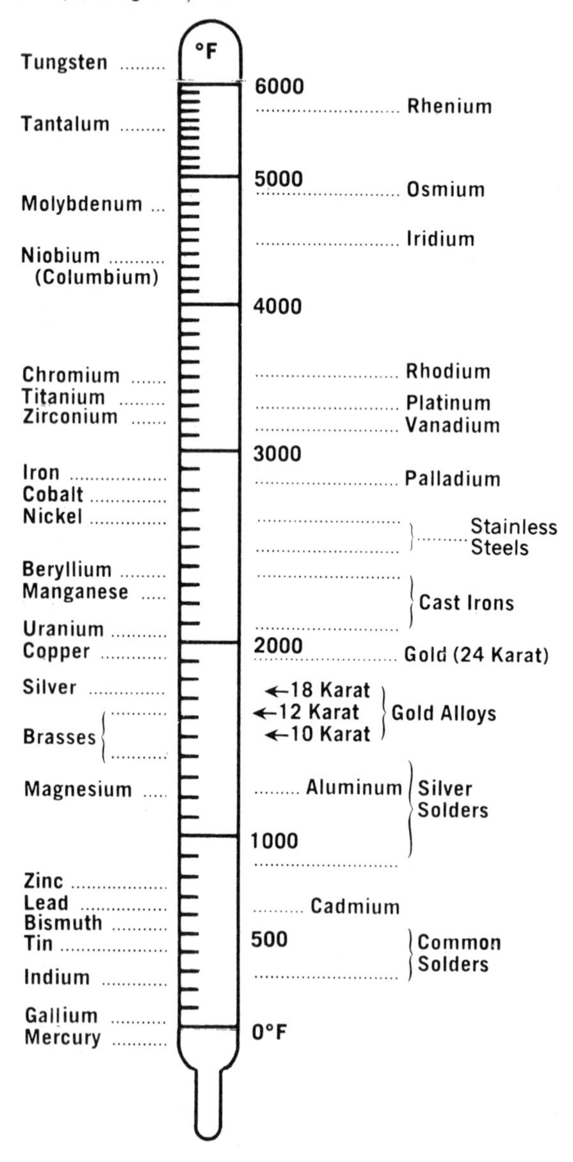

# Very High Temperature Sheath Materials

| Sheath Material | Rec. Useful Temp. | Melting Point | Environmental Conditions | | | |
|---|---|---|---|---|---|---|
| | | | Oxidizing | Hydrogen | Inert | Vacuum |
| Molybdenum | 4000°F | 4730°F | Not Rec. | Fair | Fair | Good |
| Tantalum | 4500°F | 5425°F | Not Rec. | Not Rec. | Not Rec. | Good |
| Platinum | 3050°F | 3223°F | Very Good | Poor | Poor | Poor |

# Thermometry Fixed Points

| THERMOELECTRIC FIXED POINT | MELTING POINTS FROM THE PRACTICAL INTERNATIONAL TEMPERATURE SCALE IPTS-68 | |
|---|---|---|
| Boiling point of oxygen | −183.0 °C | −297.3 °F |
| Sublimation point of carbon dioxide | − 78.5 | −109.2 |
| Freezing point of mercury | − 38.9 | − 38 |
| Ice Point | 0 | 32 |
| Triple point of water | 0.01 | 32 |
| Boiling point of water | 100.0 | 212 |
| Triple point of benzoic acid | 122.4 | 252.3 |
| Boiling point of naphthalene | 218 | 424.4 |
| Freezing point of tin | 231.9 | 449.4 |
| Boiling point of benzophenone | 305.9 | 582.6 |
| Freezing point of cadmium | 321.1 | 610 |
| Freezing point of lead | 327.5 | 621.5 |
| Freezing point of zinc | 419.6 | 787.2 |
| Boiling point of sulfur | 444.7 | 832.4 |
| Freezing point of antimony | 630.7 | 1167.3 |
| Freezing point of aluminum | 660.4 | 1220.7 |
| Freezing point of silver | 961.9 | 1763.5 |
| Freezing point of gold | 1064.4 | 1948 |
| Freezing point of copper | 1084.5 | 1984.1 |
| Freezing point of palladium | 1554 | 2829 |
| Freezing point of platinum | 1772 | 3222 |

# Identification

## Color Coding

Standard ANSI color coding is used on all insulated thermocouples wire and extension grade wire when the insulation permits. For some insulations a colored tracer is used to distinguish the calibration.

## ANSI Letter Designations

Thermocouple and extension grade wires are specified by ANSI letter designations. Positive and negative legs are identified by the letter suffixes P and N respectively. Example: JP designates the positive leg (iron) of the Iron-Constantan pair.

### CHARACTERISTICS TABLE

| ANSI T/C | Symbol Single | Generic and Trade Names | Color Coding: Single | Color Coding: Overall T/C Wire | Color Coding: Overall Extension Grade Wire | Magnetic Yes | Magnetic No | Maximum Useful Temp. Range | EMF (MV) Over Useful Temp. Range | Average Sensitivity μV/°C | Environment (Bare Wire) |
|---|---|---|---|---|---|---|---|---|---|---|---|
| T | TP<br>TN | Copper<br>Constantan, Cupron, Advance | Blue<br>Red | Brown | Blue | | X<br>X | °F -328 to 662<br>°C -200 to 350 | -5.602 to 17.816 | 40.5 | Mild Oxidizing, Reducing. Vacuum or Inert. Good where moisture is present. |
| J | JP<br>JN | Iron<br>Constantan, Cupron, Advance | White<br>Red | Brown | Black | X | <br>X | °F 32 to 1382<br>°C 0 to 750 | 0 to 42.283 | 52.6 | Reducing. Vacuum, Inert. Limited use in oxidizing at High Temperatures. Not recommended for low temps. |
| E | EP<br>EN | Chromel, Tophel, T[1] Thermokanthal KP<br>Constantan, Cupron, Advance | Purple<br>Red | Brown | Purple | | X<br>X | °F -328 to 1652<br>°C -200 to 900 | -8.824 to 68.783 | 67.9 | Oxidizing or Inert. Limited use in Vacuum or Reducing. |
| K | KP<br>KN | Chromel, Tophel, T[1] Thermokanthal KP<br>Alumel, Nial, T[2] Thermokanthal KN | Yellow<br>Red | Brown | Yellow | <br>X | X | °F -328 to 2282<br>°C -200 to 1250 | -5.973 to 50.633 | 38.8 | Clean Oxidizing and Inert. Limited use in Vacuum or Reducing. |
| S | SP<br>SN | Platinum 10% Rhodium<br>Pure Platinum | Black<br>Red | | Green | | X<br>X | °F 32 to 2642<br>°C 0 to 1450 | 0 to 14.973 | 10.6 | Oxidizing or Inert Atmos. Do not insert in metal tubes. Beware of contamination. |
| R | RP<br>RN | Platinum 13% Rhodium<br>Pure Platinum | Black<br>Red | | Green | | X<br>X | °F 32 to 2642<br>°C 0 to 1450 | 0 to 16.741 | 12.0 | |
| B | BP<br>BN | Platinum 30% Rhodium<br>Platinum 6% Rhodium | Gray<br>Red | | Gray | | X<br>X | °F 32 to 3092<br>°C 0 to 1700 | 0 to 12.426 | 7.6 | |
| C* | CP*<br>CN* | Tunsten 5% Rhenium<br>Tungsten 26% Rhenium | White/Red Trace<br>Red | | White/Red Trace | | X<br>X | °F 32 to 4208<br>°C 0 to 2320 | 0 to 37.066 | 16.6 | Vacuum, Inert, Hydrogen Atmospheres. Beware of embrittlement. |
| G* | GP*<br>GN* | Tungsten<br>Tungsten 26% Rhenium | White/Blue Trace<br>Red | | White/Blue Trace | | X<br>X | °F 32 to 4208<br>°C 0 to 2320 | 0 to 38.564 | 16.0 | |
| D* | DP*<br>DN* | Tungsten 3% Rhenium<br>Tungsten 25% Rhenium | White/Yellow Trace<br>Red | | White/Yellow Trace | | X<br>X | °F 32 to 4208<br>°C 0 to 2320 | 0 to 39.506 | 17.0 | |

*Not ANSI Symbol.

# Temperature Dependent Data

OMEGACLAD™ is a 3 part system composed of compacted ceramic insulation, thermocouple wire and metal sheath. Therefore, three factors determine the useful service temperature for OMEGACLAD™.

- Range for the thermocouple wire (see table—limits of error)
- Maximum service temperature of insulation. In the case of MgO, this is in excess of 3000°F (1650°C)
- Properties of the sheath material:

| Material | Melting Point | Continuous Max. Temp. | Tensile (PSI) Strength | |
|---|---|---|---|---|
| | | | at 200°F | at 1000°F |
| 304 S.S. | 2560°F | 1650°F | 68,000 | 15,000 |
| 316 S.S. | 2500°F | 1650°F | 75,000 | 23,000 |
| Inconel* | 2550°F | 2100°F | 93,000 | 5,000 |

*Vacuum or Inert atmosphere only.

## MAXIMUM* LONG-TERM SERVICE TEMPERATURE FOR OMEGACLAD™

| Thermocouple Type | Nominal Diameter | Sheath Material | |
|---|---|---|---|
| | | Stainless Steel | Inconel |
| T | ¼″ | 700°F (370°C) | 700°F ( 370°C) |
| J | ¼″ | 1400°F (770°C) | 1400°F ( 770°C) |
| E | ¼″ | 1650°F (900°C) | 1650°F ( 900°C) |
| K | ¼″ | 1650°F (900°C) | 2100°F (1150°C) |

*Based on limitations imposed by sheath or thermocouple wire, whichever is lower

## MAXIMUM LONG-TERM SERVICE TEMPERATURE FOR PROTECTED BARE WIRE THERMOCOUPLES OF VARIOUS WIRE DIAMETERS.

| symbol | Alloy Combination | 0.005″ | 0.015″ | 0.020″ | 0.032″ |
|---|---|---|---|---|---|
| T | Copper-Constantan | 400°F | 400°F | 400°F | 500°F |
| J | Iron-Constantan | 600°F | 700°F | 700°F | 900°F |
| E | Chromel-Constantan | 700°F | 800°F | 800°F | 1000°F |
| K | Chromel-Alumel | 1100°F | 1600°F | 1600°F | 1800°F |
| R,S | Platinum-Rhodium | | | 2700°F | |
| B | Platinum-Rhodium | | | 3100°F | |
| C, G, D† | Tungsten-Rhenium* | 3600°F | 4200°F | 4200°F | |

†Not ANSI Symbols

Trade Names: Cupron Nial and Tophel-AMAX. / Advance, T1 and T2-Driver-Harris Co. / Chromel and Alumel-Hoskins Mfg. Co./Thermokanthal KP and Thermokanthal KN-Kanthal Co.

**NOTE:** The table above is only meant as a guide. Wire may be used to higher temperatures within overall range with subsequent decrease in lifetime. (ie: type K wire serviceable to 2300°F).

# Glossary

**Active filter.** A filter that uses amplification to improve its characteristics.
**A/D converter.** A circuit designed to convert an analog voltage into a binary code, which can be read by a computer.
**Address decoder.** A circuit that detects the presence of a specific code on the address bus of a computer and outputs a single signal or pulse when the code is present.
**Apparent current.** The vector sum of effective and reactive currents.
**Armature.** The rotating part of a motor.

**Band pass filter.** A filter that passes all frequencies between two specified frequencies.
**Band stop filter.** *Also* band reject filter. A filter that blocks all frequencies between two specified frequencies. When the bandwidth is very narrow, this may be called a notch filter.
**Bandwidth.** 1. The range of frequencies amplified by an amplifier or the range of frequencies passed by a filter; $BW = f_{cu} - f_{cl}$. 2. The width of the pass or stop band of a filter. The difference between the upper cutoff frequency and the lower cutoff frequency.
**Bellows pressure sensor.** A pressure measuring device.
**Bimetal thermostat.** Temperature sensor that depends upon the different rates of expansion of different metals. Two metal strips are bonded together — as the temperature changes, the bonded metal strip bends in response.

**Bonded strain gage.** A strain measuring device where fine wire or foil is bonded to a plastic or paper support. Reports strain as a small resistance change.
**Bourdon tube pressure sensor.** A pressure measuring device made of a flexible tube formed into a *C* shape. Increasing pressure causes the *C* to straighten.
**Bridge inverter.** An ac motor speed controller that supplies power at different frequencies for motor speed control.

**Capacitive level detector.** Level sensor that uses the difference in dielectric constant between air and the fluid being measured. Change in level of fluid causes change in capacitance between the two dip rods.
**Capacitor start motor.** An induction motor that uses a capacitor in series with the start winding to create the phase shift needed for starting the motor.
**Capsule pressure sensor.** A pressure measuring device composed of two back-to-back diaphragms.
**CEMF (counter EMF).** The reverse EMF generated in the armature of a running dc motor. The motor speed stabilizes when the CEMF equals the applied armature voltage.
**Check valve.** A valve that permits fluid to flow in only one direction. The fluid analog of a diode.
**Closed loop gain.** The gain of an amplifier with feedback. Negative feedback reduces amplifier gain. Positive feedback increases amplifier gain.
**CMOS (complementary metal oxide semiconductor logic).** A logic family that uses complementary pair switches as its output. Lower current draw than TTL is leading this family into predominance, especially in portable equipment.
**CMRR (common mode rejection ratio).** The ratio of output voltage to input voltage of an op amp whose inputs are *common mode*, or connected together. A measure of how well the op amp will reject noise induced in the input lines.
**Commutator.** Sectored ring on the armature of a motor which connects the power source to the armature coils.
**Compound motor.** A dc motor in which a portion of the field is series connected and a portion of the field is shunt connected.
**Compression strain.** Strain caused by weight sitting on a member.
**Control setpoint.** The desired condition of the controlled variable.
**Controlled variable.** That part of an industrial process that is controlled by an automation system.
**Critically damped.** The degree of damping that provides the best compromise between the undamped response and the overdamped response. The overshoot is at a minimum for a fast response time.
**Cutoff frequency.** The "corner frequency" of a filter. That frequency at which the signal level is 3 dB less than it is in the passband.
**Cycloconverter.** A system for supplying an ac motor with power at frequencies below the line frequency for motor speed control.

**D/A converter.** A circuit that converts a binary code into an analog voltage or current.
**Damping.** The tailoring of the response of a system to meet system tolerances. Damping is used to limit the oscillatory nature of a control system.
**Degree of freedom.** The axes of motion available to a robotic device.
**Delta connection.** A connection scheme for three-phase ac. The load is connected in a triangular pattern with one corner to each phase.
**Derivative control.** A control system that uses a differentiator. The output is proportional to the rate of change of the error signal. Used to correct for transient disturbances. Always used with proportional control.
**Detent torque.** The torque of a permanent magnet stepper motor when the power has been removed from its windings.
**DIAC.** Four-layer breakover device used to extend the range of control in a TRIAC circuit.
**Diaphragm pressure sensor.** A pressure measuring device.
**Difference amplifier.** An operational amplifier circuit that finds the difference between two input voltages.
**Differential pressure flowmeter.** A device for measuring the flow of fluids. Depends on the drop in pressure created when a fluid flows past an obstruction or around a bend.
**Differentiator.** An amplifier that performs the calculus operation of differentiation. The output signal is proportional to the rate of change of the input signal. When used in a controller, it is often called derivative control.
**Direction control valve.** A valve that allows control of the path of fluids. The fluid analog of a selector switch.
**Drum controller.** A controller made up of one or more cam-operated limit switches. The cam is usually driven by a clock motor.
**Dynamic braking.** Braking a motor by connecting a resistive load across its armature windings after power is disconnected from the armature while the field is still connected.

**Effective current.** The in-phase portion of the current in a reactive circuit.
**Effective power.** The power consumed by the resistive portion in an ac circuit.
**Electronic ice.** A system of reference junction compensation used in thermocouple circuits. This is an electronic means of creating the thermocouple reference junction.
**EMI (electromagnetic interference).** Interference caused by stray magnetic fields from fluorescent light ballasts, motors, relays, and other electromagnetic devices.
**End effector.** The special tooling at the end of a robotic arm. Usually designed to enable the robotic arm to handle a special task.
**Events-per-unit-of-time counter.** A counter that measures the number of events in a fixed period of time. The most familiar example is the frequency counter.

**Face plate starter.** A manual motor starter in which series resistance is gradually switched out of a motor circuit. Used to limit the starting current.

**Feedback.** 1. Returning information about the result of a process to the process controller. Allows the controller to adjust the process in response to disturbances or changes in the control setpoint. 2. Returning a portion of an amplifier's output signal to the input.
**Filter.** A circuit that passes only signals of a desired frequency or band of frequencies. May be high pass, low pass, or band pass.
**Float switch.** Level-sensing limit switch, actuated by a float on the surface of a liquid.
**Force.** A directed effort that changes the motion of a body.
**Forward breakover voltage.** The voltage between anode and cathode of an SCR at which forward-bias conduction will begin.
**Four-wire resistance measurement.** A system for measuring resistance without the problems of lead resistance. Requires a constant current source and a high-impedance voltmeter circuit.

**Gage factor.** The ratio of change in resistance to the change in length of a strain gage. Approximately two for a bonded foil strain gage, 200 for a semiconductor strain gage.
**Gain (amplification).** The ratio of increase in signal level between the input and output of an amplifier. May be an increase in signal voltage, signal current, or signal power.

**High pass filter.** A filter that passes all frequencies above a specified frequency.
**Holding current.** The current from anode to cathode of an SCR, which will maintain conduction without applied gate current.
**Holding torque.** The torque of a stationary stepper motor with power applied to the windings.

**Ideal amplifier.** A theoretically perfect amplifier. Characterized by infinite input impedance, zero output impedance, infinite gain, and infinite bandwidth.
**Immersion heater.** A heating element designed to be operated in a liquid.
**Induction motor.** An ac motor in which the current in the armature windings is caused by induction from the field windings.
**Input port.** A pathway for allowing a computer to access data from external devices.
**Instrumentation amplifier.** 1. A difference amplifier with very high input impedances at both inputs. 2. An operational amplifier circuit that offers high gain for low-level signals. Basically, a difference amplifier with high-impedance inputs.
**Integrator.** An amplifier circuit that performs the calculus function of integration. The output is proportional to the length of time an input signal has been applied. When used in a controller, it is often called reset control.
**Interbase resistance.** The resistance between the two bases of a UJT.
**Interface.** To connect two devices together so that they work harmoniously.
**Intrinsic standoff ratio.** The ratio between the base-emitter pairs in a UJT.
**Inverting amplifier.** An amplifier whose output is 180 degrees out of phase with its input.
**Isothermal block.** A connecting block used with thermocouples. The temperature at all parts of the connecting block is equal. Used to prevent the creation of additional thermocouple junctions where the thermocouple is connected to the system.

**Ladder diagram.** A type of schematic diagram often used in control circuits.
**Limit switch.** A switch that is arranged to be actuated by a workpiece. The switch that turns on your car's dome light is a limit switch.
**Load cell.** A device for measuring weight. Weight resting on the device causes compression strain. Weight suspended from the device causes tensile strain. Strain is reported as a change in resistance by a coupled strain gage.
**Low pass filter.** A filter that passes all frequencies below a specific frequency.
**LVDT (linear variable differential transformer).** A device used for position detection. Consists of an ac excited primary, two secondaries, and a movable core. Voltage and phase-angle differences are determined by the position of the core.

**Magnetic overload protector.** A device used to protect equipment from short-circuit overloads.
**Magnetic reed switch.** A magnetically operated switch. Made of two or three magnetic leaves in a glass tube. Proximity of a magnet causes the switch to close.
**Mass flowmeter.** A fluid-flow measuring device that measures the mass of the fluid instead of its velocity. Used when great accuracy is required.
**Memory-mapped interface.** An interface system in which the input/output ports are addressed as memory locations.
**Microprocessor.** A single IC that contains most of the logic needed for a small computer system.

**Negative feedback.** Feedback that is 180 degrees out of phase with the input signal. Reduces overall gain.
**Noninverting amplifier.** An amplifier whose output is in phase with its input.
**Notch filter.** A band stop filter with a very narrow bandwidth.

**Open-loop gain.** The gain of an amplifier without feedback.
**Optical isolator.** A device that isolates a low-voltage device from the high-power device it controls by using light as a data transfer medium.
**Output port.** A pathway that allows a computer to send data to external devices.
**Overdamped response.** The response of a system that has a great deal of damping. Overdamping is used when overshoot is not acceptable.
**Overshoot.** The amount that a system moves beyond the control setpoint in response to a disturbance.

**Passband.** The range of frequencies that are passed by a filter.
**Passive filter.** A filter made of passive components: resistors, capacitors, and inductors.
**Peripheral tooling.** Special tooling that is used to make a task more easily adapted to automation or robotics.
**Permanent magnet dc motor.** A motor in which the field is supplied by a permanent magnet.
**Photodiode.** A light-sensitive diode that is operated in reverse bias. When light strikes the junction the diode goes into reverse breakdown and conducts.

**Phototransistor.** A light-sensitive transistor that is usually operated in the cutoff region. Light striking the base-to-emitter junction causes the transistor to become saturated.
**Photovoltaic.** The property of responding to light with an electrical current. Photovoltaic cells are used in generating electricity from solar energy.
**PID controller.** A controller that uses proportional, integral, and derivative control in one unit.
**Piezoresistance.** The property of semiconductors that causes a change in resistance in response to strain.
**Plugging.** Braking a dc motor by reversing the polarity of the armature voltage.
**Positive displacement flowmeter.** A device that measures fluid flow by passing the fluid in measured increments. Usually accomplished by alternately filling and emptying a chamber.
**Positive feedback.** Feedback that is in phase with the input signal. Increases overall gain.
**Power.** The rate of doing work. Measured in horsepower or watts.
**Power factor.** The ratio of effective power to reactive power in an ac circuit.
**Power factor correction.** The technique of adding capacitive reactance to the circuit of an induction motor to reduce the power factor. Required by most utilities for industrial users.
**Proportional control.** A control system in which the output signal is proportional to the error signal.
**Pull-out torque.** The "starting torque" of a stepper motor.
**Purge bubbler.** A level-sensing system that measures the variation in air pressure required to maintain a constant stream of bubbles flowing in a tank. Increased depth requires increased pressure.

**Ramping.** Gradually increasing or decreasing the speed of a motor instead of abrupt on-off control, especially in stepper motors.
**Reactive current.** The out-of-phase portion of the current in a reactive circuit.
**Reference junction.** The junction of a thermocouple whose temperature is known. In laboratory situations, this is an ice bath. *See also* electronic ice.
**Register I/O interface.** An interface system in which the input and output ports have unique addresses and control signals that are separate from memory operations.
**Relay.** An electromagnetically operated switch.
**Reset control.** A control system that uses integration. The output signal is proportional to the duration of the error signal. Used to correct for steady-state error. Always used along with proportional control.
**Resistance heating element.** A heating element based on the power dissipation of current flowing through a resistance.
**Resistance level detector.** Level-sensing device that relies on variation of resistance in conductive fluid. Change in fluid level causes change in resistance between the two dip rods.

**Resistive start motor.** An induction motor in which the start winding is made of many turns of small wire. The added resistance of the smaller wire creates the phase shift needed to start the motor.
**Resolver.** A rotary variable transformer used to report the angular position of a motor shaft. The primary is in a fixed position, and the secondary rotates with the shaft. The phase angle between the input and output indicates shaft rotation angle.
**Response time.** The time required for a system to return to within system tolerance following a disturbance. Also called settling time.
**RFI (radio frequency interference).** Interference caused by radio frequency emissions from arc welding, motor brushes, computers, and other high-frequency sources.
**Rolloff.** The rate at which the output signal of a filter circuit is reduced beyond the cutoff frequency. Expressed in dB per decade or dB per octave.
**Rotor.** The moving part of an ac motor.
**RS232.** A serial data communications standard.
**RTD (resistive temperature detector).** A wire-wound resistance, usually using platinum wire, whose resistance varies in direct proportion to its temperature.

**Schmitt trigger.** A digital device that is used to convert slowly changing analog signals into sharply rising and falling signals that are useful for digital inputs. Triggers low to high at a different input voltage from high-to-low trigger.
**SCR (silicon controlled rectifier).** A four-layer semiconductor device used in switching circuits.
**Self-heating.** The tendency of a sensor to heat up when a measuring or an operating current is passed through it. Can be overcome to some degree by limiting currents and using physically larger sensors.
**Semiconductor strain gage.** A strain measuring device that is based on the change in resistance of a semiconductor material under strain. More sensitive than wire or foil strain gages.
**Series wound motor.** A dc motor in which the field coils are connected in series with the armature windings.
**Settling time.** The time required for a system to return to within the system's tolerance following a disturbance. Also called response time.
**Shaft encoder.** A device used to report the angular position of a motor shaft.
**Shunt wound motor.** A dc wound field motor in which the field windings are connected in parallel to the armature windings.
**Slip.** The difference in speed between the rotor of an induction motor and the synchronous speed. Usually expressed as a percentage.
**Speed regulation.** The measure of how well a motor maintains its rpm when the load is changed.
**Squirrel cage rotor.** The rotor of an induction motor.
**Start winding.** An extra winding that is added to a single-phase motor to get the motor started.

**Starting torque.** The torque generated by a motor when power is first applied. Also called breakaway torque.
**Stator.** The fixed part of an ac motor.
**Steady-state error.** That degree of error that is inherent in a system. This is caused by tolerances in components, among other things.
**Stepper motor.** A motor in which the armature is rotated in increments. Used for accurate positional control.
**Strain.** Change in dimension of a material when force is applied.
**Strain gage.** A device used to measure strain.
**Summing amplifier.** 1. An amplifier whose output is proportional to the sum of two or more input signals. 2. An amplifier used to add (sum) its input voltages.
**Summing point.** The point in a control system where the control setpoint and the measurement of the controlled variable are compared. The inputs to a comparator or a summing amplifier.
**Synchronous motor.** An ac motor with a permanent magnet rotor. The rotor speed will equal the synchronous speed.
**Synchronous speed.** The speed of the rotating magnetic field in an ac motor.

**Tach generator.** A device used to measure motor speed. The output is a voltage or a frequency that is proportional to motor speed.
**Tensile strain.** Strain caused by force pulling on a member.
**Thermal overload protector.** A device that protects a motor from long-term overloads.
**Thermistor.** A semiconductor temperature sensor. Behaves like a resistor with a negative temperature coefficient.
**Thermocouple.** A temperature-sensing device. Output is a current or voltage that is proportional to the difference between the temperatures at two junctions of dissimilar metals.
**Thermopile.** A system of several thermocouples in a series-aiding configuration. This configuration increases the sensitivity of the thermocouple.
**Three-wire bridge.** A Wheatstone bridge with a third "sense" lead added to the measuring circuit. Helps overcome problems caused by lead resistance.
**Time-delay relay.** A relay that creates a time delay, either when power is first applied (delay on make) or when power is removed (delay on break).
**Torque.** Twisting or rotary force, such as that delivered by a motor shaft.
**Transducer.** A device that converts mechanical or physical quantities into electrical quantities, or a device that converts electrical quantities into physical quantities.
**TRIAC.** A semiconductor switching device used in ac circuits.
**TTL (transistor-transistor logic).** The most common type of IC logic in use today. The 74xx and 54xx series of chips are examples.

**UJT (unijunction transistor).** A type of transistor used as a relaxation oscillator in SCR control circuits.
**Unbonded strain gage.** Strain gage made of fine wire stretched tightly between support posts. As wire is stretched, its resistance increases.
**Undamped response.** The response of a control system that has no damping. The system usually oscillates above and below the control setpoint following a disturbance.
**Underdamped response.** The response of a system with only a small amount of damping. The system is still oscillatory, but the amount of both the oscillation and the overshoot is reduced.

**VA (volt amps).** The apparent power in an ac circuit. The vector sum of reactive and effective powers in the circuit.
**VAR (volt amps reactive).** The power consumed by the reactive portion of an ac circuit. This power is first consumed and then returned to the line.
**Velocity flowmeter.** A device that measures fluid flow directly. The most common is the turbine flowmeter.
**Voltage difference amplifier.** An amplifier whose output is proportional to the difference between two input voltages.
**Voltage follower.** An amplifier with a gain of one. Used as a buffer amplifier.

**Watt density.** The ratio of power (in watts) per square inch of heater surface. High watt density causes a greater concentration of heat. Low watt density heaters generally have a longer operating lifetime.
**Wheatstone bridge.** A network of resistors used to measure very small changes in resistance.
**Work.** The effect of force over distance. Measured in foot pounds, or kilogram meters.
**Work envelope.** The volume that can be reached by the end effector of a robotic arm.
**Wound field motor.** A dc motor in which the field magnets are electromagnets.
**Wye connection.** A connection scheme for three-phase ac. The load is connected in a star pattern.

# Index